AF530522

FRUITS FOR THE FUTURE

Volume 1 : Well Versed Arid and Semi Arid Fruits

Fruits For The Future

Volume 1 : Well Versed Arid and Semi Arid Fruits

Vishal Nath
Dinesh Kumar
V. Pandey

2008

403, Express Tower, Commercial Complex, Azadpur, Delhi - 110033 (INDIA)
Phone : 011-27672852 Fax : 91-11-27672046
E-mail : info@satishserial.com, hkjain1975@yahoo.com
Website : www.satishserial.com

Published by :

SATISH SERIAL PUBLISHING HOUSE
403, Express Tower, Commercial Complex, Azadpur, Delhi - 110033 (INDIA)
Phone : 011-27672852 Fax : 91-11-27672046
E-mail : info@satishserial.com, hkjain1975@yahoo.com

ISBN : 81-89304-43-7

Printed at
SALASAR IMAGING SYSTEMS, DELHI-35

Prof. R. K. Pathak
Chief Consultant
National Horticulture Mission
Krishi Bhawan
New Delhi - 110001

भारत सरकार
कृषि मंत्रालय
(कृषि एवं सहकारिता विभाग)
कृषि भवन, नई दिल्ली – 110001
GOVERNMENT OF INDIA
MINISTRY OF AGRICULTURE
(DEPARTMENT OF AGRICULTURE & COOPERATION)
KRISHI BHAWAN, NEW DELHI-110001

Foreward

Fruit production has been recognized as a remunerative and sustainable enterprise in the country. On one hand, it has great potential for generating employment and on the other; it improves the quality of diet. Now a day, many fruits are being used as source of nutrition energy and antioxidants and there consumption has been advocated world wide. The medicinal properties of fruit have also been understood clearly.

Land, being non-renewable and limited natural resource, plays a crucial role in spread of area under any commodity. For the benefit of human kind, the arable lands are being tapped to their full potential for the production of crops of major economic importance. However, a sizable area is still either untapped or under utilized for the one or the other reason. In India alone approximately 175 million hectare is degraded or waste land which is seldom used for crop production. Mostly the fruit crops, being deep rooted and perennial in nature, hold promise in crop diversification particularly under adverse soils and climatic conditions.

There are a number of fruit yielding multipurpose tree species which are very much adaptive to degraded land forms and adverse ecological conditions across the country. It is high time that due importance is given for these crops for amelioration of degraded land and increased availability of nutritive foods to the local masses.

I am happy to know that such fruit crops have been identified and a series of book entitled "Fruits for the Future" is being written by a team of devoted scientists for the popularization of these fruit crops in different regions. The major research results and information generated on these crops in India as well as abroad, has been compiled and put together. This compilation will ensure a better understanding of production and utilization techniques of these crops. This is a timely and valuable publication, which I am sure, will be of much use to the researchers, teachers, policy makers, development of officials, students and growers who are engaged in development of horticulture in the country.

(Prof. R. K. Pathak)

New Delhi
21st December, 2006

Preface

Fruits are an important source of nutritional security and income generation. The country ranks second in terms of fruit production after China with an annual production of nearly 50 million tons from an area of about 4.0 million hectares. However, we are short in supply the required quantity of fruits as per dietary recommendation for our increasing population. The major reason for non-availability of required quantity of fruits is, lack of sufficient production from the existing area under cultivation and post harvest management in different fruit crops. The possible solution to mitigate the shortfall could be either increase in productivity with improved management practices and varieties of existing crops and post harvest handling techniques or increase in area under fruit production. The former is being taken care off by different organizations and institutions engaged in development of technologies to enhance productivity and reduce post harvest losses where as the later needs much attention by the means of introducing newer crops and increased area coverage under existing crops. Through various developmental schemes launched by Government of India and the State Governments such as Integrated Development of Horticulture (IDH), National Horticulture Mission (NHM), Macro Management Projects, Chief Minister's Horticulture Plan etc., the emphasis has been laid for area expansion, but due to the utilization of arable lands to their full potential, the required momentum and growth has not been achieved. Presently, approximately 175 million hectare area in the country is under degraded or waste land which is not suitable for agronomical crops by enlarge. These lands are however, suitable for deep rooted, hardy perennial fruit yielding species, referred as **"Fruits for the Future"** in the book. These crops can withstand extremes of moisture stress, biophysical constraints and fragile ecological conditions and can produce quality fruits.

In the present book, the information pertaining to importance and uses, taxonomy, plant genetic resources and varieties, plant propagation and nursery management, production technologies and processing and value addition techniques of 21 such well versed fruit crops mainly concentrated in arid and semi arid areas of the country has been compiled. The updated information on these crops have been reviewed and suitably placed with the help of tables and figures. The long working experience in these crops has enable to collate and synthesize the useful informations for the better understanding. The main purpose of writing this book is to assemble information on various aspects of these underutilized and lesser known fruit crops in order to identify the constraints and research gaps to work upon and develop sustainable technology for the future so that these crops can find place in Indian Horticultural Scenario and help to nutrition security and poverty elevation programmes of the country. I hope the information presented in this book will be a useful tool for teachers, students, extension functionaries, development personals, policy makers, fruit growers and traders. I am sure that the information furnished in the book will help increase further production of these crops through area expansion with vegetatively multiplied quality plants of improved varieties and adoption of improved agro-techniques vis-à-vis minimization of post harvest losses through proper handling and increased availability of consumer acceptable value added products from these lesser known fruits.

For developing this book, I have constant guidance and encouragement from our senior colleagues and teachers. In this endeavor, I would like to express my sincere gratitude to Dr. O. P. Pareek, Ex-Director, Central Insitutte of Arid Horticulture, Bikaner and Prof R.K.Pathak, Ex-Director, Central Institute for Sub-tropical Horticulture, Lucknow, who have inspired me to bring this book on the crops which are often used but most of them are not cultivated. I am also thankful to my Co-authors who has really put a lot of efforts to compile the informaitons and various crops with their vast experience. Thanks are also due to Dr. S. D. Shikhamany, Director, IIHR, Bangalore, Dr. R. S. Singh, Ex-Director, CIAH, Bikaner, Dr. Bikash Das, Sr. Scientist (Fruit crops), HARP, Ranchi and Dr. V.K. Rao, Sr. Scientist (Argil. Chemistry) IIHR, Bangalore for their support and help in supplying of informations on various topics in the book. Sincere thanks are also due to Dr. H. S. Singh, Sr. Scientist (Entomology) and Sri. K. Gouri Sankra Rao, T-3 (Computer) for their valuable help in bringing out the manuscript in the present form.

Bhubaneswar
30th June, 2007

(Prof. Vishal Nath)

Content

1

INTRODUCTION

Land, a non-renewable and one of the limited natural resource is central to all primary production systems. For benefit of man kind, the land resources are being utilized in various ways. Focusing to agriculture as whole and crop production in particular, the most suitable arable lands have already being tapped to their full potential, baring the few examples where efforts are on to make potential use of available land resource. Besides, a sizable portion of total geographical area is under utilized or not utilized at all owing to various limiting factors. These areas mainly face different abiotic pressures which restrict their utilization for crop production. These areas are mainly comprised of the waste lands, upland plateau, hills and mountains and water bodies. However, the waste lands at greater extent and hillocks and plateau lands at limited scale are being used for production of some hardy perennial plant species including fruit species. Of these fruit species, some are either naturalized or adopted to these conditions owing to their tolerance capacity for these abiotic stresses. Be it the soil and environmental moisture stress under desertic condition or the stress of salinity and sodicity under tropical and subtropical conditions, these species have been found to yield satisfactorily. These fruit plants are virtually hardy in nature which not only can withstand the abiotic stresses but also produce sustainable yield of antioxident rich nutritive fruits. Some of the these species have been adapted to the cultivation level while others need to be recognized and adopted. The group of such fruit crops has been referred as "Fruits for the Future".

India has 329 million hectare total geographical area, which is 2.4 per cent of the world's geographical area. Over the years, the country's landmass has suffered from different degradation processes and forces of biotic and abiotic nature. These pressures have led to drastic changes in the proportion of land utilized for agricultural activities, urbanization and industrial development. At present, the country has about 173.64 million hectares degraded land of which more than 70 per cent is due to water and wind erosion (Table-1.1).

Table 1.1 : Causes of Land Degradation and their share in total degraded land in India

Causes of Degradation	Area (Million hectares)	Percentage of total area
Water erosion	107.12	61.7
Wind erosion	17.79	10.24
Ravines	3.97	2.28
Salt-affected	7.61	4.36
Water logging	8.52	4.90
Mines and quarry wastes	-	-
Degraded land due to shifting cultivation	4.91	2.82
Degraded forest lands	19.49	11.22
Special problems	2.73	1.57
Coastal sandy areas	1.46	0.84
TOTAL	173.64	100.00

Source : Status reports of degraded land (1985). Ministry of Agriculture, Government of India

Intensive agricultural practices that rely heavily on water, pesticide and synthetic fertilizers have caused water logging and salinity in many parts of the country leading to more area under degraded categories. The quest for increased total agricultural production, has also led to extensive cultivation on the marginal lands causing their degradation. As a result, around 69 per cent of our total geographical area has been converted to dry zones. According to land use statistics (2002), published by Department of Agriculture, Government of India, 13.9 million ha of the total waste land are infact culturable waste land. Reports from National Remote Sensing Agency (NRSA) also depict 63.85 million ha area under waste land in the country. Reports from forest survey of India (1999), Government of India, the country has 23 per cent of the total geographical area as forest cover on record which has recently reduced to only 19.39 per cent under actual forest cover. The remaining (3.61%) area is under degradation which may be the part of total waste land mentioned by NRSA. Apart from the waste land indicated by NRSA, other areas such as deserts, drought and flood prone areas and bushes / jungles in tribal areas have been subjected to severe terms of degradation. If we consider only the arid and semi arid areas of the country, it is spread over nearly 10 and 27.7 per cent area, respectively in the states of Rajsthan, Gujarat, Andhra Pradesh, Punjab, Haryana, Karnataka, Maharashtra, Uttar Pradesh, Madhra Pradesh and Tamil Nadu (Table 1.2).

Table 1.2 : Area under arid and semi arid region in India

State	Total area (km^2)		Per cent area	
	Arid zone	Semi arid zone	Arid zone	Semi arid zone
Rajasthan	1,96,150	1,25,000	57.42	36.67
Gujarat	62,180	93,103	33.72	47.50
Punjab	14,510	5,333	28.60	75.4
Haryana	12,840	26,274	29.32	59.77
Maharashtra	1,290	1,88,028	0.42	61.17
Andhra Pradesh	21,550	1,23,809	7.18	44.66
Karnataka	8,570	1,39,623	4.27	72.60
Utter Pradesh	–	63,939	–	21.73
Madhya Pradesh	–	59,048	–	13.38
Tamil Nadu	–	85,274	–	65.54

Source: 25 years of Arid Zone Research (1952-77), CAZRI Jodhpur.

These regions pose biophysical constraints for crop production such as poor soil conditions, meagre irrigation water resources, extremely high and low temperatures, hot and speedy dry winds, high solar radiation and very low and scanty rainfall resulting in conditions of low moisture index values which cause high potential evapo-transpiration (Table-1.3). While these conditions make the environment much fragile for crop production but help development of high quality in some fruits. Therefore, the success of fruit industry in these regions depends on efficient management practices to harness the available resources for increased harvest of high quality produce from these areas.

Table 1.3: Mean vapour pressure, relative humidity and rainfall in different ecosystems in India

Parameters	Month	Arid	Semiarid	Sub humid
Vapour pressure (mb)	July August	28.7 28.1	28.7 28.7	27.1 27.0
Relative humidity (%)	July August	65.0 70.0	69.0 75.0	79.0 82.0
Annual rainfall (mm)		310	841.0	1209.0

Source : Pareek, O.P. and Vishal Nath (2006). Arid and minor fruits. Text book of Horticulture (Ed. K.L.Chadha.), MPH, New Delhi.

Arid horticulture for example, encompasses the production of horticultural crops under arid environment with or without the use of irrigation water. Under rainfed conditions, production is dependent on the quantum and pattern of rainfall incidences. Thus, rainfall coupled with other environmental factors would govern the pattern of crop production. Under arid and semi-arid environment, rainfed horticultural production would obviously require special technologies mainly related to efficient water use. Of course, under arid environment efficient water use is important even for irrigated production. In arid region, rain spells are not only few (19-21) but confirmed to a short period from July to September with a mean annual rainfall from 100-450 mm in Rajasthan to 300-500 mm in Gujarat and 200-400 mm in Haryana and Punjab. In the peninsular region, the annual rainfall varies from 520 mm in Bellary (Karnataka) to 748 mm in Cuddapah (AP). The solar radiation during the winter and summer months is respectively, 414 and 612 cal./cm^2/day. The lowest and the highest temperature is -40 and 50^0C, respectively in cold arid and hot arid conditions in the country. The average wind speed in these regions are 12-16 km/hr and the moisture index values range from -60 to -91 (Table-1.4).

Table 1.4 : Moisture balance scenario in arid region

State	Rainfall (mm)	PE (mm)	Moisture Index
Rajasthan	100-450	2063-1503	-71 to -91
Gujarat	300-500	2144-1700	-70 to -82
Punjab and Haryana	200-400	1656-1362	-59 to -73
Peninsular India	520-750	1859-1738	-60 to -71

Source : Pareek, O.P. and Vishal Nath (2006). Arid and minor fruits. Text book of Horticulture (Ed. K.L.Chadha.), MPH, New Delhi.

The soils of the region are poor not only in fertility but also in water holding capacity. These conditions cause extreme degree of moisture stress. Similar is the situation in other ecological conditions where many of the unrecognized and under utilized fruits so called 'fruits of future importance' are being grown. While selecting fruit species for these regions, the following factors must be taken into consideration for their sustainable production.

1. Avoidance or adaptive mechanisms against drought, e.g. deep root system as in *ber, aonla,* Lasora, jamun, wood apple, bael, etc.
2. Tendency towards summer dormancy as in *ber,* mulberry, karonda and custard apple for tolerance to high temperature condition.
3. High water holding capacity in leaf cells as in fig and plant water conservation by morphological structures such as waxy coating on leaf surface and pubesence on lower leaf surface as in *ber,* sunken stomata as in *gonad* (lasora), modified stem and leaf as in cactus pear and ker.
4. Maximum vegetative growth and reproductive period of the plant coinciding the monsoon period as in *ber* and *aonla* or amenability of the plants for such coincidence by cultural manipulations, e.g.. *bahar* treatment in pomegranate and guava, pruning in ber, etc.
5. Fruit plants or types which can utilize conditions of high temperature and solar radiation for development of desired quality in fruits e.g. date palm and pilu.
6. Fruit plants or types which can make use of warm day and cool night conditions and sharp fluctuations in diurnal temperatures, e.g. development of colour and sweetness in pomegranate, date palm, rose apple, etc.

Based on these criteria, fruit crops have been identified and put under four categories. The first category includes fruit yielding plants like *ker, ber, gonda,* etc. which possess extremely drought hardy characteristics such as deep root system, scanty foliage, muscilagenous sap and leaf fall during summer. Under the second category, the fruit plants have their growth and flowering phase synchronous to the period of moisture abundance in soil and atmosphere, e.g. *karonda, aonla, guava, custard apple, jamun,* etc. Fruit crops such as pomegranate and sour lime, have been put under the third category which have continuous flowering with more than one flush in a year and can be made to synchronize their reproductive phase during the period of moisture abundance. The fourth category includes the fruits such as *phalsa*, mulberry, fig, date palm, sweet orange, etc. which develop excellent fruit quality under dry climate but will require assured irrigation and moisture conservation practices. In dates, *pilu* and sweet orange best quality is attained under the low rainfall arid environment.

In view of the growing population of the country and their needs for nutritional security as per the dietry standards, the expansion of area under fruit production is of utmost importance. Since, most of the fertile land is utilized for production of cereals, pulses and oil seeds in the country, the fruit crops do not find appropriate place in the existing production system especially in fertile arable lands. Also, on account of compulsion of producing more food for growing population, it is difficult to divert fertile land for the expansion of area under perennial fruits in general and to lesser known but other wise nutritive fruits in particular. On the other hand, a vaste area of the country is under waste land, which invites attention, for rehabilitation and can be put to efficient use by planting perennial fruit species. All the fruit species can not do better under these resource poor conditions however, some of the lesser known but hardy fruit species have been found most suitable for these land forms. Taking land utilization and nutrition security into account some of the identified lesser known fruits holds promise for the future.

These fruits species are on one hand drought hardy and widely adaptable owing to their phenological features and on the other hand highly nutritive and rich in anti-oxidant properties.

Under Utilized Fruits as Source of Antioxidant

Antioxidants are basically phytochemicals required in human diet, which neutralize harmful free radicals generated in the body. Fruits in general and underutilized minor fruits in particular are the important source of antioxidants in the human diet. These phytochemicals can be called as dietary antioxidants.

Free radicals are produced in the body by normal metabolic and immune processes and also generated from external toxins such as pollutants, cigarette smoke and ozones. High level of free radicals such as Reactive Oxygen Species (ROS) and Reactive Nitrogen Species (RNS) cause damage to cells and thus contribute to cellular disfunction and diseases. ROS and RNS also damage other body components that do not reside in cell such as blood and other fluids like sinovial and cerebrospinal fluids. Basic defense mechanism to limit the damage include enzymes such as glutathione peroxidase, which reduces oxygen radicals to harmless water and proteins which chelate transition metals that could otherwise react with free radicals. Because the protective mechanisms are not 100% efficient, dietary antioxidants are potentially important in reducing oxidative damage.

A dietary antioxidant is a substance in foods that significantly decreases the adverse effects of reactive oxygen species or reactive nitrogen species or both on normal physiological function in humans. Various compounds like catacholamines in the human body generate free radicals in their metabolism. In addition, activated phagocytes produce ROS as one of the defense mechanism against infection. Oxidative damage caused by increased levels of production of ROS and RNS may contribute to the development of many chronic diseases, including age related mascular degeneration, atherosclerosis, cancer, coronary heart disease, diabetes, neuro degenerative disease and rheumatoid arthritis.

Medical evidence suggests that many chronic diseases can be prevented or delayed by dietary intake of antioxidants which play a role in preventing cancer, cardio-vascular disease and many other degenerative disease associated with oxidative stress. There is also evidence that antioxidants such as β-carotene, ascorbic acid, Y-tocopherols (vitamin-E), flavoids and selenium (Se) are significantly associated with reduced cancer risks. Antioxidants such as vitamin-E, carotenoids and vitamin-C are able to neutralize oxygen free radicals, inhibit LDL oxidation and potentially reduce the risk of coronary artery disease.Thus diets rich in fruits are associated with protection from lings cancer, gastric cancer and colorectal cancer.

Fruits serve as major sources of natural antioxidants in human diet. Some of the important antioxidants nutrients of fruits are 1) Phenols, Flavonoids and Polyphenols 2) Carotenoids, 3) Vitamin-C, 4) Tocoferols, 5) Selenium.

Phenols, Flavonoids and Polyphenols

Aromatic components with hydroxyl groups are collectively referred to as phenolics or phenols. Phenolics include wide range of compounds such as catechol, cyanidine, caffeic acid, ferulic acid, chlorogenic acid, lignin etc. Large molecules with many sub units of phenolics are called polyphenols eg. tannins. Ellagic acid, which is particularly plentiful in pomegranates, prevents carcinogen oxidation of cellular membranes. Coumarine and chlorogenic acids prevents the formation of cancer causing nitrosamines and are found in a wide variety of minor fruits.

Carotenoids

Carotenoids are tetraterpenoid (C 40) compounds widely distributed in lesser known fruits like bael, phalsa, jamun and cactus pear. They are fat soluble yellow or red pigments with long conjugated chain. Carotenoids protect against photosensitized oxidation by quenching singlet oxygen. These carotenoids are associated with 1) protection against malignancies at multiple organ sites. 2) Immune response 3) Protection of skin against UV radiation. They also protect detoxification enzyme in the liver and are thus involved in elimination of pollutants and toxins from the body.

Vitamin-C

Vitamin-C is the cheapest, safest and best antioxidant found in under utilized fruits like aonla, barbados cherry, phalsa, mulberry and wood apple. It is a water soluble antioxidant (reducing agents) which is oxidized to form dehydro-ascorbic acid. Ascorbate is a most powerful anti oxidant in plasma. A significant inverse correlation was found between plasma as corbonate and coronary disease mortality. The effect of vitamin-C is inhanced in the presence of Vitamin-E and Selenium. The other roles of Vitamin C are in collagent formation, absorption of inorganic iron and metabolism of folic acid and some other amino acids.

Tocoferol

It is the major lipid soluble antioxidant responsible for protecting the polyunsaturated fatty acid in membranes against lipid peroxidation by free radicals and singlet oxygen species. It enhances the immune system and helps to reduce short-term memory loss. Alpha lipase and co-enzymes Q10 help regenerate. Mango, papaya, pomegranate, *Inga* and guava fruits are relatively rich in tocoferols.

Selenium

It apparently plays a role in T3 (thyroid) production. Helps prevents various cancers including prostrate, colorectal and lung. Deficiency is associated with low levels of glutathione. Sea food products are rich source of Selenium. Antioxident properties of some of the under utilized and lesser known (fruits for the future) fruits have been enumerated.

Aonla

The ascorbic acid and other constituents are well retained in dried aonla fruits. The stability of ascorbic acid in dried aonla fruits and their astringency could be due to presence of polyphenols in the fruits. Gallic acid present in the fruit has marked antioxidant activity. Use of dry aonla is an ancient practice.

Custard apple

The fruit of custard apple are sweet and non-acedic. Carbohydrate are present to the extant of 20-23%,which are mainly sugar. It is very rich in crude fibre (0.9-6.6%), Vitamin C (37mg), calcium (17-27 mg) and phosphorus (14-32 mg/100g). However, it lacks vitamin-A. The ripe fruits is rich in tannin, which is dried, pulverized and employed against diarrhoea and dysentery.

Jamun

The jamun fruit are small with decp purple to bluish black colour. They are rich in antioxidant compounds like anthocyanins and phenolics.The fruits are indigenous to India and are good source of sugars (15.6%) and Vitamin C (14.3 mg/100g). They act as effective medicine against diabetes, heart and liver disease. The seed powder from unripe reduces sugar very quickly.

Phalsa

The fruits are red or purple in colour when ripe. These fruits are acidic in taste with high amount of vitamin-A (419 µg), vitamin-C (22mg) and calcium (129mg), phosphorus (39 mg), Mg (72mg), iron (3.1mg) and K (351 mg/100g). The fruits are some what astringent. Fruits are eaten fresh when ripe and used for tasty deserts. The fruit help to cure inflammation heart and blood disorder, fever and constipation.

Fig

Dried fig are high energy yielding food (269 calories per 100 g edible portion). Important nutrient in dry fig (per 100 g edible portion) are protein-4g, carbohydrate (6.9g), fat (1g), Ca (200mg), iron (4mg) and vitamin-A(100 IU). Fig is valued for its laxative properties and is used in the treatment of skin infection.

Ber

Ber fruits are richer than apple in protein, phosphorous, calcium, carotene and vitamin C. The fruits are employed in pulmonary ailments and fevers. They are mixed with salt and pepper and given to treat indigenous and biliousness. The dried ripe fruit is a mild laxative. The seeds are sedative and are taken sometimes with buttermilk to halt nausea, vomiting and abdominal pains in pregnancy and check diarrhoea.

These fruits are now considered as fruits for future owing to their hardy nature, wider adaptability, good productivity under degraded and marginal lands and high nutritive value and antioxident properties and therefore, efforts are being made to develop varieties, agro-techniques, plant protection measures for these crops by the National and International research institutions. Nevertheless, it is very much clear that these crops need special attention particularly with respect to ensured orchard establishment, risk minimization, resource conservation and utilization, plant protection and post harvest management for sustainable production. Some of the important aspects for making efficient use of waste lands through plantation of hardy, under utilized but nutritious fruit species in holistic manner are as follow:

Ensured Orchard Establishment

Raising of new orchard in fragile region is comparatively difficult. Nursery plants raised either in polythene bags or in nursery beds generally give poor survival in the field due to coiling or damage of roots. When raised in poly-tubes in the nursery or by *in situ* sowing in the field, the plants develop straight and well distributed root system. In the nursery, seeds of desired rootstocks are sown in polythene tubes (open at both the ends) of 300 guage (25 cm length and 10 cm diameter) filled with a mixture of sand, clay and FYM in equal proportion. The size of tube may vary depending upon the species, e.g. for *aonla*, it should be 35 x 15 cm. On the tube-raised rootstocks of desired thickness, budding or grafting operation should be done at appropriate time. The tubes train the roots to grow straight downwards which helps in their successful establishment in the field. Under extreme agro climate, high moisture loss from newly planted saplings often leads to a very low establishment in the field. Under such conditions, *in situ* budding/ grafting should be preferred. This would ensure deeper, well distributed and extensive root system. The technique is useful in fruits like *ber, aonla, bael* and *custard apple* in the arid and semi arid parts of the country.

Risk Minimization through Alternative Land Use

Even after successful orchard establishment, the productivity of fruit plants under such agro-climatic conditions are often poor. This calls for creating microclimate conducive for good productivity by plantation of shelter belts around the orchard especially on the western and northern sides. For this, 2-3 rows of fast growing tree species are planted in staggered manner. For temporary shelter, fast growing short duration bushes such as castor, jatropha, etc. can be planted around individual blocks. Damage by wild animals is another serious problem in these regions because their natural habitat exist near by. Planting of thorny bushes (*wild ber species, karonda, jharber, palmyra palm* etc.) between the shelter belt tree rows would not only protect the orchard from wild animals but also provide additional income.

Under the great uncertainty in production owing to high environmental insecurity mono-crop culture in such region is highly risk prone. Crop failures are common sometimes leading to great economic losses. Therefore, multi-tree-crop combinations are advisable in these region. During young age of the orchard, short duration fruits crops can be grown as filler crop or legumes / oil seeds / grasses / coase cereals / medicinal and aromatic plants can be raised as inter crop. In semi-arid region at Godhara (Gujarat), for example; horti-silvi system incorporating *soobabool* and *Eucalyptus* and at Faizabad growing of *aonla* and *guava* between *ber* trees, in arid region at Jodhpur, horti-pastoral system incorporating buffel grass and at Bikaner, horti-horti system using vegetables have been found remunerative. While selecting plant species, its suitability in the region with respect to vegetative growth period, flowering and fruiting season, market demand must be taken into consideration, e.g. at Faizabad, *aonla* has been found better in *ber* based system, because fruiting in both the crops coincides which saves the cost of watch and ward.

Resource Conservation for Enhanced Production

To reduce the water losses through deep percolation and evapo-transpiration in such regions, some special techniques have been found effective. For examples, provision of 2-3 cm thick clay and bentonite layers at the bottom and sides of the pit during its preparation allows the penetration of roots and minimizes the percolation and seepage losses of water in sandy soil. In dune lands, to reduce the water loss through percolation and seepage, a special pot comprising of two walls joining with each other at bottom leaving a hollow inner cylinder like structure (*Jal Tripti*) proved useful. The sapling is planted in inner space of the double walled pot and water is applied in the space between two walls. The pot allows roots to penetrate deeper in soil but controls rapid percolation of water. Painting of outer wall of the pot with enamel paint reduces the seepage loss.

Evaporation loss of water from soil can be effectively reduced by mulching. For this, black polythene has been found much effective. Locally available weeds, crop residues and other organic wastes can also be used as mulch material. *Bui* (*Arvea* sp.) in date palm at Bikaner, *sarkanda* (*Saccharum* sp.) in *ber* at Bawal, paddy straw in *aonla* at Faizabad and paddy husk and groundnut shells in pomegranate at Anantapur have been found effective. The organic mulches, besides conserving moisture, also add organic matter to the soil. Owing to intense radiation and high wind velocity in the region, considerable amount of water is lost from the leaf surface. Spray of radiation reflectant (Kaolin, phenyle mercuric acetate, power oil, waxol, etc.) reduces the transpiration losses by lowering the leaf temperature. Spray of 1.5 per cent power oil, 7.5 per cent Kaolin and 10^{-4} M phenyl mercuric acetate (PMA) increased fruit yield in ber and 10 per cent Kaolin and 10^{-5} M PMA proved beneficial in improving pomegranate production under semi-arid conditions.

In the absence of irrigation facility, the production from these fruit trees depends upon the extent and pattern of rainfall which is usually scanty and erratic. Besides, loss of rainwater occurs in the light textured soil having various gradients through evaporation and surface runoff and as a result of high solar radiation through evapo-transpiration. Rain water can be concentrated around active root zone of plants by provision of slopy catchments area on two sides or all around the plant. The catchments area would vary according to the water requirement of the fruit crop. While determining the size of micro-catchments for a perennial fruit plant factors such as natural rainfall of the region, water requirement of the fruit crop, characters of soil and soil surface, nature of root growth and spread, etc. are kept in view. In any case, the water requirement of the species for its optimum production must be met by rainfall and the runoff from the micro-catchments. The root characteristic (depth and spread) of the species will determine the volume of soil profile where water should be stored by rainfall runoff. Thus, runoff from the micro-catchments at one point of time should not exceed the moisture storage capacity of this soil profile.

Weeds compete with the orchard trees for water and nutrients. This causes extreme stress to the plants in arid and semi-arid region where water is scares commodity and soil fertility is poor. For effective control of monocot and dicot weeds, although application of chemicals (weedicides) are effective, mechanical weeding and hoeing helps to break the hard crust and provide aeration to feeder roots. Post-emergent application of 3.5 kg diuron per hectare followed 2.75 lit./ha glyphosate at 20 days interval provides effective control of monocot and dicot weeds in date palm orchard at Abohar (Punjab). Mulching also suppresses the weed population near tree base. Effective weed management conserves the available natural resources like water and nutrient in tree basins and enhances the fruit production.

To enable full utilization of the rich radiation resources in the region, optimization of canopy architecture of the tree is important. It is better to train the plants to a lower stature to avoid damage by high wind velocity and to reduce cost of orchard operation. In bushy crops like pomegranate where multi stem training is generally practiced, not more than 4 stems should be allowed from ground level to reduce possibility of the pest and disease infestation on these stems. In tall growing crops such as aonla, jamun, bael, wood apple, palmyra palm which also require wider spacing, provisions of filler and inter crops should be essentially made to harness the solar radiation at different strata vertically. Under plateau region of Ranchi, a successful multi-storied model with aonla crop has been developed and demonstrated. The model gives 2-3 times higher income than sole crops of aonla besides other environmental benefits.

Resource Optimization through Canopy Management

Pruning allows maintaining optimum canopy which help increased available resource utilization on one hand and improved production on the other hand. In some fruit trees annual pruning is necessary to maintain a balance between vegetative growth and fruit production. Fruits like ber, guava and phalsa bear on current season's growth. Therefore annual pruning would induce more fruiting area in such crops. Pruning is done preferably during dormant period. In fruit trees like aonla and gonda extension growth takes place during spring and monsoon and thus do not require regular pruning. During the initial stage of plant establishment, light pruning helps to develop better framework. The ideal time of pruning depends upon the prevalent climatic conditions at the location , e.g., in ber, pruning is done during January-March at Aruppukottai in Tamil Nadu, during April at Rahuri in Maharashtra and during May at Bawal in Haryana. The severity of pruning also varies at different locations. In drier regions lighter pruning is done than in the moderate and humid areas e.g., pruning is done at 25 buds at Jobner (Rajasthan), at 20 buds at Bawal (Haryana), at 15 buds at Rahuri (Maharashtra) and Sardarkrushinagar (Gujarat), by removal of 50 per cent past season's wood at Faizabad (U.P.) and at 20 buds coupled with pre-pruning spray of 3 per cent thiourea or potassium nitrate at Bangalore (Karnataka). At Bangalore, pruning of 50 per cent shoots to 20 buds and the remaining to 1-2 buds helped in avoiding the development of long unfruitful shoots in ber.

In phalsa, also timing and severity of pruning depends upon agroclimatic conditions and on the age of the bushes. Established phalsa bushes should be pruned at 150cm height once a year during January in north India and twice in a year (December and June) in south India. Pruning from ground level is done either to rejuvenate old bushes or to train young plants into bush form.

Nutrient Management for Optimum Yield

Nutritional requirement of a plant varies with soil characteristics, rainfall and age of the plant. In a well grownup plant of ber, application of 750g N per tree has been recommended in sandy soils at Jobner whereas 250g N per tree has been found enough in loamy soils at Rahuri. In pomegranate, application of 720g N, 230g P_2O_5 and 220g K_2O per tree under semi-arid conditions at Udaipur (Rajasthan), 625g N, 250g P_2O_5 and 250g K_2O at Rahuri (Maharashtra) have been recommended for optimum production. In fig, application of 900g N, 250g P_2O_5 and 250g K_2O per tree per year to Poona Fig improved productivity whereas in phalsa, application of 40g N and 10g K_2O per bush gave the highest fruit yield. Judicious use of nutrients on the basis of tissue analysis (leaf sampling) should be done at an appropriate time. Micronutrient deficiencies are common in the orchards in these regions owing to high soil erosion and nutrient leaching and therefore correction of deficiencies will determine their productivity. Deficiency of zinc in pomegranate orchards can be corrected by two foliar sprays of 0.06 per cent $ZnSo_4$, once at flowering and another during fruiting. Necrosis in aonla can be prevented by the spray of micronutrient mixture containing 0.4 per cent $ZnSO_4$, 0.4 per cent boric acid and 0.2 per cent $CuSO_4$. Foliar sprays of 0.2 per cent $ZnSO_4$ and 2 per cent urea to bael and phalsa controls the problem of yellowing and intervenial leaf chlorosis. In highly sodic soils, filling of pits with 25 kg FYM along with 120 kg pond soil per pit improves the soil condition around root zone and thereby growth and productivity of fruit trees. In gravelly lands pits should be filled with pond soil mixed with locally available cakes for better establishment, growth and production from these fruits of future importance.

Use of Plant Growth Regulators for Beneficial Manipulation

Some of these fruit species bear flowers and fruits throughout the year. To harvest an optimum fruit yield of good quality, fruiting from only a particular flush should to be taken, which would depend upon the agroclimatic condition e.g. in pomegranate *mrig bahar* (June -July flowering) in dry regions of Rajasthan and *ambia* and *hasth bahar* in moderate climate of Maharashtra and Karnataka are preferred. The selection of crop is regulated by cultural, mechanical or chemical treatments. Ploughing of orchards exposing plant roots and spray of 200 ppm NAA before the unwanted *bahar* induce dropping of flowers and small fruits in pomegranate. Then crop load per tree can be regulated depending upon agroclimatic condition and cop management practices. A load of 50 fruits per tree in pomegranate gave the maximum number of medium size fruits and the highest gross income per tree at Rahuri in Maharashtra. In date palm, removal of one-third strands from the centre of the bunch at the time of pollination increased the size and weight of berries without reducing yield per palm. Application of 1000 ppm ethephon at colour turning stage induces early ripening by about one week in Halawy date berries. Under moderate climatic conditions at Bangalore, early induction of new growth in custard apple by inducing early defoliation either manually or by two sprays with 0.5 per cent potassium iodide, at one week interval, increased fruit yield. Application of 750 ppm ethephon at colour turning stage induced early ripening in ber and reduced number and cost of picking.

Plant Protection for Managing Enemies

Based on the occurrence and epidemiology of diseases in these under studied fruit species, chemical and cultural control measures needs to be standardized and tested against the major diseases and pests of such fruits. Efforts are being made to develop plant protection schedule for these crops of future importance. For the control of ber powdery mildew, fungicides such as 0.1% dinocap or carbendazim or triadeomorph or thiophenate methyl and 0.2% wettable sulphur have been found to be the most effective when they are sprayed 2 to 4 times at 15-20 days interval starting from the initiation of the disease. One spray of the fungicide should be done at the start of new growth after pruning as a prophylactic treatment. Black leaf spot (*Isariopsis indica*) which is common under more humid conditions, can be controlled by 2-3 sprays of 0.2% captafol or copper oxychloride or mancozeb at 15 days interval. For the control of leaf and fruit spot in pomegranate, four sprays with 0.3% ziram or mancozeb and 1% Bordeaux mixture at 15 days interval are most effective. Since, the intensity of the disease is more under humid conditions more sprays at shorter interval should be done to manage the disease. For the control of ring rust in aonla, 4 sprays of 0.2 per cent chlorothalonil at 15 days interval after initiation of disease symptoms give the best control. Leaf spot in date palm can be kept under check by 4 sprays of 0.4 per cent copper oxychloride at 15 days interval.

Fruitfly (*Carpomyia vesuviana*) causes serious damage to ber fruits. To control the infestation, a chemical spray schedule consisting of first spray at pea stage with 0.3% monocrotophos, second spray after 15 days with 0.03% fenthion and third spray after another 15 days with 0.1% carbaryl should be followed. The same schedule can also be used effectively against fruit borer (*Meridarchis scyrodes*) which causes serious damage in southern and western India. Collection and destruction of the fallen fruits and digging the soil under the tree canopy also decreases its infestation. Pomegranate butterfly (*Virachola isocrates*) causes considerable damage to pomegranate fruits. Bagging of fruits with butter paper gives good protection. For control, sprays of 0.002% deltamethrin and 0.2% carbaryl in rotation of 21 days starting form fruit set are the most cost effective. Scale insect (*Parlatoria blanchardi*) causes damage to date palm against which *Chilocorus nigritus* can be used as an effective predator. Midge and mites have also been found to cause damage to fig trees particularly in humid locations against which 0.05% monocrotophos provides the most effective control. Birds cause considerable damage to ber, pomegranate, jamun and date palm. Bird damage in date palm can be prevented by covering the bunches either with 600 guage low density polythene bags or with reusable wire gauge covers whereas in pomegranate nylon netting of individual fruits should be done. Since most of the fruits are under utilized and lesser known and are grown in their natural abode, efforts should be made to manage their disease and pest problems through natural enemies and organic means to save them from pesticide residue.

Post harvest Management and Value Addition

Owing to harsh climatic conditions and lack of transport network in these regions, the highly perishable fruits remain in farmer's field for longer duration which some times leads to a great loss. Unorganized marketing sector in the region aggravate the problem and the producer often faces great difficulty in disposing off their produce. Very little efforts have been made towards the post harvest management of such fruits. Fruits such as custard apple, ber, aonla, date palm, etc. do not have long shelf life at ambient temperature. To store them for longer time, only low temperature storage is effective. Post harvest dip of fruits with non-hazardous chemicals can also be done, e.g. 0.1 per cent Bavistin for ber and 4 per cent borax for aonla. Standard grades for pomegranate, aonla and ber are available for export while processing techniques for date palm (*chhuhara, pind*), aonla (preserve, jam, pickle), bael (squash, nectar), tamarind (paste), etc. are now available. Other minor fruits like woodapple, lasora, karonda, phalsa, ker, etc. also have immense potential to be used in processing industry (Table-1.5).

The group of fruit crops attempted in this book have potential in Indian context. The information on these crops are rather scattered and fragmented on the vital aspects of orchard establishment and management and fruit production and utilization. The updated information on plant genetic resources, suitable cultivar, improved production, propagation and protection techniques and post harvest management and value addition methods in some of the well versed minor fruits which have potential in resource poor arid, semi arid and waste land areas has been compiled in this book.

Table 1.5 : List of possible Processed products from the targeted fruits

Fruits	Processed products
Ber	Jam, preserve, candy, wine, dehydrated product, canned product
Pomegranate	Juice, sharbet, squash, dehydrated product, chutney, pickle
Aonla	Jam, preserve, candy, dehydrated product, chutney, pickle
Date palm	Jam, syrup, chhuhara, pind khajoor
Custard apple	Ice cream, mixed jam, toffee, RTS
Bael	Preserve, syrup, sharbet, squash, dehydrated product, candy, cider
Fig	Wine, candy, dehydrated product
Karonda	Jam, jelly, preserve, candy, wine, syrup, dehydrated product, chutney, squash, pickle
Wood apple	Jam, jelly, syrup
Gonda	Pickle
Khirni	Beverage, dehydrated fruits
Khejri	Dehydrated immature green pod, cookies from ripe pods
Jamun	Jam, jelly, syrup, squash, sauce, vinegar
Palmyra palm	Wine, candy (endosperm), canned product (endosperm)
Ker	Dehydrated product, pickle
Phalsa	Juice, syrup, squash, dehydrated product
Pilu	Juice, syrup
Cactus pear	Jam, squash, syrup, canned product, wine
Tamarind	Jam, juice, candy, syrup, sharbat, glazed fruit, sauce, frozen puree
Rose apple	Beverage, canned product
Mulberry	Beverage, dehydrated product

The available literatures, facts and figures of these "fruits for the future" alongwith working experiences with these crops has been tried to put together for better explanation and under standing of the content of the book. The information furnished in the book will certainly be helpful in acquiring and improving knowledge and understanding on these crops among the researches, teachers, students, extension functioning and the farming community to utilize the available wasteland of the country.

—x—

2

BER

Ber (*Ziziphus mauritiana* Lam.) is cultivated all over the drier parts of the Indian subcontinent for its fresh fruits, which are rich in vitamins (C, A and B-complex) and minerals. It can be successfully cultivated even in the most marginal ecosystems of the subtropics and tropics. Since its cultivation requires comparatively less care, the fruit tree is suitable to rehabilitate resource poor areas at extensive. It also exists in wild groves which are widespread in the warmer parts of India, Pakistan, Bangladesh, Sri Lanka, and Central to Southern Africa and in the northern parts of Australia. In addition to fruits, the wild trees yield timber of marginal value, brushwood fuel wood and leaf fodder. The tree regenerates freely and greatly resists recurrent drought. It is thus an important tree suitable for integration into agro forestry systems of warm desert eco-regions. The tree can help in economic sustenance and insurance against ecological degradation.

The composition of ber fruits has been quite widely reported and much of the information is cultivar specific. Pulp content of fresh nature fruits varies from 81 to 97% (Pareek, 1983) and from 91.6 to 92.9% (Jawanda *et al.,* 1980).

The pulp is very rich in nutritive substances (Randhawa and Biswas, 1966; Singh *et al.,* 1967; Chadha *et al.,* 1972; Singh *et al.,* 1973 a; Dhingra *et al.,* 1973; Teaotia *et al.,* 1974; Chundawat and Srivastava, 1978; Jawanda *et al.,* 1980 a, b, 1981; Bal *et al.,* 1978, 1979, 1988 b; Bal, 1981 a; Jawanda and Bal, 1978). Pareek (1983) summarized the nutritional constituents of the fruit (Table 2.1). There are however, no definitive values for composition. Ber is richer than apple in protein, phosphorus, calcium, carotene and vitamin C (Bakhshi and Singh, 1974) and excels oranges in phosphorus, iron, vitamin C, calorific value and carbohydrates. Ripe fruits provide 20.9 K calories per 100g pulp (Singh *et al.*, 1973 a).

Table 2.1 : Nutritional constituents in fruit pulp of 4 ber cultivars

Constituents	Gola	Kaithli	Banarsi Karaka	Umran
Moisture (%)	81	-	81	-
Starch (%)	0.95	-	0.86	-
TSS (^{0}Brix)	17-20	16-18	13-17	18-20
Total sugars (%)	8.3-12.1	4.9-10	5.4-12.4	7.2-7.4
Reducing sugars (%)	3.3-5.8	1.95-2.7	3.3-3.7	2.6-2.9
Non-reducing sugars (%)	2.4-8.4	2.2-8.0	3.3-8.4	4.8-4.9
Acidity (%)	0.37-0.75	0.16-0.51	0.13-0.48	0.19-0.35
Protein (%)	-	1.18	-	1.03
Total ash (%)	-	0.45	-	0.34
Ca O (%)	-	0.04	-	0.03
P_2O_5 (%)	-	0.02	-	0.01
Fe_2O_3 (mg/100g)	-	0.5	-	1.00
Ascorbic acid (mg/100g)	70	89-133	66-110	73-103

Source : Pareek, 1983

Ber pulp contains 13.6% carbohydrates out of which 5.6% is sucrose, 1.5% glucose, 2.1% fructose and 1.0% starch (Jawanda *et al.*, 1981). Fruits of cultivars Umran and Gola had 8.1 and 8.05 % total sugars (Singh *et al.*, 1983 a). Ber pulp is a rich source of vitamin C also. Jawanda and Bal (1978) reported that ascorbic acid content in different ber cultivars ranged from 70-165 mg/100g of pulp. Bal *et al.*, (1978) reported that β-carotene content in cultivar Umran and Sanaur-2 was 81.2 μg and 76.8 μg per 100g respectively. The fruit contains about 70 IU vitamin A in 100g pulp. Chinese jujube (Z. *jujube*) fruits have been reported to contain 20 to 28% sugars, 0.3 to 2.5% acids, 2.9% protein, 500-600 mg/100g vitamin C, 3.5% vitamin P (Tasmatov, 1963) and minerals such as Fe, P, Ca, etc.(Ming and Sun, 1986). Chinese jujube fruits are very rich in vitamin C and vitamin B-complex (Kuliev and

Guseinova, 1974). In some fruits, Baratov *et al.*, (1974) reported 188-544 mg vitamin C per 100g pulp. Dried fruit of jujube contains several volatile substances, which have significance in imparting the typical flavour to the fruit. As many as seventy eight such compounds were indentified in Z. *jujube* var. *inermis* Rehd. among which aliphatic acids and arbonyl compounds accounted for 62.97% and 29.56% of the total volatiles, respectively (Wong *et al.*, 1996). The major components were decanoic acid (19.98%) and dodecanoic acid (15.64%). Succinic and malic acids have also been identified (Ahmedov and Halmatov, 1969).

Ziziphus-pectin A has been isolated from the fruits of Z. *jujube* var. *inermis*, which contains 2,3,6-tri-o-acetyl D-galactose units in the side chain (Tomoda *et al.*, 1985). Thus it binds bile acid, lowers plasma cholesterol and has antidiarrhoeal properties. Mature jujube fruits also contain polyphenols and rutin (Kriventsov and Karakhanova, 1970). Isoquinoline alkaloids isolated from the fruits of Chinese jujube, have been identified as steppharin, N-nor-nuciferine and asimilokine (Khokhar, 1978). A pyrrolidine alkaloid obtained from the dried fruits of Chinese jujube showed sedative activity in laboratory trials with mice (Han *et al.*, 1987). Cyclic adenosine monophosphate (Cyong and Hanabusa, 1980) and guanosine 3′, 5′-monophosphate (AMP) (Cyong and Takahashi, 1982) have been isolated from the fruits of Chinese jujube, which contained 100 to 600 n moles /g AMP in dried fruit and from 100 to 150 n moles /g in fresh fruit. Both the percarp and the seeds of Chinese jujube cultivar Ta-yan-tszao contains the phospholipids of the two main classes, viz. phosphatidylcholines and phosphatidylglyerols among which oleic acid was present in the fatty oil of the seed while the neutral lipid of the pericarp was palmitoleic (Goncharova *et. al.*, 1990)

It is known that ber has been used in India for almost 4000 years (Mehra, 1967). At present, nearly 90% of its production is consumed as fresh fruit. Although there seems to be good potential for use of the fruit in the processing industry and of different parts of the tree in pharamacology, these have still not been fully exploited.

Fruits of ber are commonly used in Indian households as fresh fruit and dehydrated for later use. However, the fruits of wild seedling trees are mostly used for sub drying. In Zimbabwe, fruit is consumed fresh from June to September and after dehydration throughout the year (Funkhouser, 1998). Its powder is used for baking and to prepare jam (Maposa and Chisuro, 1998). Cakes from the dried and fermented pulp are made in western Sudan (Dalziel, 1937) and in Zmbia (Kalikiti, 1998). The Toureq nomads prepare a flat rounded loaf of bread from dry fruit pulp (Chevalier, 1947). The mature green fruits are also occasionally used in India to prepare chutney, pickle and jelly. Powder is made from the ripe fruits. Chinese jujube fruits are eaten either fresh or after dehydration. In Azerbaijan, jujube is grown for fresh fruit consumption, dry fruit and canning (Tagiev, 1992). In USA, various confectionery recepies such as Graham bread, cake, cake filling, butter pudding and sweet pickle have been recommended from dried jujube fruits (Lanham, 1926). *Z. nummularia* fruits are dried and powdered with spices in western India to prepare a product called "churan" (Pareek, 1983).

A number of products such as *murabba* (preserve), candy and dehydrated *ber* are prepared from the fruits. Pulp made from ripe fruits is used to prepare products such as ready-to-serve beverage, squash, etc.

Fully ripe fruis are dried to prepare a dehydrated product similar to dry dates. The fruits of wild and cultivated ber are dried in the sun. The dried fruit is relished as a dessert (Cheema *et al.*, 1954). Improved methods to prepare good quality dehydrated products have been developed both in India (Khurdiya and Singh, 1975;) and China (Kim *et al.*, 1981; Yang and Niu, 1992). A dehydrated product prepared from Chinese jujube fruits is exported from China under the name "Chinese dates". Very good quality preserves and candy can be prepared from hard mature ber fruits for which techniques have been standardized in India (Khurdiya and Singh, 1975; Pareek, 1983; Sanjeev Kumar *et al.*, 1992). Pulp obtained from fully ripe ber fruits can be used for the preparation of ready-to-serve beverage and squash (Khurdiya and Singh, 1975). Ready-to-serve beverage prepared on a small scale is already popular in some locations in India. Other products such as jam, nectar and fruit leather can be prepared from ber pulp.

Different plant parts of *Ziziphus* species have medicinal value. Use of ber, Z. *nummularia*, Z. *rugosa* and Z. *jujube* in *Ayurvedic* and *Yunani* (Bhandari, 1969; Kirtikar and Basu, 1975), Z. *jujube* in Chinese (Tanaka and Sanada, 1991) and Z. *spina-christi* in African, Egyptian and Saudi Arabian (A1-Akeely, 1985; Depommier, 1990; Abdul-Galil and E1-Jissary, 1991) systems of medicine have been reported. Such utilization, however, seems to be sporadic and not in common practice. Detailed pharamacological investigations are required on the medicinal value of *Ziziphus* species.

Ber leaf is a valuable source of fodder (Dastur, 1952 and Dinesh Kumar, 2001) as it is very nutritious. It is a popular fodder tree in arid regions since it grows and regenerates very quickly even under stress conditions. One Z. *nummularia* bush easily provides 2.5 to 3.0 kg air-dried leaves per year (Bohra and Ghosh, 1981). Even under extremely arid conditions (100-125 mm annual rainfall), a ber tree yields 3 to 5 kg green leaf fodder (Vahishtha, 1997 b) which is palatabile to camel, sheep and goat. A six year old tree of ber produces a biomass of 11.6 kg in the arid northwest India (Toky and Bisht, 1993) which is used as fuel wood. Reddy (1988) reported that, depending upon pruning severity, the weight of air-dry pruned wood vary from 4 to 6 tons per hectare from 8 year old ber trees at Bangalore whereas Bajwa *et al.*, (1986) mentioned that the weight of pruned wood per tree ranged from 19.5 to 37.4 kg from a 13 year old tree. Ber trees yield 8-10 kg air dried fuelwood frôm annual pruning in north-west India (Vashishtha, 1997 b).

Ber trees are considered amongst the best for rearing lac insects (Hussian and Khan, 1962; Anon., 1996). A lac yield of 1.5 kg per tree per year has been obtained by collection during October-November at Ranchi (Anon., 1996). Ber is a chief host plant of *Kerria lacca* and *K. sindica* (Li and Hu, 1994). An yield of 3-6 kg raw lac can be obtained in 3 years by allowing 6-8 shoots of 2-3 m length and 2-3 cm thickness on a stump for inoculation of lac insects. It is notable that if the plant is used for lac production, fruit production is not viable. Besides, the ber plant has other uses like fencing of agricultural fields, improvement of degraded lands by plantations, minor timber, preparation of agricultural implements and rearing silk worm.

Origin and Taxonomy

Ber (*Ziziphus mauritiana* Lamk) is native of India. Liu and Cheng (1995), based on their investigations, believe that the Indo-Malaysia region, (South and Southeast Asia) is the centre of both evolution and distribution of the genus *Ziziphus*. De Candolle (1886) guided by its abundance in a wild state, states that the original abode of the ber (*Z. mauritiana* Lam.) seems to be Myanmar (Burma) and India. This assumption is based on the specimens gathered by Wallich from Burma and its mention by Kurz in the Flora of Burma and Beddone, in Forest flora of India, stating it to be present there in a wild state. Chinese ber (Z. *jujube* Mill) is originated in Northern and Western China (Vavilov, 1951).

Ber or Indian jujube belongs to the genus *Ziziphus* of the family Rhamnaceae and Order Rhamnales. The Order Rhamnales is characterized by having the stamens opposite to the petals. The stamens are in one whorl in a bisexual, actionomorphic flower, which often has a tendency to unisexuality and perigyny or epigyny. Calyx lobes are very small and the calyx is entire. An intrastaminal disk is often well developed. Rhamnaceae is a large plant family of spiny trees, shrubs or climbers. Leaves are stipulate, simple alternate or rarely opposite. Flowers are small regular, bisexual, rarely polygamous-dioecious or unisexual by abortion of one of the essential whorls in cymose or paniculate inflorescences. An intrastaminal disk is often present. Sepals 4-5, connate in tubular 4-5 lobed calyx and the lobes are valvate, usually ridged internally. Petals small, 4-5 or rarely absent, inserted on the calyx tube, free, usually clawed and hooded. The family is characterized by the 4-5 stamens, which are opposite to another often embraced by the petals; anthers 2-locular, opening lengthwise; disk mostly present, perigynous, sometimes liming the calyx tube. The carpels form a 3- celled ovary. Ovary is sessile, free or sunk in the disk; 2-4 locular, style shortly lobed; ovules solitary in each chamber, rarely paired, erect from the base, anatropous with two integuments. Fruits are of various types but often drupes. Seeds mostly have copious endosperm and a large straight embryo. According to Bailey (1947), the family has about 50 genera and more than 600 species. Rendle (1971) stated that the family contains about 45 genera and 500 species and the most widespread genera are *Rahmnus, Ziziphus* and *Gouania*.

Sussenguth (1953) divided the family Rhamnaceae into the five tribes viz. Rhamneae (Fruit berry or schizocarp; testa thick), Zizipheae (Fruit drupe; testa papery, thin), Ventilagineae (Fruit winged, indehiscent; or capsule), Colletieae (Fruit capsule or schizocarp; testa thick) and Gouanieae (Fruit with wing-like expansions of carpellary portions; testa massive). The genera *Zizphus* belongs to tribe Zizipheae.

The generic name *Ziziphus* has been derived from 'Zizouf', the Arabic name of *Z. lotus* (Bailey, 1947). The genus contains erect, small trees or shrubs having spitules modified into straight or curved spines. Spines solitary or in pairs, usually one straight, the other curved. The leaves are alternate, or sub-opposite (Plate 2.1) with stipules extrapetiolar (Rendle, 1971). The leaves are coriaceous, strongly nerved (3-5) from the base, branchlets have stipular spines. The flowers are cymose. Calyx, triangular acute lobes, valvate, keeled inside, tube obconical. Petals 5, small, cucullate, deflexed, or absent. Stamens 5, slightly adnate to the base of petals, usually longer than petals. Ovary sunk in flat or pitted disk; disk 5 or 10 lobed with a free margin, and confluent at base, styles 2-4, distinct, more or less connate, stigma small, papillose. Fruit subglobose, drupaceous and 1 to 4 seeded; embryo large, endosperm present or absent (Trimen, 1893; Ridley, 1922). Hooker (1875) recorded 18 species and 5 indeterminate species in the genus *Ziziphus.* Watt (1893) mentioned 50 species, Rdiley (1922) 60 species and Rendle (1959) 40 species. According to Evreinoff (1964), the genus comprises 80 species. Johnston (1972) reported that the genus has about 86 species in the temperate and subtropical parts of the world. Bhansali (1975) stated that the genus contains as many as 135 species of which nearly 90 are found in the Old World and 45 are in the New World. Liu and Cheng (1995) reported that the number of *Ziziphus* species published until 1994 is 271, and that considering the synonyms and homonyms, the actual number of *Ziziphus* species would be about 170.

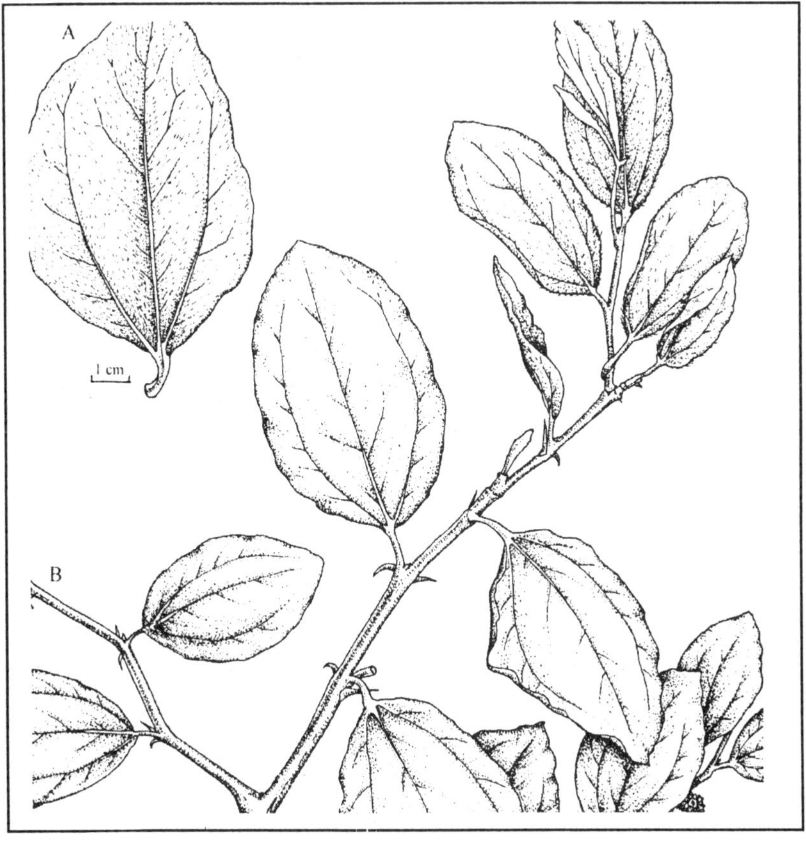

Plate 2.1. A Typical Ber twig

Don (1831) divided 39 species of *Ziziphus* into two groups, i.e., those having leaves smooth on both surfaces and those having leaves downy beneath. Hooker (1875) divided the 18 Indian species into three groups; i) having flowers in sessile cymes, ii) having flowers in predunculated axillary cymes, and iii) having flowers in predunculated cymes, which are disposed in leafless simple or compound spikes. Brandis (1906) classified 14 Indian *Ziziphus* species into two groups: 1. Cymes axillary, sessile or shortly predunculate (*Z. jujube* Lam., *Z. nummularia* Weight., *Z. oenoplia* Mill., *Z. incurve* Roxb., *Z. trinervia* Roxb. *Syn., Z. glabrata* Heyne, *Z. glabra* Roxb., *Z. hosfieldii* Mill., *Z. vulgaris* Lam., *Z. oxyphylla* Edgew., *Z. xylopyrus* Willd., *Z. horrida* Roth., *Z. apetala* Hook.) and 2. Cymes pedunculate, arranged in terminal or lateral panicles (*Z. rugosa* Lam., *Z. funiculosa* Ham.). Sussenguth (1953) divided *Ziziphus* into four groups : i. Cymes simple, axillary, ii. Cymes large, disposed in terminal or lateral panicles, spines short, iii. Inflorescene disposed in leafy panicles, and iv. Cymes in leafy spur. Liu and Cheng (1995) did not regard these as valid subdivision since they considered that the characters used were quite unstable. They have given the following classification based on morphological characters, growth habit and geographical distribution : i) plants glabrous with deciduous fruiting branches and distributed in temperate regions, and ii) plants having at least two kinds of pilose organs, without deciduous fruiting branchlets and distributed in tropical and subtropical regions. The second group has two series. The section of the genus is enumerated (Pareek, 2001) below:

Section 1. *Ziziphus*

Plants glabrous with deciduous fruiting branchlets and mainly distributed in temperate zones

Z. jujube Mill

Z. acidojujuba C.Y.

Section 2. *Perdurans*

Plants with at least two kinds of pilose organs and without deciduous fruiting branchlets; distributed in tropical and subtropical areas

Series-*Cymosiflorae*

Axillary cymes, glabrous and fruit, thick and hard endocarp, widely distributed in tropical and subtropical area

Z. *mauritiana* Lam.

Z. *oenoplia* Mill.

Z. *cambodiana* Pierre

Z. *spina-christi* (L.) Willd.

Z. *nummularia* (Burm.f.) Wight & Arn.

Z. *horrida* Roth.

Z. *incurve* Roxb.

Z. *lenticellaria* Merr.

Z. *lotus* Lam.

Z. *apetala* Hook.

Z. *glabrata* Heyne ex Roth

Z. *abyssinica* Hochst. Ex. A. Rich

Z. *mucronata* Willd.

Z. *truncate* Blatt. and Hall.

Z. *guatemalensis* Hemsl.

Z.*xiangchengensis*. Chen and Chou

Z. *mairei* Dod.

Z. *Montana* W.W. Smith

Z. *publinervis* Rehd.

Z. *laui* Merr. and others

Series-*Thyrsiflorae*

Terminal or axillary thyrse, pilose young fruit and/or ovary, thin and fragile endocarp. Concentrated in South and South East Asia

Z. rugosa Lam.

Z. *calophylla* Wall.

Z. *attopensis* Pierre

Z. *hoaensis* Pierre

Z. *kunstleri* King

Z. *borneensis* E.D. Merrill

Z. *havanensis* BL.

Z. *fungi* Merr. and others

Trimen (1893) and Cooke (1902) had earlier prepared keys for the classification of *Ziziphus* species based on whether 5 petals were present or absent and whether the styles were connate or not as well as on leaf morphology. However, they did not take into account the important species such as Z. *sativa* and Z. *spina-christi*. Kirtikar and Basu (1975) also considered fruit size (whether more or less than 13 mm in diameter) and included Z. *sativa* as a species having smaller fruits, but stated that its cultivated forms had larger fruits.

Botanical description of ber (Z. *mauritiana* Lamk.) has been made by many workers. Trimen (1893) has described that it is a small, much-branched tree or a large shrub with dense spreading crown, bark dark grey with deep, longitudinal fissures, brachlets elongated, flexuose, woolly-pubescent; leaves, 2.5-3.75 cm, broadly oblong-oval or rotundate, rounded at both ends, faintly and irregularly denticulate, base oblique and 3-nerved, nerves depressed on the glabrous shining upper surface, densely covered beneath with whitish or buff coloured tomentum, petiole short, woolly, paired spines short, very sharp, one usually curved, the other straight, often absent; flower on hairy pedicel, in small axillary clusters or very shortly pedunculate, paniculate cymes (Plate 2.2 and 2.3); calyx woolly outside, segments very acute; petals very small, spathulate, recurved; disk 10-lobed, 10-grooved; styles 2,

Plate 2.2. Inflorescence of Ber

Plate 2.3. Ber flowers with small fruits

connate for half the length, trifid; drupe about 1.25 cm, globose, fleshy, smooth, yellow, stone 2-celled, brown, excavated on surface. Flower greenish-white. The fruit of the cultivated types (Portuguese name 'Masau') is larger and more elongated and stone generally one celled (Trimen, 1893). Troup (1921) reported that the tree sometimes reaches large dimensions (24m high, and 7m girth). Kirtikar and Basu (1975) have given a similar description and mentioned that it is a sub-deciduous tree having furrows about 1.2cm apart on the bark, blade 9-13 mm, short fibre, pink with or without polar steaks, the juice turning purplish black on the blade of knife; leaves 3-6.3 x 2.5 x 5cm, petiole 2.5 to 10mm long. Bailey (1947) also gave similar account and stated that the trees may be 9-15m in height, flowering in March-June and berries ripen in November-December. The trees of *Ziziphus* species according to Johnston (1972) have densely but minutely pubescent branchlets, leaf blades elliptic to ovate to nearly orbicular, 3-8cm long, 1.5-5cm broad with rounded symmetrical or nearly symmetrical base, obtuse, densely tomentose beneath; petioles 1-2cm long and broad, few to many flowered cymes; peduncles 1-4mm long, tomentose; pedicels 2-4mm long in flower, 3-6mm in fruit, tomentose; sepals 1.5-2.0mm long, dorsally tomentulose; petals 1-1.5mm long; ovary cells and seeds 2; drupes globose to ellipsoidal, 1-2cm wide (Plate 2.4).

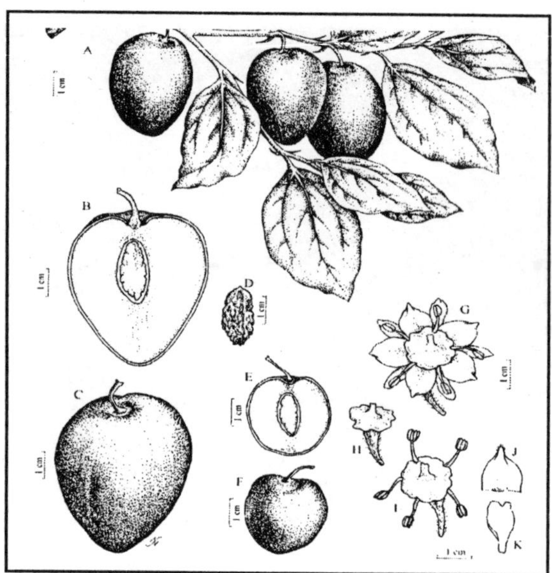

Plate 2.4. Sketch of Ber fruits and flowers

Area and Distribution

The major production area of ber is in the semi-arid and arid parts of the country. From 1984 to 1995, the area under orchards of large-fruited budded ber varieties has been almoset doubled. The total area under ber during 1994-95 reported to be 88,000 ha, producing about 0.9 million tons fruits (Pareek, 2001). Marketable produce is also harvested from wild seedling groves, which yield fruits of variable quality. Precise data of production from these groves are not available but it is estimated to be around 0.2-0.3 million tons. Ber is also grown in other south Asian countries particularly in Pakistan and Bangladesh.

In African countries, fruits are harvested from natural seedling groves of ber. Regular orchards of identified varieties are rare. Some plantations have recently been established in Israel and also in some Middle Eastern countries. Chinese jujube is produced in the cold desert areas of China (450 thousand tons), including Taiwan (13.5 thousand tons) and in Thailand (31.6 thousand tons).

Ber productivity from scattered plants are mainly for domestic use. Where as large plantations are for multiple use and distribution to local or distant markets. All such areas have potential for enhanced production. The spread of plantations should originate from around current production centres where management techniques and technical skills are available and marketing is in place. Production techniques need to become increasingly intensive and competitive with emphasis on specialized quality for different uses. Cultivation can be expanded in arid and semi-arid areas in, i) watersheds near the town, ii) resource-poor areas where crop failures are common, and iii) wastelands where no other crops are possible.

Eighteen species are found in the Indian subcontinent, from the Himalayas to Cape Camorin, and from the western desert region to the eastern wet tropics. Ber is found throughout the greater part of India, either wild or naturalized, ascending to 1500m in the Himalayas. It is common in dry regions where it is often gregarious either in the tree form or in the bushy from in grasslands (Troup, 1921). It is found wild in the southern districts of Rajasthan (Sebastian and Bhandari, 1990). It is widely cultivated in Punjab, Haryana, Uttar Pradesh, Rajasthan, Gujarat, Madhya Pradesh, Maharashtra and to some extent in several other states of India. Brandis (1906) reported this species to be wild in Siwalik forests, basins of the Ganges and in forests of Central India.

Z. xylopyra is found in the sub-Himalayan tract and outer hills upto 900m from Sutlej eastwards. It is also present in Bihar, Chotanagpur, Rajasthan, Central and Southern India, the Deccan and the Peninsula (Brandis, 1906; Troup, 1921). *Z. nummularia* is common throughout north-west India and in the drier parts of south India (Brandis, 1906, Bhandari, 1978; Sebastian and Bhandari, 1990). It is mainly found wild in the dry arid regions of Punjab, Rajasthan, Gujarat, Sindh and Baluchistan and the Indian Peninsula. It is found in the Mukundara ranges of the Vindhyan series in Madhya Pradesh where it is regarded as a weather indicator plant (Sharma, 1992). *Z. oenoplia* is common throughout the greater part of India extending as far as Burma. Other *Ziziphus* species are also available in India but the commercial cultivation is limited only to *Z. mauritiana* in drier parts of the country.

Soil and Climate

The ber grows on a wide variety of soils from gravelly, shallow soils to deep aridisols and to some extent in entisols, however deep sandy to clay soils are considered best. It is also found in drought prone, low rainfall areas (Pareek, 1983). Even in soils underlain with murram (calcic subhorizon) within one metre depth, roots were found to penetrate upto 4.5m (Pareek, 1977). Data from field trials in India and Zimbabawe suggest that in *Ziziphus* rooting depth is critical for the maintenance of high rates of assimilation and conductance throughout the day making it tolerant to drought conditions (Clifford *et al.,* 1996).

Neutral or slightly alkaline soils are considered most suitable for good growth and production of ber. Seed germination was reduced in salinised soil and was almost inhibited at 14.2 dSm^{-1} (Dahiya *et al.,* 1981), while the growth of seedlings was considerably reduced above 14.2 dSm^{-1} (Dahiya and Dhankhar, 1982). Growth of ber trees was reduced at 5 dSm^{-1}, fruit set was adversely affected at 15 dSm^{-1} resulting in 70% yield reduction but at 20 dSm^{-1} salinity level, not a single plant survived in an orchard on sandy loam soil (Hooda *et al.,* 1990). With increasing soil salinity, the contents of Na^{+}, Ca^{++}, Mg^{++}, Cl and SO_4 increased in the leaves (Dahiya and Dhankhar, 1982; Hooda *et. al.,* 1990 and Pandey *et al.,* 1993) and the nitrogen, phosphorous and potash contents decreased (Pandey *et. al.,* 1993). Increasing salinity levels also increased total chlorophylls, total free amino acids and praline content (Pandey *et al.,* 1991).

Ber is known to survive even in soils having pH as high as 9.2 (Jawanda *et al.,* 1981). No leaf scorching in trees growing on alkaline soils was observed while those of mango and guava were seriously damaged (Samra, 1985). Young ber seedlings survived when transplanted in soils upto 66 ESP (Exchangeable Sodium Percentage) in a greenhouse (Singh, *et al.,* 1983b). Increasing ESP levels, however, reduced growth, decreased Ca and Mn but increased Na and Fe contents in the leaves. Kumar *et al.,* (1990) also observed a decrease in leaf micronutrient content with an increase in ESP levels from 5 to 45.3. The total chlorophyll, free amino acid and proline contents increased with increasing sodicity (Pandey *et al.,* 1991). Awasthi *et al.,* (1994) reported that 62.5% ber seedlings survived at 60.5 ESP but none of the grafted plants of Umran and Gola survived. Shoot and root growth of Umran ber decreased with increasing ESP (Patil *et al.,* 1981).

Ber grows well under varying climatic conditions from sea level upto an elevation of 1000m (Singh *et al.,* 1973 a). Commercial ber plantations are found growing in areas having a minimum temperature of 4 to 12°C and occasionally reaching as low as -2°C for short periods. Freezing temperatures and frost, however, cause damage to young twigs and developing fruits and result in considerable crop loss and decline in tree growth. Ber trees can withstand extremely high summer temperatures and are found to grow well in regions having maximum temperatures of 39 to 42°C, and can tolerate temperatures as high as 49-50°C. However, fruit set is adversely affected at temperatures above 35°C. The trees shed leaves and enter into dormancy during the extremely hot summers in the subtropics of northern India. Ber is highly drought tolerant. Its deep taproot enables it to survive under extended moisture stress even when the surface soil completely dries out. Lower loss of water through leaf transpiration makes ber trees more adapted to arid microclimates (Singh *et al.*, 1998). Drought-stressed *Ziziphus* trees effectively control water loss by stomatal closure and even by reduction in leaf area which results in higher intrinsic water use efficiency and assimilation / leaf conductance ration than in unstressed trees (Jones, 1992).

Natural groves of ber grow and produce well in subtropical and tropical regions receiving over 400mm annual rainfall. Fruit yields are higher during higher rainfall years (Singh *et al.*, 1998). Management of groves in low rainfall areas necessitates some irrigation. Conditions of high atmospheric vapour pressure deficit result in the production of better quality fruits than in humid areas. The incidence of pests and disease is also lower in drier regions. *Z. nummularia* is widespread even in regions receiving as little as 100mm annual rainfall.

Plant Genetic Resource Management and Varieties

Genetic variability of ber is high in India. Several other countries in which *Ziziphus* has now spread also have secondary diversity. It is known by the several vernacular names in Indian languages (Table 2.2).

Several horticultural varieties of ber are in cultivation in India. These varieties developed as a result of selection in different ecorgions from progenies emanating from cross pollination between different *Ziziphus* species and/or between types /cultivars of ber. Over 180 named cultivars have been mentioned in the literature (Ram, 1940; Dutta, 1954; Chandra, 1964; Randhawa and Biswas, 1966; Singh and Khanna, 1968; Singh *et al.*, 1971, 1972a, 1973c; Bajwa *et al.*, 1972; Chadha *et al.*, 1972; Chopra and Srivastava, 1973; Dhingra *et al.*, 1973; Singh and Singh, 1973; Teaotia *et al.*, 1974; Khera and Singh, 1976; Gupta, 1977; Madelgiri *et al.*, 1977; Chundawat *et al.*, 1978; Bisla *et al.*, 1980; Godara *et al.*, 1980; Singh and Jindal, 1980; Pareek and Vashishtha, 1983, 1986; Nehra *et al.*, 1984; Yamdagni *et al.*, 1985 b; Babu and Kumar, 1986, 1988; Kumar and Babu, 1987; Tomar and

Singh, 1987; Singh and Tomar, 1988; Kundi *et al.*, 1989 a b; Vashishtha and Pareek, 1989; Vashishtha *et al.*, 1989 a, b; Bal, 1992; Chovatia *et al.*, 1992, 1993; Reddy *et al.*, 1995). A large number of germplasm collections have been conserved in field genebanks and documented by different institutions in the country (Table 2.3).

Table 2.2 : Names of ber in different languages

Language	Name
Assamese	Bogori
Bengali	Kul
Gujarati	Ber, Bor, Bordi, Boyed
Hindi	Ber, Beri
Kannada	Badari, Bogari, Bore, Egasi, Elasi, Ilanji, Ilisi, Jelachi, Karkhandhu, Yalachi
Marathi	Ber, Bhor, Baher, Bor, Bordi, Bora
Malayalam	Badaram, Badari, Kolam, Lanta, Perintutali
Punjabi	Ber, Beri, Unab
Sanskrit	Ajapriya, Badari, Balastha, Dridhabija, Dviparni, Ghonta, Gudaphala, Kantaki, Karkaramadhu, Koli, Kuvali, Madhuraphala, Madadebara, Nakhi, Nripabadari, Nripeshta, Prithukoli, Phalashayashira, Rajabadari, Rajakoli, Rajavallabha, Sukshmaphala, Sukshmapatrika, Srigalakoli, Svacha, Sukorapriya, Suphala, Tanubija, Ubhayakantaka
Sindhi	Ber, Jangri
Tamil	Adidaram, Attiram, Ilandai, Iradi, Iratti, Koli, Kondai, Kullari, Kulvali, Padari, Sivagam, Vadari, Vettiram, Veyam
Telugu	Badaramu, Badari, Gangaregu, Gangarenu, Karkhanduvu, Regu, Renu
Urdu	Ber
Oriya	Barkoli, Bodokoli, Bodori, Koli

Cultivars are often named according to the locality of origin, morphology (e.g. shape, style end, etc.), quality characters (e.g. texture) of the fruit, or after the name of an individual person. Considerable confusion has arisen as a result of, i) different names in different localities for the same cultivar (e.g. Umran, Katha, Ajmeri, Madhuri, Chameli) and ii) the same name for different cultivars, e.g.

Chhuhara (Chadha *et al.*, 1972) and Rashmi (Chundawat and Srivastava, 1978).

Table 2.3 : Indian Institutions holding ber germplasms

Institution	*Ziziphus* sp	Ber Varieties
Central Institute for Arid Horticulture, Bikaner-334 006, Rajasthan, India	7	162
Central Arid Zone Research Institute, Jodhpur-342003, Rajasthan, India	-	68
Indian Institute of Horticultural Research, Bangalore-560089, Karnataka	3	32
Central Horticultural Research Station, Godhra-389 001, Gujrat, India	-	22
Haryana Agricultural University, Hisar-125004, Harayana, India	-	74
Narendra Deo University of Agriculture and Technology, Kumarganj, Faizabad-224 229, U.P., India	-	32
Mahatma Phule Agricultural University, Rahuri-413 722, Maharashtra, India	-	87
Gujarat Agricultural University, Sardar Krishunagar-385 506, Gujarat, India	-	75
Fruit Research Station, Punjab Agricultural University, Bahadurgarh, Patiala-147001, Punjab, India	-	42
Indian Agricultural Research Institute, New Delhi-110012, India	4	52
Dryland Agriculture Research Station Haryana Agricultural University, Bawal-123501, Haryana, India	-	36

Source : Pareek, 1988; Pareek and Sharma, 1993

The ability of *Ziziphus* species and the different varieties / types within ber to cross freely has allowed the build up of a rich genepool. Vegetative growth (tree form, leaf shape, apex, base and pubescene, petiole length, colour and pubescence, etc.), flowering (flower emergence, anthesis, etc.), fruiting (set, maturity), and fruit characters (shape, size, style end, skin colour, stone and pulp content, pulp colour and sweetness, etc.) have been described by different authors. Vashishtha (1983) has described the morphological characteristics of a large number of ber cultivars at Jodhpur in northwest India.

The cultivars show different growth forms ranging from an erect branching habit to those with vine-lime, spreading branches (Singh *et al.*, 1971; Vashishtha and Pareek, 1989). A wide range of variation also exists with respect to length, width, shape, apex, base of leaf, petiole length and other leaf characters. Ber cultivars can be assigned to two groups on the basis of anthesis time, those in which anthesis takes place in the morning and those with anthesis in the afternoon (Teaotia and Chauhan, 1963; Prasad, 1964; Desai and Patil, 1978; Vashishtha and Pareek, 1979). The fruit maturity period varies in different ber cultivars according to location depending on the agroclimatic conditions (Table 2.4). Thus, some cultivars mature early and others mature in mid- or late-season (Chadha *et al.*, 1972; Pareek, 1983; Singh *et al.*, 1983a).

Table 2.4: Fruit maturity periods in ber cultivars at different locations

Location	Early	Mid Season	Late
Tamil Nadu (8.5-14^{0}N 77-80^{0}E)	–	Banarsi, Kaithli	–
Andhra Pradesh (13-20^{0}N 77-80^{0}E)	Gola, Mundia	Kaithli, Umran,	Banarsi Seb
Maharashtra (16-22^{0}N 73-79^{0}E)	Shamber, Badami, Gola	Maharun, Chhuhara	Umran
Gujarat (21-24.5^{0}N 69-74^{0}E)	Gola, Mundia	Banarsi Karaka, Kaithli	Vikas (Randeri) Umran (Chameli)
Rajasthan (23-30^{0}N 70-78^{0}E)	Gola, Mundia	Jogia, Seb, Banarsi Karaka Kaithli, Banarsi Peondi	Umran (Katha) Maharwali Bagwadi, ZG-3 Kali
Uttar Pradesh (24-29^{0}N 77-84^{0}E)	Gola	Banarsi Karaka, Kaithli, Banarsi Peondi	Umran
Haryana (28-31^{0}N 75-77^{0}E)	Gola, Safeda Selected, Sandhura Narnaul, Chonchal	Kaithli, Sanaur-5 Muria Mahrara (Mundia), Banarsi Karaka	Umran, Illaichi
Punjab (29.5-32.5^{0}N 74-77^{0}E)	Nazuk, Gola, Safeda Selected, Sandhura Narnaul	Banarsi, Dandan Kaithli, Sanaur-2	Umran, Illaichi ZG-3

Based on ber and Chinese jujube it would seem that *Ziziphus* is a monobasic genus with x = 12 (Morinaga *et al.*, 1929; Bowden, 1945; Srinivasan, 1952; Khoshoo and Singh, 1963; Nehra *et al.*, 1983). However, reports of 2n=40 in Chinese jujube and 2n=20 in *Z. oenoplia* by Srinivasachar (1940) and 2n=26 in *Z. sativa* by Chiarugi (1930) indicate that the genus could be tribasic (x=10, 12, 13) as also assumed by Darlington and Wylie (1955). Unlike Chinese jujube, which is diploid, *Z. oenoplia*, which is tetraploid and *Z. rotundifolia*, which is tetraploid and hexaploid, ber ranges from tetra-to octaploid. Khoshoo and Singh (1963) observe that i) the chromosome number in 33 ber cultivars (Banarsi, Banarsi Pewandi, Chhuhara, Dandan, Desi Alwar, Golar, Gorva, Kaithli, Kala Gola, Katha Bombay, Kathaphal, Laddu, Mirchia, Nalagarhi, Narikeli, Nazuk, Noki, Pathani, Sandhura Narnaul, Sanaur, Sanaur-2, Sanuar-3, Sanaur-4, Sanaur-6, Seo, Safeda slected, 28/1, Umran, Wilayti, ZG-2, ZG-3, Hoshiarpur) was n=24 in 2 cultivars (Illaichi and Mithianwali) it was n=48. In one cultivar (Kalianwali), it was n=30 ii) in a wild ber rootstock as well as in 3 types of *Z. rotundifolia*, the chromosome number was n=24 and in 2 types of *Z. rotundifolia* it was n=36. Nehra *et al.*, (1983) confirmed that the chromosome number in cultivar Illaichi and two wild ber types was n=48, in Umran n=24 and in *Z. rotundifolia*, n=24, 36. Ber cultivars show wide variation in fruit characteristics (Table 2.5).

Table 2.5 : Characteristic feature of fruit in some ber genotypes/cultivars

Characteristics	Range
Shape	Round-oval-ovate-obovate-oblong-oblate
Style end	Round-round with depression-blunty tapering-obliquely beaked-obliquely pointed-straight pointed-flat
Length (cm)	1.82 to 5.80
Breadth (cm)	1.1 to 4.7
Size (length x width) (cm)	2.2-1.7 x 5.7-3.5
Weight (g)	3.8 to 39.5
Stem end	Round-round with shallow cavity-obtuse or flattened-slightly ridged-faintly grooved-distinctly ridged-necked-tapering
Cavity in pericarp	At stem end-at styler end-at both ends
Colour of developing fruit	Light green-green-dark green-green with red pigment-brownish red-half red half green

Contd...

Colour of ripe fruit	Green-greenish-yellow-light yellow-yellow-bright yellow-golden yellow-brownish-cream yellow rutning reddish brown with stripes-lemon yellow turning brownish faintly striped-yellow with brownish tinge-greenish yellow with brown blush

Chemical composition of the fruits of different ber varieties also varies considerably with respect to constituents such as sugars, ascorbic acid, acidity and minerals (Table 2.6).

Table 2.6 : Variability in morphology and chemical components of ber fruits

Characteristics	Range
Pulp (%)	81 to 97
Stone (%)	2.7 to 12.0
Pulp / stone ratio	7.6 to 32.1
Starch (%)	0.72 to 1.15
Total Soluble Solids (^{0}Brix)	12 to 23
Total sugars (%)	3.1 to 14.5
Reducing sugars (%)	1.4 to 9.7
Non-reducing sugars (%)	1.3 to 9.7
Acidity (%)	0.13 to 1.42
Ascrobic acid (mg/100g)	39 to 166
Fe_2O_3 (mg/100g)	0.01 to 2.0

Singh *et al.*, (1971) classified 39 ber cultivars into three groups on the basis of primary leaf characteristics such as length : width ratio and type of apex, i.e. broad-leaved, long-leaved with sub-acute to acute apex and long-leaved with obtuse apex. The cultivars of each group were further divided on the basis of branching habit (erect, semi-erect and spreading) as secondary characters and shape and type of leaf base including leaf habit, colour, petiole length, type of margin and lobes on apex as tertiary characters. Later, Singh *et al.*, (1972 a) classified 40 ber cultivars on the basis of fruit characters only. The cultivars were divided into 8 sections based on fruit shape alone or in combination with apex type as primary characters. Absence and presence of cavities in the pericarp and the type of base were taken as secondary characters. Presence or absence of styler scar and its colour, size of fruit, colour of developing fruit and markings on the base of the apex were taken as tertiary characters. Ber cultivars could, thus, be identified either during fruiting or non-fruiting periods. Vashishtha and Pareek (1989) prepared a key using both fruit and vegetative characters simultaneously to classify 59 ber cultivars. Easily seen qualitative characters like tree growth habit and fruit and leaf shape were used to divide the cultivars into 3 groups and 28 sub-groups. Three groups (erect, semi-erect and spreading) were divided into 9, 8 and 11 sub-groups respectively (Pareek, 2001).

The important charactesrstics of some of the commercial ber cultivars have been enumerated.

Bagwadi

Spreadings trees, Leaf oval, broad (L/W : 1.3); apex obtuse, flat; base broad. Petiole light green on both surfaces, densely hariy; length 1.9cm. Flowering starts late in August with a peak in early September and end in early October. Anthesis occurs during 13.00 to 14.30 hrs. Fruit set starts late in August with a peak in early September and end in late October. Fruits mature in mid-seasson (January 22-February 4). Fruit oval; style end round; stem end round with shallow notch and prominent ridges, skin yellowish green, smooth, cavity in percap at both ends; pulp white, soft 3.8 x 3.0cm; weight 19.5g; pulp/stone ratio 27; TSS 14.8^0 Brix.

Banarsi Kadaka

Semi-erect trees. Leaf oblong, long (L/W : 1.5); apex subacute; curved margin with tapering base. Petiole light green with purple tinge on dorsal and light green on ventral surface; length 1.4cm. Floweing starts in fourth week of August with peak in frist week of September and ends in frist week of October. Anthesis occurs during 13.00 to 14.30 hrs. Fruit set starts in first week of September with peak in fourth week of September an ends in second week of October. Fruits mature in mid season (January 27-Febraury, 12). Fuit oval; styler end obliquely pointed; stem end round, narrow, slightly ridged; skin light yellow, smooth glossy with streaks; cavity in pericarp at stem end; pulp light cream, soft; size 4.9 x 3.1 cm (Plate 2.5); pulp /stone ration 31. TSS 16.7^0 Brix.

Plate 2.5. Ber cv. Banarasi Kadaka

Banarsi Pebandi

Semi-erect trees. Leaf oval, broad (L/W : 1.3); spex sub-acute. Curved; base broad. Petiole light green with purple tinge on the dorsal and light green on ventral surface; length 1.4cm. Flowering starts late in Aguust with a peak early in September and ends late in September. Anthesis occurs during 13.00 to 14.30 hrs. Fruit set starts late in August with a peak one month later and ends early in October. Fruits mature in mid season (January 25-February 10). Fruit oblong; styler end obliquely pointed, necked; slightly ridged stem end; skin greenish yellow, smooth; cavity in pericarp at both ends; pulp white, semi-soft; weight 33g; size 5.2 x 3.4cm; pulp/stone ration 32, TSS 13.7^0 Brix.

Chonchal

Semi-erect trees. Leaf oblong, broad (L/W : 1.4); apex acute, flat with tapering base. Petiole light green with purple tinge on dorsal surface, light green on ventral surface, length 2.7cm. Flowering starts early in September with a peak in mid September and ends late in October. Anthesis occurs during 07.00 to 08.30hrs. Fruit set starts late in September with a peak early in October and ends late in October. Fruit mature in mid season (January 28-February 12). Fruit oblong;

styler and obliquely beaked; stem end narrow, necked; skin light yellow, smooth, streake; cavity in pericarp at stem end; pulp white, soft; size 5.3 x 3.2cm; weight 24.9g; pulp / stone ration 19; TSS 19.8^0 Brix.

Dandan

Semi erect trees. Leaf oval, broad (L/W : 1.2); apex abtuse, slightly curved; base broad. Petiole light green with purple tinge on dorsal and light green on ventral surface; length 1.5cm. Fowering starts late in August with a peak early in September and ends late in October. Anthesis occurs during 07.00 to 08.30 hrs. Fruit set starts late in September with a peak early in October and ends late in October. Fruits mature in mid season (January 28-Febraury 12). Fruit oblong; styler end bluntly tapering with ridges; stem end slightly ridged; skin yellowish green, smooth, streaked; cavity in pericarp at both ends; pulp light cream, soft, granular; size 5.7 x 3.5g; weight 27.6g; pulp / stone ration 17.4; TSS 17.2^0 Brix.

Gola

Spreading tree. Leaf cordate, broad (L/W : 1.1); apex obtuse; old leaves curved towards ventral surface; base broad. Petiole light green on both surfaces; length 1.2cm. Flowering starts late in August with a peak in early September and end in late September. Anthesis occurs during 13.00 to 14.30 hrs. Fruit set starts early in September with a peak late in September and ends early in October. Fruits mature early (January 7-25). Fruit round; style end flat; stem end round; skin bright yellow, smooth and glossy; cavity in pericarp at stem end; pulp white, soft; size 3.8 x 3.5 cm; pulp / stone ratio 14.3; weight 21.4 g; TSS 21^0 Brix. Chovatia *et al.,* (1992) have reported 160 mg / 100g vitamin C content in fruit pulp.

Goma Kirti

It is an early maturing clonal selection from a population of trees of Umran cultivar (Hiwale and Raturi, 1999). Growth and vegetative characters of trees and physico-chemical characters and keeping quality of the fruit are almost similar to those in Umran. Flowering and fruit maturity occurs 3 weeks earlier and the fruit yield, during the first 5 years after planting, has been observed to be over 25% higher owing to higher fruit retention (62%) than in Umran (31%).

Illaichi

Erect trees. Leaf cordate, broad (L/W : 1.2); apex obtuse; base flat, broad. Petiole light green on both the surfaces; length 1.8cm. Flowering starts late in fourth week of August with a peak in second week of September and ends in fourth week of October. Anthesis occurs during 07.00 to 08.30 hrs. Fruit set starts in third week of September with a peak in fourth week of September and ends in fourth week of October. Fruits mature in mid season (January 25-February 10). Fruit oblate; styler end flat, ridged; stem end rounded with prominent stripes, ridged; skin yellowish green (Plate 2.6), smooth, streaked; cavity in pericarp at stem end; pulp white, soft; size 2.2 x 2.2cm; weight 8.5g; pulp / stone ration 18.4; TSS 20^0 Brix. Vitamin C content of 138 mg / 100g pulp has been reported by Yamadagni *et al.*, (1985b).

Plate 2.6. Ber cv. Illaichi

Jogia

Semi-erect trees. Leaf oblong, broad (L/W : 1.3); apex sub-acute, slightly curved at margin and apex; base broad. Petiole light green on both surfaces; length 1.4cm. Flowering starts in fourth week of August with a peak in third week of September and ends in second week of

October. Anthesis occurs during 07.00 to 08.30 hrs. Fruit set starts in fourth week of August with peak in fourth week of September and ends in fourth week of October. Fruits mature early (January 12-30). Fruit oblong; styler end blunty tapering; stem end ridged; skin yellowish green, surface coarse with ridges; cavity in pericarp at stem end; pulp white, semi-soft; size 4.9 x 3.3cm; weight 24g; pulp /stone ratio 24; TSS 14^0 Brix.

Kaithli

Erect trees. Leaf oblong, long (L/W : 2.2); apex acute, slightly curved; base tapering. Petiole light green with purple tinge on dorsal and light green on ventral surface; length 1.5cm. Flowering starts in first week of September with a peak in fourth week of September and ends in fourth week of October. Anthesis occurs during 13.00 to 14.30 hrs. Fruit set starts in third week of September with a peak in third week of October and ends in fourth week of October. Fruits mature in mid-season (January 26-February 15). Fruit is oblong; styler end round, stem end shallow; skin yellowish green, smooth; cavity in pericarp at both ends; pulp white, soft; size 4.7 x 3.4cm; weight 29.5g; pulp/stone ratio 32.1; TSS 15.3^0 Brix. Yamdagni *et al.*, (1985b) have recorded per tree fruit yield of 123.45kg and Gupta (1977) reported ascorbic acid content of 125 mg / 100g pulp.

Mehraun

Spreading trees. Leaf oval, broad (L/W : 1.2); apex acute, flat, base broad. Petiole light green on both the surfaces; length 1.1cm. Flowering starts early in September with a peak one month later and ends late in October. Anthesis occurs during 07.00 to 08.30 hrs. Fruit set starts early in September with a peak one month later and ends late in October. Fruits mature late (February 21 to March 3). Fruit oval; styler end round; stem end flat; skin yellow, smooth; cavity in pericarp absent; pulp light cream, soft; size 2.2 x 1.7cm; weight 3.5g; pulp/stone ratio 7.9; TSS 22.9^0 Brix.

Mundia

Semi-erect trees. Leaf oblong, long (L/W : 1.5); apex sub-acute; margin curved; base tapering. Petiole light green with purple tinge on dorsal and light green on ventral surface; length 1.6cm. Flowering starts late in August with a peak one month later and ends late in October. Anthesis occurs during 13.00 to 14.30 hrs. Fruit set starts late in August with a peak one month later and ends late in October. Fruits mature early (January 10-30). Fruit oblong; styler end round; stem end narrow with shallow notch; skin yellowish green, smooth with some depressions; cavity in pericarp at stem end; pulp white, soft; size 5.7 x 3.5cm; weight 39.5g; pulp/stone ratio 23.8; TSS 19.9^0 Brix. Godara *et al.*, (1980) has reported 125 kg fruit yield per tree and TSS of 22.8^0 Brix.

Poplar Gola

Spreading trees. Leaf oblong; broad (L/W : 1.3); apex acute; curved at margin; base broad. Petiole light green on both surfaces; length 1.1cm. Flowering starts late in August with a peak one month later and ends late in October. Anthesis occurs during 13.00 to 14.30 hrs. Fruit set starts early in September with a peak late in October and ends early in November. Fruits mature in mid season (January 23-February 5). Fruit round; styler end flat with depression; stem end round depressed; skin yellow; cavity at stem end; pulp; white, soft; size 3.8 x 3.7cm; weight 24.5 g; pulp/stone ratio 14.1; TSS 20^0 Brix.

Seb

Erect trees. Leaf oval, broad (L/W : 1.2); apex sub-acute, slightly curved; base broad. Petiole light green with purple tinge on dorsal and light green on ventral surface; length 1.8cm. Flowering starts late in August reaching a peak one month later and ends early in October. Anthesis occurs during 07.00 to 08.30 hrs. Fruit set starts early in September with a peak late in September and ends one month later. Fruits mature early (January 14-February 10). Fruit is round, resembling crab apple; styler end round with depression at the centre; stem end broad, deeply grooved; skin yellowish green with brown specks (Plate 2.7), slightly rough; cavity in pericarp at stem end; pulp white, hard; size 3.6 x 3.3cm; weight 24.6g; pulp/stone ratio 14.4; TSS 16.8^0 Brix.

Plate 2.7. Ber cv. Seb

Umran

Semi-erect trees. Leaf oval, broad (L/W : 1.3); apex sub-acute; curved margin; base broad. Petiole light green with purple tinge on both surfaces; length 1.5cm. Flowering starts late in August with a peak one month later and ends late in October. Anthesis occurs during 13.00 to 14.30 hrs. Fruit set starts early in September with a peak late in September and ends one month later. Fruits mature late (February 20-March 2). Fruit oval; styler end round; stem end round and ridged; skin yellow, smooth and glossy (Plate 2.8); cavity in pericarp at stem end; pulp white, hard, brittle; size 4.7 x 3.4cm; weigh 33g; pulp/stone ratio 25.2, TSS 17.9^0 Brix. Gupta (1977) has reported per tree fruit yield of 195 kg and Singh and Jindal (1980) have reported 39.8g fruit size. Fruits have long storage life.

Plate 2.8. Ber cv. Umran

Vikas-2

Semi-erect trees. Leaf oblong, broad (L/W : 1.3); apex acute; base flat, broad. Petiole light green with brown tinge on both surfaces; length 1.3cm. Flowering starts late in August with a peak one month later and ends late in October. Anthesis occurs during 07.00 to 08.30 hrs. Fruit set starts early in September with a peak early in October and ends late in October. Fruits mature late (February 17-March 3). Fruit round; styler end flat; stem end round; skin yellow with light brown streaks on surface exposed to sun, smooth; cavity in pericarp at stem end; pulp white, soft; size 2.0 x 2.2cm; weight 6g; pulp / stone ratio 8.5; TSS 21.5^0 Brix. The fruits are suitable for dehydration.

ZG-3

Semi-erect trees. Leaf oval, broad (L/W : 1.3); apex obtuse, base flat, broad. Petiole light purple on the dorsal surface, ventral surface light green, hairy, length 1 cm. Flowering starts late in August with peak early in September and ends late in October. Anthesis occurs during 07.00 to 08.30 hrs. Fruit set starts early in September with a peak in mid October and ends late in October. Fruits mature late (February 17-March 3). Fruit ovate; styler end round; stem end broad and shallow with ridges; skin smooth, greenish yellow; cavity in pericarp at stem end; pulp white, semi-soft; size 3.8 x 3.1 cm; weight 18.6g; pulp/stone ratio 14.5; TSS 15.8^0 Brix.

Varietal Improvement

The existence of polyploidy and entomophily in addition to a range of self and cross incompatibilities in *Ziziphus* species has resulted in wide hybridization. Current ber cultivars have been evolved through selection of promising plant types from this widespread natural variability followed by their vegetative propagation. Over several generations, this has helped the development of acceptable ecospecific genotypes having characteristic quality attributes and productivity potential. Some of these identified genotypes attained widespread popularity and became established as cultivars suitable for orchard planting. However these cultivars tend to lack one or more quality and / or productivity traits since these have developed through single plant selections from the natural population. The need for improvement is realized particularly when commercial ber cultivation has been undertaken under agro-ecologically different environments and under more intensively managed orchard practices. The main objectives for breeding in ber are to develop:

- types resistant to i) diseases, particularly powdery mildew (*Oidium erysiphoides* f. *ziziphi*) and *Isariopsis*, ii) pests, particularly fruit fly (*Carpomyia vesuviana*) and fruit borer (*Meridarchis scyrodes*), iii) soil salinity / alkalinity conditions, and iv) damage by freezing temperatures.
- types giving good fruit set under high temperature conditions.

- types producing fruits with i) a long shelf life, and ii) quality attributes suitable to prepare dehydrated and other processed products.
- rootstocks suitable for high density orchards.

The available information on desired traits has been given in Table 2.7 and 2.8. The Breeding should aim at the ideotypes mentioned in Table 2.9.

Table 2.7 : Traits identified in some ber cultivars

Traits	Cultivars	Author/Source
Fruit maturity	Early (Gola, Mundia), Mid season (Banarsi, Kaithli), Late (Umran)	Vashishtha, 1983
Sweetness (high TSS)	Reshmi, Umran	Vashishtha, 1983
Pulp texture	Coconut-like (Umran), Juicy (Gola, Aliganj), Melting (Illaichi)	Vashishtha, 1983
Fruit size	Very large (Ponda), Large (Umran), Medium (Mundia, Banarsi, Gola), Small (Illaichi)	Vashishtha, 1983
Fruit shape	Apple like (Seb), Cardamom-shaped (Illaichi), Bell shaped (Mundia), Round (Gola), Oblong (Umran)	Vashishtha, 1983
Fruit colour at maturity	Bright golden (Sanaur), Bright yellow (Gola), Greenish yellow with brown blush (Kathaphal) (Plate 2.9)	Vashishtha, 1983
Acidity	Very low (Umran), Low (Gola), Moderate acidic (Sanaur), Acidic (Kathaphal)	Vashishtha, 1983
Shelf life of fruits	Good (Umran, Maharwali), Poor (Gola)	Vashishtha, 1983
Processing uses	Dehydration (Vikas, Raja, Babu, Jeevan, Chinense cultivars, Umran, Bagwari, Chhuhara) Preserve (Umran, Banarsi Karaka, Kaithli) Candy (Illaichi, Umran, Kathaphal, Kaithli) Beverage (Gola, Mundia)	Gopani, 1976 a, Khurdiya, 1980 Gupta *et al.*, 1981b Gupta *et al.*, 1981a Khurdiya and Singh, 1975

Contd...

Resistance to fruitfly	Tikadi, Meharun, Illaichi	Singh and Vashishtha, 1984
Resistance to fruit borer	Banarsi Pewandi, Gola Gurgaon, Jhajjar Selection	Pareek and Vashishtha, 1986
Tolerance to powdery mildew	Illaichi, Jhajjar, Sanaur-5, Safed Rohtak, Kathaphal, Gola, Seb, Meharun	Pareek and Vishal Nath, 1996
Field resistance to powdery mildew	Dharkhi-1, Dharkhi-2, Guli, Villaiti, Seedless	Pareek and Vishal Nath, 1996
Tolerance to *Isariopsis*	Safed Rohtak, Sanaur-1, Seo Bahadugarhia, Jhajjar Selection	Pareek and Vishal Nath, 1996

Table 2.8 : Wild species as source for some exploitable attributes

Wild Relatives	Exploitable attributes
Z. nummularia	i) Drought tolerance ii) Dwarf tree stature and extensive root system iii) Early fruit maturity
Z. jujube	i) Resistance to low temperature damage ii) Excellent dehydration quality of fruits iii) High vitamin C and P contents in fruits
Z. mistol	Resistance to low temperature damage
Z. mauritiana var. *rotundifolia*	i) Vigorous tree frame ii) Wood of marginal timber value

Table 2.9 : Ideotype of ber

Character	Ideotype	Character	Ideotype
Root system	Deep (> 2m)	Fruit size	>30g
Canopy	Spreading, low headed and thornless	TSS in pulp	>15^0 Brix
Leaf area	High	Acidity in pulp	<0.3%
Internodal length	Short (<5cm)	Fruit shape	Round to oval
Leaf colour	Dark green (chlorophyll >0.04 mg/kg fresh wt.)	Fruit colour	Bright golden
Leaf surface	Glossy dorsal and tomentose ventral	Fruit surface	Glossy and smooth
Leaf size	Large (>50 cm^2)	Pulp / stone ratio	>20

Plate 2.9. Ber cv. Khataphal

Work for improvement in *Ziziphus mauritiana* Lamk initiated during the 1960s with major emphasis on collecting and evaluating germplasm which led to the identification of location specific cultivars for commercial cultivation. Clonal selections, Kakrola Gola and CAZRI Gola, were made from vegetatively propagated plantations of cultivar Gola, which yield fruit with good quality attributes. The fruits of a clone of cultivar Umran, named Goma Kirti mature earlier than those of Umran itself. Vashishtha *et al.*, (1999) used Tikadi as a parent in a crossing programme to induce fruitfly resistance in the popular commercial cultivars Gola and Seb and obtained in F1 hybrid (Seb x Tikadi) which was resistant to fruitfly but had inferior fruit quality. The other crosses were incompatible. Subsequently, the back cross Seb x F1 regained the desired fruit quality whilst retaining the fruitfly resistant trait. At central Institute of Arid Horticulture, Bikaner, new hybrid of ber Thar Sevika (Seb x Katha) for high yield has been developed. Thar Bhubaraj, a clonal selection has also been released for Rajasthan and Gujarat region (Personal Communication). Hoang and Tuynth (1989) selected some mutants in Vietnam. The pre-germinated seeds of ber cultivars were treated with 0.02 to 0.04% N-methyl-N-nitrosurea for 12h. The mutant named as Ma Hong maintains the early fruit maturity character of its parental cultivar Gia Loc allowing two harvests per year with good quality.

Future improvement strategies in ber should be focused on selection from the existing genepool with a continuous stress for the identification of genotypes giving high productivity and having the desired traits. Plant types showing the least mid day depression in photosynthetic efficiency should be identified. Screening for tolerance to frost can be done by evaluating the tissue moisture retention capacity under dehydration stress. Variability can be induced by use of chemical and radioactive mutagens to obtain useful mutants.

Hybridization seems to be the most practical approach in ber to introduce resistances to diseases and insect pests. Hybridisation also seems to be an appropriate method to incorporate cold and drought tolerances and dwarf tree stature. The genotypes of ber are heteropolyploid woody perennials having a long juvenile phase in their sexually propagated progenies and consequently require a long time to raise inbreds for the desired traits. Moreover, the nature of the resultant inbreds would still be uncertain with respect to homozygosity for the desired character due to the polygenic nature of many of them. Therefore, making crosses between the carefully selected donor parents and the recipient parents, keeping in view their reproductive biology, followed by perpetuation of selected desirable hybrids by vegetative propagation should be the best approach. Interspecific hybridization may be needed in this endever (Pareek, 2001). *Z. nummularia* could be an excellent donor parent to introduce drought tolerance, early fruit maturity and dwarf tree stature. To introduce cold tolerance, high vitamin content and excellent dehydration quality of the fruit in cultivars of ber, the most appropriate donor parents would be cultivars of Chinese jujube. Since the species flower at different times, making crosses between their cultivars would necessitate pollen collection, storage and effective pollination.

Making use of DNA finger printing to identify genes for resistance to fruitfly, powdery mildew, salinity /alkalinity, drought and high ascorbic acid content and their transfer in the recipient parent through somatic hybridization could be another approach. Standardization of callus culture and organogenesis from mature as well as juvenile plant parts of *Z. mauritiana* (Mathur *et al.*, 1993) have been attempted and found successful. This line of work can further be intensified.

Plant Propagation and Nursery Management

In India, ber is propagated by grafting the scion buds of the desired cultivar into seedling rootstocks during the summer monsoon. The Chinese jujube is propagated by whip or tongue grafting onto rootstocks during the dormancy period or by budding during the active growth period. The rootstock seedlings are raised by sowing seeds of wild *Ziziphus* species of their native. No graft incompatibility has been observed between the scions of the commercial ber cultivars and the rootstock seedlings of ber (*Z. mauritiana), boradi (Z. mauritiana* var. *rotundifolia), Z. nummularia, Z. oenoplia, Z. rugosa, Z. spina-christi* or *Z. xylopyrus* (Pareek, 1983). However, vigour and growth pattern of the combinations differ greatly. The tree form of *Z. nummularia* tends to be dwarf with corresponding reduction of fruit production and on boradi the tree takes a vigorous growth form giving higher fruit yield (Pareek and Vishal Nath, 1996). This was found to be related to higher stomatal density (61753/cm^2 in ber compared to 8099/cm^2 in *Z. nummularia*) as reported by Banker and Prasad, (1992). Budding on *Z. nummularia* results in an inverted-bottle graft union, since the rootstock remains thinner than the scion portion (Vashishtha, 1987 a, b). Budding success on *Z. nummularia* is also lower (50%) compared to 70% on boradi rootstocks (Anon., 1981). In Israel, over 95% success has been obtained when budding onto ber and *Z. abyssinica* rootstocks compared to less than 35% success on *Z. spina-christi* (Nerd and Mizrahi, 1998). The grafts on *Z. spina-christi* developed a bottleneck form. Thus, the rootstock seedling plants of different species, although from variable progenies, influence the vigour, growth form and productivity of the scion and presumably also its adaptation to adverse climatic and soil conditions. The best choice, at present, seems to be to raise rootstocks from seeds collected from vigorous trees of ber (*Z. mauritiana*) and boradi (*Z. mauritiana* var. *rotundifolia*) or other suitable wild species such as *Z. abyssinica.*

Usually seed stones are sown to raise seedlings. Each stone contains as many as 3 seeds embedded in the endocarp of the drupe. The quickness and extent of germination depends upon viability and after ripening status of the seed, presence of endogenous inhibitors, weathering of the stoney endocarp and environmental conditions such as temperature, moisture, salinity and alkalinity of the growing medium. Thus, it is always advisable to extract seed kernels from stone for assured germination. Storage of seed kernels at reduced temperatures e.g. at 4.5 ± 0.5^{0}C in perforated polythene bags, result in retention of viability for a reasonable period (Kulkarni *et al.*, 1992). Ber seeds also has after ripening period therefore, seeds should be stored for 4-5 months before sowing. One or two months after extraction, germination increases (Ghosh and Sen, 1988; Reddy and Murthy, 1990) and one year old seeds germinate better (Kajal and Singhrot, 1986). Water soaking of kernels for 24 hrs improves germination. Such seed germinates within a week. Treatment of seed kernels by soaking in 100 ppm (Shanmugavelu, 1970), 200 ppm (Hore and Sen, 1994) or 500 ppm GA (Krishnan and Kulasekaran, 1984) or in 0.1% thiourea (Murthy and Reddy, 1989 b; 1990 a) enhanced the germination percentage.

To raise rootstock seedlings, seeds should be sown at an appropriate time, when the temperature conditions (30^{0}C) are favourable. The best period has been found to be spring through summer (April to July) in tropical and subtropical climates (Pareek, 1978 b; Singhrot and Makhija, 1979 b; Anon., 1981; Kajal and Singhrot, 1986). Sand, silt or clay and organic manure in the growing medium gives the best results in terms of seed germination and top and root growth of the seedlings. A mixture composed of equal parts of sand, clay and manure has been used to raise seedlings (Pareek, 1978 b; Bisla *et al.*, 1984). Salinity in the growing medium delays or inhibits germination of ber seeds (Dhankar *et al.*, 1980, Dahiya *et al.*, 1981; Dahiya and Dhankar, 1982; Mehta, 1982; Gupta *et al.*, 1989 a). Rootstock seedlings may be raised directly in the field or in the nursery by sowing healthy and viable seeds of the selected *Ziziphus* species during the spring to monsoon period at 2 cm depth (Kajal and Singhrot, 1986; Chattopadhyay and Dey, 1992). Seedlings raised in the nursery may be used to prepare budded plants or can be planted in the field for budding *in situ*. For raising seedlings in the field, sowing is done during the monsoon period at demarcated locations directly in the field. The seedlings become ready for *in situ* budding during the following spring and summer, i.e. 7-12 months after sowing. Since no transplanting is involved, the plants retain the deep tap roots and are thus more drought hardy and vigorous. For raising seedlings in nursery beds, sowing is done at 30 x 30 cm spacing (Bakhshi and Singh, 1974) any time from spring through the monsoon period. The seedlings can either be budded in the nursery 90-100 days after sowing (Singhrot and Makhija, 1979 a; Chattopadhyay and Dey, 1992) or can be transplanted in the field with 20 x 15cm earth ball (Singh *et al.*, 1982 b) for *in situ* budding. Nursery bed raised seedlings have to be carefully handled while lifting from the beds, packing, transporting and transplanting in the field to obtain good survival. Even then, mortality is higher than in plants raised in alkhathene bags (Singhrot *et al.*, 1970) because of root injury. By treatment of roots with 250 ppm IBA, transplanting success can be improved to 78.8%. To obtain better survival, Singhrot and Makhija (1979 a) shifted the seedlings within the nursery one month after sowing and then transplanted them in the field after 7 months, i.e. during the spring. Seedlings can be transplanted bare-rooted (without earth ball) when these are not growing actively during the winter in subtropical climates. A very high survival of 95% has been obtained by treating bare rooted seedlings with 250 ppm IBA. Seedbed raised seedlings are not considered suitable for planting under rainfed conditions owing to loss of their tap roots while lifting from the seedbed. Seedlings can also be raised in pots and polythene bags in the nursery. Seedlings grown in the nursery in 28 x 23 cm pots (Singh, 1961, 1979) and polythene bags of 23 x 10cm size (Bhambota and Singh, 1971) can be transplanted in the field at any suitable time and give fairly good survival rates. Sowing time can also be adjusted depending upon the stage of growth required for transplanting in the field. An innovative technique of raising ber seedling has been developed (Pareek, 1978b). In this technique, sowing of seed kernel is done in 300 gauge polythene tubes of 25 x 10cm size during the spring. After 90-100 days, seedlings can be transplanted in the field for *in situ* budding. Polytube raised seedlings develop straight and uncoiled roots and thus make nearly as vigorous growth as seedlings raised by sowing seeds directly in the field. Alternatively, the seedlings can be budded while in the polytubes about 100 days after sowing. The polythene tubes have to be protected from disintegration by exposure to solar radiation either by keeping them buried in sunken beds or by proper covering when kept above ground.

Budding can be carried out by different methods such as I or T (shield), ring and patch (Husain, 1973; Moti *et al.*, 1976; Anon., 1981; Mawari and Singh, 1992 a), chip (Anon., 1981), flute (Mawari and Singh, 1992 a, b) and forkert (Jyotishi *et al.*, 1967); however, shield (Singh, 1952; Anon., 1953; Singh, 1957; Joshi, 1960; Singh, 1964; Jyotishi *et al.*, 1967) and patch budding are the most commonly used methods. The ring method is not popular because of the necessity for the scion stick to be of equal diameter to the rootstock limits its use on a large scale. Using patch budding, 75-85% success has been obtained (Anon., 1981). Budding should be performed at 8-10 cm from the ground level. Over 93% success has been obtained by budding at a height of 10 cm on seedlings with a stem diameter of 0.53-0.55 cm (Singh *et al.*, 1981 b; Chattopadhyay and Dey, 1992). Lopping and topping of the rootstock from 8 (Kundu, 1983; Singhrot and Kajal, 1986) to 15 days (Singh *et al.*, 1984 c) after budding in the nursery beds has been found to ensure 90 to 100% success. Topping has also been found to increase growth of the budlings (Anon., 1981). Pareek *et al.*, (1999) has recommended that just before *in situ* budding in the field, rootstock seedlings should be lopped at 30-45 cm height. Budwood becomes available during the active growth period in the summer. The bud sticks, with well swollen and recently matured buds (but still not open) (Plate 2.10) are collected. Immature and undeveloped buds from the upper part of the new shoots are unsuitable. Similarly, over mature and inactive buds should not be used. Buds collected from a flowering shoot give very poor success. Buds should be collected from juvenile shoots. Such shoots can be induced to grow by severe pruning of the mother trees.

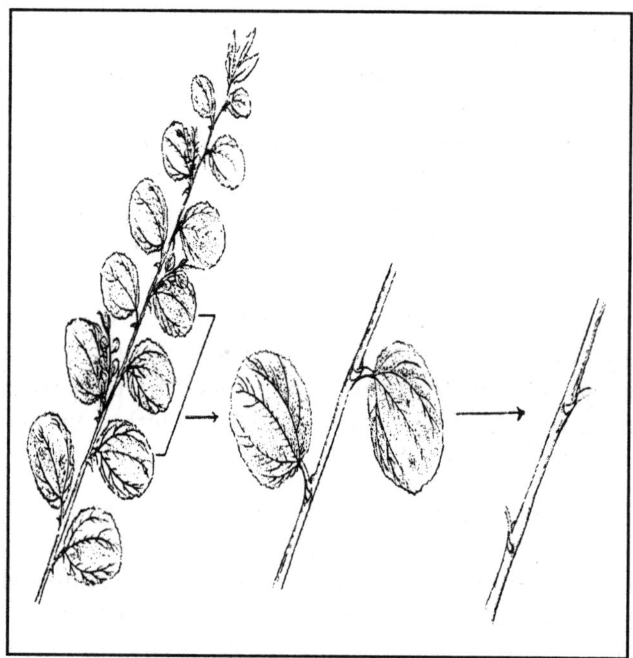

Plate 2.10. Shoot and appropriate stage of scion Suitable for taking scion bud

After collection, if the budwood is stored for some period or has to be transported, considerable loss of viability may take place. Budwood is observed to retain good viability when kept under ventilated conditions and wrapped in most jute cloth. Kanwar and Singh (1981) found that buds stored in polythene sheets, even for one day at 31-38°C and 67.68% relative humidity, failed in bud-take, however, buds kept in moist jute cloth for one day gave bud take of 68.8% and 34.2% when kept for 5 days. Singh (1988) found that budwood wrapped in polythene sheets and moist sphagnum moss gave better results than those stored in moist jute cloth. The best time for successful budding is during the active growth period. The active growth period is indicated by easy and clear separation of the bark from the wood in both scion and the rootstock. Kaundal *et al.,* (1984) observed maximum bud take (80-87%) at ambient temperature between 30 and 34^{0}C irrespective of variations in atmosphere relative humidity (RH) from 45 to 73.5%. Bud take declines below 36% RH and 18.5-20^{0}C temperatures. Favourable conditions for budding occur during the summer and monsoon period (June to September) in the arid and semi-arid subtropics of northwest India (Pareek, 1978 a; Kaundal *et al.,* 1984). Maximum sprouting of 93.2% has been recorded by budding when the maximum temperature was between 33 and 36.5^{0}C and minimum temperature was around 26^{0}C (Kaundal *et al.,*1984). Budlings prepared in polytubes become ready for transplanting at about 30 days after budding. The polytubes with the budlings are removed from the nursery and kept in shade for a week. These can then easily be transported and transplanted to the field with over 90% survival (Pareek, 1978 b) since there is little stress or root damage during lifting, transport and planting. The roots of the plants do not coil and, therefore, retain the drought hardy character and vigour almost similar to plants raised *in situ*. Ber can also be propagated through grafting, stooling, air layering, cuttings and micro propagation at a limited scale. The seedling population of ber can be converted into improved variety by top working (Plate 2.11 and 2.12). The process involves lopping of plant at 1 m height followed by budding on newly emerged 2-3 month old sprouts with suitable varieties and after ward care.

Plate 2.11. Initial stage of top working

Plate 2.12. Top working in BER

Planting

Planting of ber plants can be done either in spring or during the monsoon (Bajwa *et al.*, 1972; Singh *et al.*, 1973 a; Bakhshi and Singh, 1974). In arid regions, the best time for planting is with the onset of the monsoon (Pareek, 1983). Planting distance varies from place to place depending upon agroclimatic conditions and variety. The spacing between trees varies from 6 to 7m in arid northwest India (Pareek, 1983) and 7-8m (Bajwa *et al.*, 1972; Nijjar, 1975; Sharma *et al.*, 1980) in semi-arid regions.

The layout for planting is generally on a square pattern. After laying out, pits of 60 x 60 x 60cm dimension are dug well in advance so that they can be filled about one month before planting time. For filling the pit, top soil should be mixed with about 20 kg farm yard manure and 50 g chloropyriphos dust (10%) for protection against termites. After the soil is well settled, planting is done prefereably during the rainy season and plants immediately irrigated. Irrigation after every 4 or 5 days for the first two months after planting is essential.

Training and Pruning

Ber has a bushy and spreading growth form with long straggly branches and weak crotches. The wood of ber is weak and brittle and splits easily, particularly at the narrow crotches. A strong, upright and open frame should therefore be encouraged by proper training to obtain good fruit production. This can be done during the first two to three years after transplanting of budlings or after *in situ* budding of rootstocks in the field.

In subtropical regions, where cold winters and warm summers are common, ber trees are trained during the first three years after planting. During the first year, after plant establishment at the onset of monsoon (June-July), the plants are allowed to grow until the following spring (March) when the bush is headed back keeping 1-2 basal buds on the scion portion just above the graft union to induce development of vigorous shoots. Only one upright vigorous shoot is allowed to grow from the scion bud. The emerging shoot forms the tree trunk in due course. The trunk is kept clean up to a height of 30cm from ground level by removing all branches (Pareek, 1978a). From the trunk, 3 or 4 properly spaced and favourably placed branches are allowed to grow (Plate 2.13). The top of the trunk is headed back

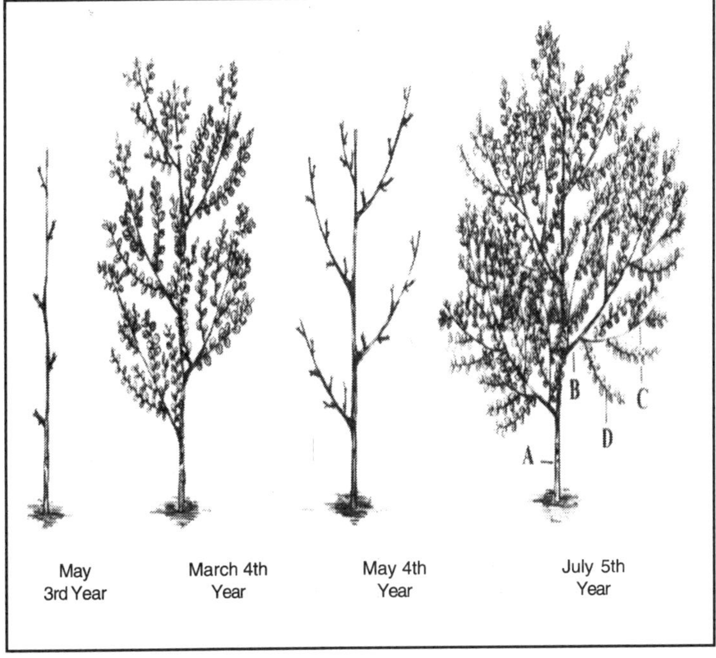

Plate 2.13. Training system for Ber after 2 yrs. of establishment to have strong frame.

during summer to encourage growth of these branches. Ber has a characteristic growth form in which production branches, usually starts from the sixth or ninth node from the base and subsequently at regular intervals of three internodes (Reddy and Chadha, 1993). These form the first order sylleptic branches or secondaries. It is observed that during spring or summer of the following year, shoots emerge from the basal buds of these secondaries and grow vigorously but the secondaries themselves either dry out or remain insignificant in growth vigour. Therefore, during the spring of the second year, the secondaries are pruned to retain their basal buds from which vigorous shoots emerge. More than one shoot may emerge from the basal bud, but only one of them is retained. These form the main branches of the tree frame. On the main branches, 3-4 upright growing and properly spaced side shoots are retained and then the top of the main branches are headed back. During the spring of the third year, these side shoots are pruned retaining their basal buds. From these buds, vigorous shoots emerge to from the sub-main branches of the tree frame.

Under the moderate climatic conditions of tropical regions, ber plants continue to grow during the winter months and thus can be trained within the first two years. About 5-6 months after planting, the scion portion is cut back retaining only the basal 1-2 buds. From the new growth emerging from the scion buds, only one upright growing vigorous shoot is retained and the others are removed. This forms the tree trunk. The tree trunk, main branches and sub-main branches are similarly developed as in the subtropical region except that the intervening growth periods between the lopping operations are shorter compared to about one year in the subtropics.

Pruning is necessary in ber to maintain productivity of the trees and to improve the quality of fruits. It is, therefore a regular yearly practice in well managed orchards. Through the pruning operation, the unproductive upper part of the past season's main shoots and its secondary branches as well as undesirable, weak, criss-crossing diseased and broken branches are removed so that the most healthy and vigorous growth is induced at the most productive nodes. Pruning also opens up the tree form for sunlight, air and ease of chemical spraying resulting in regular production of good quality fruits. Fruit bearing in ber is on the current season (Singh *et al.*, 1970; Pareek, 1983) and remains confined to its first and second order sylleptic branches, i.e. the secondary and tertiary branches (Reddy, 1983). Pruning should therefore induce the emergence of maximum number of secondaries and tertiaries on vigorous shoots. This can be done by pruning at the right time and with the right intensity depending upon location and cultivar.

The severity of pruning is determined by the length of the past season's shoot retained after the pruning operation, e.g. 20 or 25cm (very severe), 40 or 50 or 60cm (severe), 70 or 75 or 90cm (moderate), 100 or 120 or 125cm (light), 150cm (very light), (Dhaliwal and Sandhu, 1984; Bajwa *et al.*, 1986, 1987; Syamal and Rajput, 1989). Fruit yield decreases as pruning severity increases, and is greatest in unpruned trees (Dhaliwal and Sandhu, 1984; Bajwa *et al.*, 1986, 1987; Syamal and Rajput, 1989; Kundu *et al.*, 1995), but the fruit weight, size and quality increase significantly with pruning severity (Gupta and Singh, 1979; Dhaliwal and Sandhu, 1984; Bajwa *et al.*, 1986, 1987; Yadav and Godara, 1987; Gupta and Godara, 1989; Syamal and Rajput, 1989). Reddy (1983) observed that if the nodes up to the fourth or sixth secondaries (17-23 nodes) on the main axis are induced to sprout by pruning, vigorous shoots giving maximum fruit yield are produced. Also the new growth arising from the nodes on the main axis is significantly more productive than that emerging from the secondaries. The secondaries should therefore be completely removed at the time of pruning. Multilocation trials have confirmed these results (Pareek and Vishal Nath, 1996) and it has been recommended that the main axis should be pruned at 15 to 25 nodes depending upon agroclimatic conditions, i.e. at 20-25 nodes in arid areas and at 15 nodes in semi-arid or more moderate regions, along with complete removal of secondaries.

The time of pruning in ber varies in different regions. In subtropical regions, the most appropriate time for pruning is during the summer (between May and June in northern parts of India) when the trees enter are fallen dormancy their leaves into (Singh and Sandhu, 1984; Gupta and Godara, 1989; Gupta *et al.*,1990; Kundu *et al.*, 1995) before putting forth new growth. Early pruning has been observed to advance flowering and fruit maturity (Singh and Sandhu, 1984; Pareek and Vishal Nath, 1996). In tropical regions with a mild winter rainfall during December-January and early onset of summer rainfall (e.g. Tamil Nadu in southern India), pruning can be carried out any time from January to April and it is also possible to regulate fruit maturity so that bearing occurs at the desired time (Pareek and Vishal Nath, 1996). In tropical regions without winter rainfall (e.g. Andhra Pradesh of southern India), pruning during the first fortnight of April results in early flowering and therefore early harvest; however, the maximum yield of good quality fruits is obtained by pruning in the second fortnight of April. In Maharashtra, the best time for pruning is before the end of April (Pareek and Vishal Nath, 1996) as further delay causes reduction in fruit yield. Old and senile orchards of ber can be made productive by adopting rejuvenation technique owing to its high coppicing capacity and ability of latent buds to produce vigourous shoots.

Orchard Management

Generally old orchards become semile on account of general neglect and infestation of parts and diseases. These orchards can be made productive through rejuvenation. For rejuvenation, serve pruning, followed by balanced nutrient application and managemat of diseases and pest is essential. Work done at Kanpur (Pandey, 1999) on rejuvenation of old ber orchards has shown promising results. In this technique, all the branches at 1.8 m above ground level are headed back during summer. With the onset of monsoon, fertilizers and manures are applied which encouraged new shoot growth. As per health of the tree and branch orientation, 4-5 primary branches and subsequently 2-3 secondary branches are allowed to grow. On these secondary branches tertiary shoot bear flowers and fruits. However, heading of main branches should be done well above the budding height and plants should be fertilized properly.

Nutrient Management

Ber orchards are seldom manured. In the arid region of northwest India, farmers do not apply any manure or fertilizer in ber orchards other than that given at planting time. Regular manuring is, however, essential to replenish the nutrient removal by the tree through fruit harvests and annual pruning of its wood besides the losses from the soil. According to Ahlawat *et al.,* (1990) as many as 75% of the orchards in the arid subtropics were deficient in organic N and 90% in P but none in K. Under subtropical and tropical agroclimates, nitrogen deficiency is widespread and thus it plays a major role in determining the yield levels in ber trees. Deficiencies of N, P and K reduce shoot length, leaf number and size, axillary branch number, flowering and fruit set and consequently productivity of the tree (Sadhu *et al.,* 1978). On an average ber tree removes, 142 to 191 g N, 59 to 87g P and 467 to 684g K from the soil during a single growing season (Mehrotra *et. al.,* 1987). Major share of nutrients removed from the soil is apportioned alongwith the havested fruits. According to Pandey *et al.*, (1990), 43.8 g N, 7.5g P, 101.2g K, 4.5g Ca and 22.7g Mg are removed by 100 kg harvest of fruits. Judicious manuring and fertilization are necessary to replace these losses and to correct deficiencies, which may occur as a result of nutrient imbalances in the soil or plant. Nitrogen and phosphorus fertilization in the light soils of the arid and semi-arid subtropics are critical and potash may be applied to guard against its possible limitation as a result of annual removal over several years.

In the low fertility soils of arid and semi-arid regions, application of organic manures is considered essential for sustained nutrient supply. Organic manures also enable greater moisture retention for longer periods in these light soils. Similarly, fertilizers having minimum losses of nutrients, e.g. CAN (Calcium ammonium nitrate) rather than urea, are preferred. The best results have been obtained in young orchards of cultivar Umran by application of 400g N + 100g P + 200g K per tree (Dhatt *et al.*, 1993); in established orchards of Umran by application of 750g N per tree (Dahiya *et al.*, 1985); in 8 years old orchards of Kaithli by application of 500g N per tree (Yamdagni *et al.*, 1980) and in 6 year old orchards of Gola cultivar by application of 1000g N + 1000g P + 150g K per tree (Singh *et al.*, 1986). A multi-location trial indicated good results by application of 750g N per tree in deep sandy soils of the northwestern subtropics and by 250g N per tree in the shallow sandy loam soils in the southern tropics of India (Pareek and Vishal Nath, 1996). Thus, it appears that ber trees require 500 to 1000g N, 400 to 800g P and 100 to 200g K per tree depending upon age of the tree and edaphic and climatic conditions of the area. The method of placement of fertilizers depends on the root growth pattern. In an irrigated orchard, most of the active roots of 10-year old ber trees are located 2.35m from the trunk and up to 0.5m deep (Khera *et al.*, 1981). Under rainfed conditions, however, the roots make more vertical growth down to a depth of 4.5m (Pareek, 1977). Under rainfed conditions, foliar feeding is often a practical method of supplementing nutrient requirement, particularly of nitrogen and micronutrients. Nitrogen fertization of ber trees through foliar sprays of urea increased fruit set, fruit retention and yield, and improved fruit quality in cultivars Banarsi Karaka (Rajput and Singh, 1976; 1977), Umran (Chauhan and Gupta, 1985; Singh and Ahlawat, 1995) and Gola (Joon *et al.*, 1984). Dahiya *et al.*, (1985) found that application of 375 g N through foliar spray and 375g N to the soil gave a fruit yield of 128-130 kg per tree, higher than by applying all the nitrogen either through soil or foliage. The optimum concentration of urea for foliar application seems to be 1 to 2% (Chauhan and Gupta, 1985; Dahiya *et al.*, 1985; Singh and Ahlawat, 1995) and 2 sprays may be required at monthly intervals depending upon the nitrogen requirement. The best time for spray application seems to be after the fruit has set (Chauhan and Gupta, 1985; Dahiya *et al.*, 1985; Singh and Ahlawat, 1995) Supplementing micronutrients through foliar sprays of 0.2 to 0.4% $ZnSO_4$ or $FeSO_4$ or $MnSO_4$ and 0.2% boric acid has been observed to increase TSS and ascorbic acid contents (Kamble and Desai, 1996) and development of golden-yellow fruit colour. Combined sprays of micronutrinets (0.03% boric acid and 0.5% $ZnSO_4$) and growth regulator (500 ppm NAA) have been found to improve fruit quality in ber by increasing total soluble solids, total sugars and ascorbic acid and decreasing acidity (Singh *et al.*, 1989 a, b). Spray application of GA3 and NAA after fruit set improves fruit retention, yield and quality of fruits (Pandey, 1999). Microbial inoculations of soil and ber roots with Azospirillum, Azotobacter (Rao and Dass, 1989), vesicular arbuscular mycorrhizal (VAM) fungi (Mathur and Vyas, 1996) and other nitrogen fixing Actinomycetes (Tuohy *et al.*, 1991) can help in effective nutrient management.

Water Management

Water supply is the single most important factor determining productivity and quality of ber fruits, particularly in dry areas. The judicious use of water in rainfed as well as irrigated orchards is of great importance for sustained production at reasonable costs.

A large number of ber orchards found in dry areas depend only on annual rainfall for water. Productivity of these orchards depends mainly upon the quantity and distribution of rainfall and the extent of its subsequent use by the trees. A considerable quantity of rain water is lost as runoff and cannot be used by the trees unless water harvesting methods are employed. These may be *in situ* where runoff is concentrated into catchments surrounding a tree or group of trees, or *ex situ* where runoff from large rocky areas is diverted directly to nearby orchards or after collection in reservoirs or farm ponds. For *in situ* water harvesting land shaping may be done to prepare sloping catchments on both sides of the rows of ber trees (Plate 2.14). In the subtropical sandy plains, which receive 360 mm annual rainfall in north-west. India, higher runoff yield was obtained from catchments having 5% slope than from the usual and slope of 0.5% (Yadav *et al.*, 1980; Sharma *et al.*, 1982, 1986).

Plate 2.14. *In situ* Water harvesting system in Ber under Arid condition

Framers generally do not irrigate established ber orchards other than for intercrops (Verma and gujar, 1994). However, water deficits decrease the productivity of trees by reducing the assimilation rate. Productivity of ber trees therefore increases with irrigation. In the subtropical semi-arid region of north India, irrigation during the fruit development period has been observed to increase productivity by decreasing fruit crop (Bakshi and Singh, 1974). Although irrigation during the flowing period causes flower shedding in ber (Singh *et. al.*, 1973 a), watering may have to be done if there are prolonged breaks in the monsoon during the growth and flowering period. Irrigation at the time of fruit maturity delays and prolongs the maturity period. Moisture loss from ber orchard soils takes place through evaporation and runoff from the soil surface, deep percolation into the soil and transpiration through the weeds and trees. To conserve soil moisture for maximum water use efficiency, the losses must be minimized by weed control, mulching and reducing losses through transpiration from the trees. Mulching conserves soil moisture. Mulching also suppresses weed growth. Available waste materials such as grasses, castor shells, other organic wastes, polythene and even pure sand can be used as mulch to reduce moisture loss. In a multi-location trial black polythene proved to be the most effective (Pareek and Vishal Nath, 1996).

Application of radiation reflectants such as kaolin, film forming chemicals (including liquid paraffin and power oil) and chemicals causing stomatal closure without affecting gaseous flow (including phenyl mercuric acetate) have been tested to reduce water loss through transpiration, from the leaf surface of ber trees. Spraying of 7.5% kaolin and 1.5% power oil have been found to be useful (Pareek, 1998).

Orchard Floor Management

Weed growth is usually checked by a combination of mechanical and manual control measures. Ploughing the inter-row spaces immediately after summer pruning exposes the soil to solar radiation and kills hibernating insect pests. It also allows maximum penetration of rain-water into the soil and induces weed seeds to germinate during the following monsoon. Ploughing is carried out just after the weeds emerge and is followed by harrowing about one month later. These operations are combined with manual weeding of the area around the trees. Herbicides can also be used to control weeds. Mechanical control by ploughing the inter-row spaces combined with chemical weed control around the trees can be an effective alternative. Herbicides such as dalapon followed by paraquat, each at the rate of 5 kg/ha (Bajwa *et al.*, 1990) and glyphosphate at the rate of 4 kg/ha (Bajwa e*t. al.*, 1993) are very efficient for killing the weeds.

Ber trees cover the entire inter-row spaces about five years after planting. In the mean time, considerable losses occur from these vacant interspaces. By growing intercrops, the losses can be minimized and additional income generated. Intercrops such as mung bean, moth bean, cowpea, clusterbean (Plate 2.15) and sesame can be grown under rainfed conditions during the summer. Perennial grass species like Cenchrus ciliaris has been found a suitable inter crop under sandy soils of Jodhpur (Plate 2.16). The yields of both ber and the intercrops are higher than in monoculture and no adverse effects are observed for upto 5 years after planting the ber tree (Singh, 1997). Under irrigated conditions, horse gram, cumin, chillies and other vegetables are more profitable (Pareek, 1983). In the semi-arid plateau of western India, a ber + clusterbean combination gave the highest net returns per hectare followed by ber + cowpea, ber + okra, and ber + brinjal combinations (Raturi and Chadha, 1993). Dinesh Kumar and Om Prakash (2001) reported pearl millet-wheat is the best intercrop in ber plantations under Agra conditions. Cover cropping with kulthi (*Dolichos biflorus*) was found to increase water-holding capacity of light soils as a result of increased organic carbon content in arid region in the tropics (Pareek and Vishal Nath, 1996). The legume, *Stylosanthes hamata* is a good cover crop in the semi-arid regions of western India (Raturi and Chadha, 1993).

Plate 2.15. Cluster bean as Intercrop in Ber orchard

Plate 2.16. Gross as Intercrop in Ber orchard

Disease Management

Ber is infected by number of diseases which cause considerable loss to ber plantations. Notable diseases and their management strategies have been enumerated.

Powdery mildew

Powdery mildew (*Oidium erysiphoides f. sp. ziziphi*) causes great losses in ber in India, particularly in humid areas. Epidemiological studies have shown that maximum development of powdery mildew occurs when the maximum temperature is between 24-35^{0}C, minimum temperature between 4-22^{0}C, morning RH is between 63-91% and evening RH between 24-57% (Pareek and Vishal Nath, 1996). The disease occurs with increased virulence during high rainfall periods however, it may also appear earlier if conditions are favourable. The symptoms of the disease have been noticed on flowers and newly set fruits. The developing young leaves show a whitish powdery mass, which causes them to shrink and defoliate. The disease also appears in the form of white powdery spots on the surface of the fruits and later covers the whole fruit surface. The spots turn into light brown to dark brown discoloration. The infected area becomes slightly raised and rough. The affected fruits either drop off prematurely or become corky, cracked, mis-shapen and remain underdeveloped. Sometimes the whole crop is rendered unmarketable.

The disease can be controlled by spraying dinocap (0.2% Karathane WP or 0.1% Karathane EC), 0.2% sulfex (Gupta *et al.*, 1977, Yadav and Singh, 1985; Singh *et al.*, 1995), 0.2% carbendazim (Yadav *et al.*, 1980a Singh and Sidhu, 1995), 0.2% carbendazim (Yadav *et al.*,1980; Singh and Sindhu, 1985) or trideomorph (Singh *et al.*, 1995). Multi-location trials in India have shown that fungicides such as dinocap (0.1%), carbendazim (0.1%), wettable sulphur (0.2%), trideomorph (0.1%), sulphur dust (250 g/ tree) and thiophenate methyl (0.1%) were effective for the control of powdery mildew. Starting with the initiation of the disease, 2 to 4 sprays at 15-20 day intervals should be carried out depending upon disease intensity. The time of spray is critical for effective control. Gupta *et al.*, (1978) and Singh and Sindhu (1985) suggested that spraying should be started as soon as the disease appears on the fruits of peanut size and repeated at 3-week intervals. In regions with an established period of recurrence of the disease, one spray of fungicide should be applied as a prophylactic measure as soon as growth appears after pruning (Pareek and Vishal Nath, 1996). Some varieties have also been reported to the resistant under field condtions. Cultivars Illaichi Jhajjar, Gola, Seb, Safed Rohtak and Mehrun have shown comparative tolerance to the disease in northern India. Five local genotypes (Darakhi-1, Darakhi-2, Guli, Villaiti and Seedless) have been reported to the free from the disease.

Alternaria leaf spot

The disease, caused by the *Alternaria anomorph* of *Leiwa (Pleospora) infectoria*, produces small, irregular brown spots on the upper suface of the leaves and dark brown to black spots on the lower surface (Gupta and Madaan, 1977 a). The spots coalesce forming large blighted patches and the affected leaves drop. The disease develops at temperatures from 20 to 30^{0}C (Madaan and Chand, 1985), but high humidity and frequent rainfall appear to be more important. Plant debris seems to be a source of primary infection. Secondary infection is possibly caused by dissemination of spores under frequent rainfall. The disease can be controlled by spraying 0.3% mancozeb or 0.2% captafol or 0.3% copper oxychloride (Pareek and Vishal Nath, 1996).

Black leaf spot

The disease is caused by *Isariopsis indica* var. *ziziphi* and is also known as *Isariopsis* mouldy leaf spot (Gupta and Madaan, 1977 b). The disease has been observed under comparatively humid conditions. The disease is characterized by sooty tuft-like circular to irregular black spots on the underside of the leaves (Plate 2.17). Later, it covers the entire lower surface giving a sooty appearance. The leaves slow

Plate 2.17. *Isariopsis* leaf spot in Ber

yellowish and brownish discoloration on the upper surface and drop prematurely. The fungus survives in plant debris and soil, which are the primary source of infection. Secondary infection is initiated from spores present in the air. The disease can be controlled by capitol (0.2%), carbendazim (0.1%), mancozeb (0.2%) and copper oxychloride (0.2%). The sprays should be applied 2-3 times at 15 days interval starting with the initiation of symptoms (Pareek and Vishal Nath, 1996). Rawal and Saxena (1989) obtained good control by 0.1% carbendazim and 0.2% chlorothalonil. Cultivars Seo-Bahadurgarhia, ZG-3, Safed Rohtak and Sanaur-1 have been found to resist the disease infection in a multilocation trial in India (Pareek and Vishal Nath, 1996). Cultivars Mundia, Banarsi Karaka, Banarsi Pewandi and Bagwadi also showed resistance to the disease at some locations.

Other foliar diseases of ber are *Cercospora* leaf spot, *Cladosporium* leaf spot and rust which also cause considerable loss to ber yield. These diseases can be managed by timely spray of suitable fungicides.

Fruit rot in ber is caused by *Phoma hissarensis, Collectotrichum gloeosporioides, Trichothecium roseum, Alternaria* state of *Pleospora infectoria, Cladosporium hebarum* and a number of other pathogens.

Pests Management

Several insects infest ber but only fruitfly, fruit borer, bark eating caterpillar, chafer beetle and hariy caterpillar cause significant damage.

Fruit fly

Carpomyia vesuviana, Dacus correctus and D. *dorsalis* (Diptera: Tephritidae) infest ber fruits (Basha, 1952; Batra, 1953; Saen, 1986). C. *Vesuviana* has been observed to damage as much as 80% of the crop under sever infestations (Cherian and Sunderam, 1941). Infestation starts with the onset of fruit setting. The adult female lays eggs by inserting its ovipositor in the developing fruit. After 2 to 5 days, the larva hatches out and starts feeding on the pulp making galleries in it. Generally, only one larva is found in one fruit. The excreta of the larva accumulate in the galleries, which may sometimes result in rooting of the fruit. Infested fruit become deformed and their growth becomes checked. A large number of such fruits drop off. The larval stage lasts 9 to 12 days. When full grown (6mm length), the larva finds its way out by making a hole in the fruit skin and drops to the ground. The larva bores down into soil upto a depth of 2 to 12cm where it pupates. The pupal period lasts for about 2 weeks after which the adult fly (5 to

8mm long, 3 mm broad) emerges. There may be 2 or 3 generations of the pest during the active period (Batra, 1953) while the fruit matures (from November to April in north India). Cultivation of the orchard soil during spring (Singh *et al.*, 1973a), summer (Chundawat and Srivastava, 1978) and rainy season (Bakhshi and Singh, 1974) destroys the hibernating pupae by exposing them to bright sunlight and birds and thus the extent of infestation in considerably reduced. Heating the soil by burning grass and irrigation during the summer also kills the pupae. Infested fruits having maggots should be collected and buried.

To prevent infestation, prophylactic sprays should be carried out with 0.03% oxidematon or dimethoate starting from the stage when 70-80% fruit attain pea size and then repeating the spray at one-month intervals (Pareek and Vishal Nath, 1996). During the maturity of fruits, if necessary, spraying should be done with 0.5% Malathion + 0.05% sugar solution at weekly intervals. Malathion has been observed to dissipate quickly in ber fruits decreasing well below the tolerance level of 3 ppm within 2 days after spraying (Popli *et al.*, 1980). The pest has been effectively controlled by insecticidal spray schedules such as i) three applications of 0.1% fenthion or two sprays of endosulfan or malathion (Patel *et al.*, 1989), ii) 0.005% fenvalerate or 0.0015% deltamethrin or 0.05% monocrotophos or phosphamidon (Bagle, 1992), and iii) two applications with 0.1% dichlorvos, one at pea stage of fruit and another after one month (Raghumoorthi and Arumugam, 1992) and monocrotophos or phosphamidon + carbaryl treatments (Patel and Patel, 1979). Based on a multi-location trial, a control schedule consisting of 3 lots of spraying, one with 0.03% monocrotophos, second with 0.05% fenthion and third with 0.1% carbaryl XLR, has been recommended (Table-2.10).

Table 2.10 : Effect of chemical control schedule (first spray at pea stage with 0.03% monocrotophos, second spray after 15 days with 0.05% fenthion and third spray after another 15 days with 0.1% carbaryl XLR) on per cent fruit damage by fruitfly.

Treatment	Jobner			Bawal			Anantapur	Sardar Krushinagar
	First pick	Second pick	Third pick	First pick	Second pick	Third pick		
Treated trees	1.5	1.5	0.6	2.1	2.7	3.6	13.7	17.4
Untreated trees	20.7	16.7	7.7	34.8	42.1	49.4	21.3	26.1

Source : Pareek and Vishal Nath, 1996.

Almost all commercial cultivars are attracked by fruitfly but the extent of damage varies in different locations and years depending upon agroclimate and growing conditions. Mann and Bindra (1976) reported that cultivars Sanaur-1, Safeda selected, Illaichi, Mirchia, ZG-3 and Umran resisted fruitfly damage. Singh (1984b) observed that the extent of infestation varied between cultivars from 6.7% in Tikadi to 73% in Gola, Gola Gurgaon-3 and Kaithili (Singh and Vashishtha, 1984) and that the percentage of larva hatched also varied indicating varying degrees of antibiosis.

Fruit borer

Damage by fruit borer (*Meridarchis scyrodes* (Lepidoptera : Carposinidae)) has been observed mainly in southern and western India (Sonawane and Dorge, 1971; Pareek and Vishal Nath, 1996). The reddish larvae bore into the fruit and feed on the pulp. Its moth is dark brown. Based on multilocation trials in India, a chemical control schedule consisting of first spray at pea stage with monocrotophos (0.03%), second spray after 15 days with fenthion (0.05%) and a third spray 15 days after the second spray with 0.01% carbaryl has been recommended (Pareek and Vishal Nath, 1996). Cultural schedules consisting of collection and destruction of fallen fruits and digging the orchard soil under the tree canopy have also given good control. Ber cultivars Banarsi Pewandi, Ajmeri, Gola Gurgaon and Jhajjar Selection have been found to be resistant to the pest.

Bark eating caterpillar

The caterpillars of *Indarbela quadrinotata, I. watsoni* and *I. tetraonis* (Coleoptera : Cerambycidae) have been reported to make winding galleries of frassy web on the stem near the forks and angles of branches. The caterpillar reins concealed in the stem in the day-time and eats away the bark during the night. Heavy infestation by this pest retards growth of the trees and adversely affects fruit yield. The moth is pale with grey marks on the forewings. It lays eggs with the onset of the rainy season in batches of 15 to 25. A single female moth can lay 300 to 400 eggs. Incubation period lasts 8 to 10 days. After emerging, the caterpillars start eating the bark. The dark brown, full-grown caterpillar is 37-50 mm long and has a dark head; its body is covered with long and thin hairs. The caterpillars pupate during the summer for about 4 weeks and from the pupae, adult moths emerge, mate and lay eggs again.

For control of the pest, application of 0.05% monocrotophos has been found to be the most effective followed by 0.2% trichlorphon and 0.05% chlorfenvinphos (Verma and Singh, 1975). Before application of the chemical, frassy galleries should be cleaned to expose the holes. Application of the solution, made up by mixing on litre kerosene and 100g soap in 9 litres of water, to the holes has been found to effectively control the bark eating caterpillar. Alternatively, each hole should be filled with a solution of 2ml monocrotophos or 20 ml trichlorphon 50EC or 30ml endosulfan 35 EC in 10 litres of water and then the holes closed with mud. Ber cultivars have been screened to assess their suceptability to the pest (Verma and Singh, 1974; Mann and Bindra, 1977; Singh, 1984 a; Kumawat and Swaminathan, 1990). Verma and Singh (1974) reported that Desi Alwar cultivar was resistant. Mann and Bindra (1977) reported that Rohtak Gola had the least number of infective holes.

Hairy caterpillar

Hairy caterpillars (*Dasychira mendosa, Euproctis fraternal* (Lepidoptera : Noctuidae) remain gregarious and scrape the leaves and young fruits. The older caterpillars spread in all directions and decour leaves and fruits and sometimes even the tender shoots. They start eating new foliage emerged after pruning and this is continued by overlapping generations. The full-grown larva is reddish brown with dark brown head. The larva pupates on the plant in a yellowish, hairy cocoon from which a yellowish moth with pale transverse lines on the forewings emerges. It then lays flat, circular yellowish eggs in masses on the lower surface of the leaves. The females lay 92-241 eggs, which hatch in 9-14 days (Bhatnagar and Lakra 1992). The pre-oviposition, oviposition and post-oviposition periods last 2,1 and 4 days respectively, and the larval, pre-pupal, pupal and adult stages last 25-26, 2-3, 7-15 and 5-7 days respectively. As many as 6 generations each of 44-84 days duration have been observed in a year. Verma *et. al.* (1972) reported that all the instars were controlled by treatment with 0.1% carbaryl or endosulfan or trichlorphon or 0.05% methyl parathion. Bhatnagar and Lakra (1992) obtained the most effective control of 3rd and 6th instar larvae by 0.005% fenvalerate. A spray of 0.05% monocrotophos and 0.2% carbaryl (Killex carbaryl 50WP) was found most effective in controlling the pest (Pareek and Vishal Nath, 1996).

Cockchafer beetle

Chafer beetle (*Adoretus decanus, A. kanarensis, A. stoliezkae, A. pallens, A. versutus*) (Coleoptera : Scarabaeidae) devour ber leaves during the night. They become active with the onset of the rainy season when new growth starts. In severe cases, the whole tree is rendered leafless. Eggs are laid in the soil during the early part of the rainy season (May to August in north India). Larvae hatch out in one week and feed on roots and vegetation. Adults emerge with the onset of rains. There is only one generation per year. Chafer beetles can be controlled effectively by spraying 0.2% carbaryl 50 WP and 0.05% monocrotophos (Pareek and Vishal Nath, 1996). *A. pallens* can be controlled by spraying 1% lead arsenate (Trehan, 1956). The beetles can be trapped by using any source of light and killed by dropping them into water containing kerosene.

Other insect pests

The nymphs and adults of the leaf hopper (*Zyginida pakistanica*) (Lepidoptera : Zygaenidae) suck sap from the underside of the leaves (Khangura and Sandhu, 1976). *Phyllodiplosis jujubae* (Diptera : Cecidomyiidae) forms galls on the leaves of developing shoots causing them to curl irregularly and lose colour (Gangwar, 1983). Wax scale (*Drepanococcus chiton*) (Homoptera : Coccidae) infest ber and can be controlled by application of dimethoate (0.05%), phosalone (0.07%) or dichlorvos (0.1%).

Flowering, Fruiting and Yield

Budded plant of ber starts bearing in the second year but the commercial production from a tree is expected only after 5-6 years. Ber plants can produce fruits after the first year of planting from budded plants or after *in situ* budding in the field. However, during the first 2-3 years, the young plants are trained to develop into balanced trees when fruit production is of little significance. In general, peak productivity is reached at 10 years of age and continues until about 40 years, depending upon care and management. Ber bears fruit on current season's growth. The shoots emerging in the June-July after pruning in May produce flowers during August-September. In north India, the minimum and maximum temperature, respectively 9.8-25.4^{0}C and 28.2-35.7^{0}C and in south India respectively, 19.7-26.3^{0}C and 28.9-38.1^{0}C favours flowering in ber (Vishal Nath and Bhargava, 2000). The flowering in ber comes in clusters as axilary cyme borne on the axil of leaves. Anthesis of flowers is cultivar specific. In some cultivars like Seb, Banarasi, etc it takes place in morning whereas in Umran, Mundia, etc it takes place in afternoon (Pareek, 2001). Generally, the stigma remains receptive for 10-14 hours. Insects are the main pollinating agents. Prevailing temperatures during bloosming period governs the fruit set. Temperature range between 25-32^{0}C considered congenial for fruit set in ber. The fruits mature by December. Yield of ber plant vary with the varieties. In the semi-arid subtropical climate of northern India under irrigated conditions, the fruit yield per tree ranges from 80 to 200kg depending on the varieties and management practices during the price bearing age of 10 to 20 years (Bakhshi and Singh, 1974). A well grownup and nourished plant of Gola produces 80-110 kg, of Banarasi Kadaka produces 100-120 kg and of Umran produces 110-140 kg fruit every year. Verma and Gujar (1994) reported that farmers get about 100 to 150kg fruit yield per tree at the age of 20 to 25 years. In rainfed regions, the yield is lower, i.e. 80 to 100 kg per tree in semi-arid areas and 50 to 80kg in arid areas depending upon rainfall pattern during the year.

Harvesting and Post harvest Management

Immature ber fruits do not mature after picking. Such fruits do not have satisfactory sweetness and remain acrid in taste. Fully ripe fruits, on the other hand, lose their attractive yellow or golden-yellow colour and crisp and juicy texture and become red or dark brown in colour and slimy in texture. It is, therefore, essential to harvest ber fruits at the optimum stage of maturity. Maturity in ber fruits is usually judged by external colour. Fruits are harvested at mature-green and mature-golden yellow stages depending on cultivar, distance from the market and the expected post harvest use. At the fully ripe stage (red brown colour), the fruits develop poor organoleptic quality but are suitable for dehydration. Fruits of cultivar Umran develop good organoleptic quality at the mature-golden yellow stage (Singh *et al.*, 1981a). Owing to low respiration rate, Umran fruits are suitable for prolonged storage when harvested at this stage (Singh *et al.*, 1981a). Fruits of cultivar Kaithli are harvested when green-yellow and of Gola at green-mature or green-yellow or yellow stage (Siddiqui and Gupta, 1989a). For transportation, fruits of Kaithli cultivar should also be harvested at the green yellow stage (Siddiqui and Gupta, 1989a).

Precise indices of fruit maturity in ber are specific gravity, TSS, TSS/acid ratio and degree day heat unit accumulation. These values also vary in different cultivars. Specific gravity values at harvest maturity (yellow colour stage) in Kaithli, Gola and Umran have been observed to be 0.88, 0.93 and 0.81 respectively with corresponding fruit wall pressure values of 2.07, 2.52 and 2.8 by Bhatia and Gupta, (1985). Meel *et al.,* (1991) recommended that fruits of Sandhura Narnaul ber should have <1 specific gravity for harvest maturity when the 'degree-day' accumulation, counting from full bloom, was 1840. At fruit maturity, the TSS and TSS/acid ratio should be respectively 16.7-17.0% and 76-84 for Gola, 16.2-16.5% and 75-100 for Kaithli and 17.1-17.4% and 93-100 for Umran (Bhatia and Gupta, 1985). The degree day heat units, above the base 7.2^{0}C, have been observed to be 1980-2236, 2236-2566 and 2516-2920 respectively for early, mid-season and late cultivars.

Harvesting of ber fruits is carried out by manual shaking or beating of the tree branches. Sometimes, a cloth sheet is spread below the tree to facilitate collection of fruits. Harvesting of fruit by shaking or beating the branches, however, results in considerable spoilage. Fruits can also be manually picked using a ladder or using an iron loop or hook fixed to a piece of bamboo. Harvesting of fruits along with attached pedicels increases their storage life. Attached pedicels had little effect on shelf life of Kaithli ber when stored upto 4 days at room temperature (32^{0}C), but these were beneficial. Fully mature fruits are picked several times since the fruits do not attain harvest maturity at one time. In one season, 4 to 5 pickings are usually made. The number of pickings can be reduced to 2 or 3 by pre-harvest sprays of 500-750 ppm ethephon (Pareek and Vishal Nath, 1996). Morning is generally considered as the best time of harvesting because at that time the fruits are cool and turgid and can be sold, stored or processed during the rest of the day.

After picking, the fruits are collected and separated according to variety and are brought to the packaging house/ shade for cleaning, grading and packing or for post harvest treatments to extend their shelf life. The under-ripe, over-ripe, damaged and misshapen fruits are removed. The remaining fruits are graded into two or three levels on the basis of size and colour. Grading can be carried out either manually or by passing through sieves. A grading standard for ber has been suggested (Table-2.11).

Table 2.11 : Grading standards of ber fruits

Grade	Standard
A	Shining yellow, large to medium size fruits of uniform shape with no blemishes
B	Uneven yellow or yellow red, large to medium size fruits, of uniform shape with some blemishes
C	Red, large to small fruits. Uneven yellow, small fruits

Source : Pareek and Gupta (1988)

The fruits are packed either for proper storage or for safe transport to local or distant markets according to grade. Packing of fruit is done in baskets with paper lining for bulk and in nylone net bags for retail market. Ber fruits can also be packed in non organic materials like crates etc. Small packages of 1-2 kg in perforated polythene bags, nylon-net bags and sometimes in cardboard cartons can also be used. Fruits are also packed in large packages of 10-20 kg in gunny bags, cloth packages or wooden or plywood boxes with holes (Pareek and Gupta, 1988). A study using gunny bags, wooden carries, bamboo baskets and corrugated cartons for packing Kaithli and Umran ber showed that corrugated cartons of 8-9 kg (40 x 25 x 20 cm) with 6 holes of 1 cm diameter on two sides and paper cuttings as cushioning material retained good fruit quality for 9-12 days (Gupta *et al.,* 1981). At low temperatures, even non-perforated polythene bags have been used. The storage time for retaining different varieties of ber fruit varied according to temperature. Baskets made from locally available materials (e.g. mulberry sticks) can be used for packaging the fruits. Spoilage can be reduced by the use of cushioning materials, such as paper cuttings, during transport and storage, particularly when large packages are used.

For long distance transport (800 km), cardboard containers and mulberry baskets with straw as cushioning material proved the best packaging materials (Jain and Chitkara, 1980). Transport by rail resulted in lower physiological weight loss than by trucks (Anon., 1987). After a 19 hours journey of 600 km by rail, fruits of Gola ber retained good quality with the least weight loss when packed in corrugated cartons (Singh and Gupta, 1987). For transportation, corrugated cartons of about 10 kg seems to be the most suitable. However, for short distances and for local transport, any cheaper local materials (e.g. gunny bags, cloth, old boxes etc.) can be used provided care is taken to minimize damage / spoilage by making provision of cushioning material and ventilation.

Ber fruits are usually stored at ambient / room temperature (25-30°C) from harvest until their disposal, i.e. at the orchard site and then during transport to local and distant markets. Storage is often done in heaps under sheds or in rooms. But it is better to store in packages such as gunny bags, net bags, polythene bags, cloth packs and boxes. Depending on cultivar and storage conditions, the fruits can be kept for 4 to 15 days without loss of organoleptic quality. Fruits of cultivars Sanaur-2, Umran and Kaithli remained in marketable condition for about one week at room temperature of 28-32°C (Panwar, 1981; Bal, 1982). During storage, the fruits lose weight and shrivel, change colour and become red, lose acidity and ascorbic acid but gain sweetness. After 9 days of storage at room temperature (30-35°C), Umran fruits shriveled somewhat, changed colour from golden yellow to grey orange, increased in TSS from 17.5 to 19.5%, in total sugars from 10.19 to 16.38%, decreased in acidity from 0.21 to 0.12%, in ascorbic acid from 117.13 to 93.72 mg /100g and in total phenolics from 0.108 to 0.039% (Bal *et al.*, 1978).

Shelf life of fruits can be executed by storage in zero energy cool chambers. High humidity develops in the cool chamber, is condusive for spoilage but the fruits could be stored in these chambers for 6 to 10 days, depending upon cultivars (Anon., 1989 a; Siddiqui and Gupta, 1990). Under cold storage conditions, ber fruits retain acceptable organoleptic quality for 3 weeks in perforated polythene bags and baskets stored at 13°C in an incubator (Jain *et al.*, 1981) where as at 10°C fruits of different cultivars remained in good condition for 28 to 42 days (Table 2.12).

Table 2.12 : Storage life (in days) of some ber cultivars in various storage environment

Storage condition	Gola	Kaithli	Umran
Room temperature (Min. 17.8°C, Max. 28°C, RH 64%)	9	7	10
Zero energy chamber (Min. 11.6°C, Max. 20°C, RH 95%)	18	14	15
Cold storage (10°C, RH 79%)	42	28	35

Source : Anon. 1989a

It has been observed that calcium content in ber fruits influences shelf life. Therefore, pre-harvest applications of calcium compounds have been helpful in extending storage life. The fruits contain calcium as pectate, carbonate, oxalate and phosphate (Minamide *et al.*, 1986). Spraying 10 days before harvest with calcium chloride using 1.7g per litre as calcium with 1% Teepol as surfactant, gave the best results in increasing shelf life of Umran ber (Gupta *et al.*, 1987). Spraying of 1% calcium nitrate at the colour turning stage (10 days before harvest) improved the shelf life of fruits of Umran, Gola and Kaithli cultivars at room temperature when stored in wooden crates lined with a cushioning material (Siddiqui *et al.*, 1989; Siddiqui and Gupta, 1989 b; Gupta *et al.*, 1989 b;). Post-harvest treatments of the fruits before packaging, e.g. cleaning, dipping in cold water to remove field heat, treatment with chemicals such as calcium compounds, anti-oxidants, growth regulators and fungicides, influence their shelf life. Dipping of fruits in cold water for 2 hour or exposing to cold air for 4 hours immediately after harvest to remove field heat prolongs their shelf life (Gupta and Mehta, 1987). Dipping the fruits in 1 to 2% calcium chloride solution containing a surfactant, has been found to prolong their shelf life. Post harvest dipping in 150 and 300 ppm ascorbic acid solution reduced over-ripening, increased TSS but had no effect on acidity and ascorbic acid content during storage (Siddiqui and Gupta, 1995). Both physiological and pathological losses were reduced in Gola fruits by dip treatments either in 5% Virosil Agro (H_2O_2) or 1% calcium chloride even when stored at ambient temperature (28±2^{0}C) for 9 days (Anon., 1998-99). Growth regulator treatments have been helpful in delaying ripening and in maintaining high quality in ber fruits during storage. Treatments with 500 ppm cycocel for 15 minutes (Siddiqui and Gupta, 1995), 200 ppm maleic hydrazide (Bandyopadhyay and Sen, 1995), benzyladenine (BA) at 100 ppm (Sandhbhor and Desai, 1991) have been reported to result in lower weight loss and improvement in quality with respect to retention of sugars and vitamin C in ber fruits when stored in wooden boxes or polythene bags.

During post harvest period, microflora such as *Aspergillus niger, A. sydowii, Rhizopus oryzae, Penicillium chrysogenum, Alternaria tenuissima, Phoma* sp. and *Curvularia* sp. may cause spoilage (Kaisana *et al.*, 1978). Singh and Gupta (1983) observed that *Cladosporium chartarum, Phoma hissarensis and Botrodiplodia theobromae* caused decay losses more frequently in the packages. The incidence of these microbes increased when doob grass (*Cyanodon dactylon*) was used as cushioning material. Treatments such as dipping the fruits in 500 ppm TBZm captaf or Dithane M-45 (Gupta and Mehta, 1987), 0.05 and 0.1% $KMnO_4$, 500ppm cycocel (Siddiqui and Gupta, 1995), fumigating with 5g bleaching powder per kilogram of fruit (Singh and Gupta, 1987) and lining the packages by diphenyl impregnated paper (Gupta and Mehta, 1987; Siddiqui and Gupta, 1988) reduced the decay losses in the packages. Fungicidal treatments reduce decay losses in storage. Decay losses were also reduced when treated with 500 ppm thiobendazole (Gupta *et al.*, 1989b) or by 0.2% $ZnSO_4$ (Siddiqui *et al.*, 1989) or by 1% Bavistine (Vishal Nath and Bhargava 1998).

Processing and Value Addition

Ber fruits can be processed in to various value added products. Important among are:

Dehydrated ber

Traditionally, the fruits are washed, drained and sun-dried. The product quality will be often poor, if care is not taken to keep the fruits of different varieties and maturity grades in separate lots and use the necessary pre-drying treatments. The techniques for preparation of dehydrated ber (Pareek, 2001) has been given in Fig. 2.1. The quality of product from different varieties differs. Fruits of the cultivars Katha or Umran, Bagwadi, Chhuhara, Mehrun, Sanaur-2, Illaichi and Karaka give a good dehydrated product (Khurdiya, 1980; Aggarwal *et al.*, 1997). In general, product quality from Umran or Katha and Chhuhara is considered better, but local taste preferences may demand a product from the slightly acidic fruits of Sanaur-2, Sanaur-3, Sanaur-4 and Mehrun. Pre-treatments such as blanching and sulphuring before dehydration improve product quality. Blanching induces development of a uniform yellow colour, softens the fruit for uniform dehydration and checks the activity of enzymes and microorganisms. The optimum time required for blanching varies with softness of the pulp and size of the fruit in different cultivars. After blanching, the fruit should be exposed for 3 hours to sulphur-dioxide fumes by burning sulphur powder at the rate of 3.5 to 10g per kg fruit in a sulphur box (Khurdiya and Singh, 1975). Sulphuring helps in the retention of ascorbic acid content and colour in the dried fruit.The product quality also varies due to different drying methods and local weather and sanitation particularly when dried in open. Drying is commonly done by spreading fruits on the floor, on mats or on polythene sheets in the sun for 7 to 10 days. Usually, drying process is completed in 4 to 5 days and the product quality is superior because the sulphured fruits retain a higher SO_2 content with very little changes in reducing and total sugars (Khurdiya and Roy, 1986).

Cleaning and **sorting** fruits

(Select ripe fruits)

Blanching

(Dip in boiling water for 2-6 minutes)

Sulphuring

(Expose to SO_2 fumes in a box for 3 hours by burning sulphur powder at the rate of 3.5-10 g/kg fruit)

Dehydration

(Retain 10-15% moisture in the product)

a. Under the bright sun for 7-10 days

b. In solar drier for 4-5 days

c. In cabinet drier at 60-65^{0}C for 20-35 hours

↓

Packing

(Polypacks/Tins)

Fig. 2.1 : Flow chart for the preparation of dehydrated ber

Murabba and candy

Delicious preserve and candy can be made from mature ber fruits. Fruits of Umran, Banarasi Karaka and Kaithli cultivars are suitable for the preparation of preserve (Gupta *et al.*, 1981b). Sanjeev Kumar *et al.*, (1992) observed that candy prepared from Umran ber was of the best organoleptic quality since its fruits had the highest total soluble solids, the lowest acidity and average size. Fully mature fruits at the hard stage are used for making preserve and candy. Flowsheet for preparation of candy and preserve has been shown in Fig. 2.2. The fruits are pricked and softened so that sugar impregnation is uniform and effective.

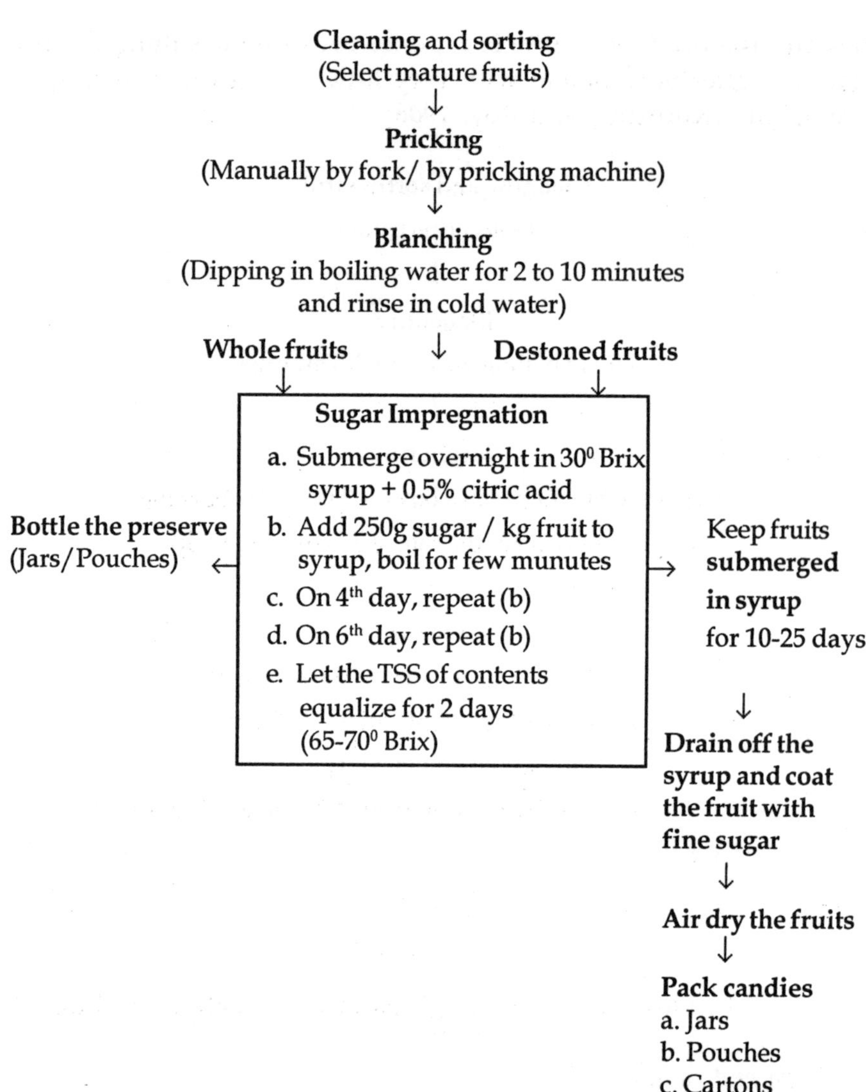

Fig. 2.2 : Flow chart for the preparation of preserve and candy

Pricking of the fruits can be done manually with a fork or a pricking board or by using a pricking machine. Fruits are then blanched for 3-10 minutes depending upon the cultivar (Gupta, 1983). The prepared fruits are kept submerged overnight in sugar syrup of 50^0 Brix (Gupta, 1983). Addition of citric acid at 0.5% (Khurdiya and Singh, 1975) to the syrup imparts a bright colour and slightly acidic taste to the product. The strength of the sugar syrup is then slowly increased (5^0 Brix per day) by repeatedly adding sugar (Pareek, 2001).

Candy is a sugar impregnated and surface-dried fruit. Sugar impregnation is carried out in the same way as preparing preserve. After equalization of TSS at 65^0 to 70^0 Brix, the fruits are drained from the syrup by spreading onto wire trays. The drained fruits are dried in the sun or in a warm room until they attain 10 to 15% moisture. The candies are packed in jars, tins and cardboard cartons or in polythene pouches. The air-dried, sugar-implemented fruits can be rolled over powdered sugar to prepare a crystallized product (Pareek, 1983). Various other products like, RTS, nectar, wine, jam, squash, etc. can be prepared from pulp of ripe ber fruits.

Future Research Needs

Although ber cultivation is established in India, improvement in technologies with respect to both production and utilization are required for economically sustained growth. The following technology gaps still exist:

1. Success of ber production often depends upon effective management of insect pests (mainly fruitfly and fruit borer) and diseases (mainly powdery mildew and *Isariopsis*) which is being managed by chemical means. An integrated pest and disease management (IPM) system using biocontrol agents should be developed.
2. Physiological adaptation of ber genotypes to stress conditions like drought, high temperature, low temperature, salinity and sodicity.
3. Unfavourable climatic conditions result in loss of production in the absence of:
    a. Efficient rain water management techniques to ensure production under low rainfall regions
    b. Efficient irrigation methods to provide the maximum water use efficiency from scarce water resources
    c. Genotypes / cultivars having frost tolerance vis-à-vis can set fruits at high temperature
4. Absence of market information network and organized marketing channels.
5. Absence of standard products acceptable in the known markets.

REFERENCES

Abdul-Galil, F.M. and El-Jissary, M. A. (1991) Cyclopeptide alkaloids from *Zizyphus spina-christi*. *Phytochemistry*, 30(4): 1348-1349.

Aggarwal, P., Bhupindeer, K. and Bal, J. S. (1997) Studies on dehydration of different ber cultivars for making ber chhuharas. *Journal of Food Science Technology* (Mysore), 34(6): 534-536.

Ahlawat, V.P., Sindhu, S.S., Gupta, O.P. and Thareja, R.K. (1990) Nutrient element stauts of soil and leaves of ber (*Ziziphus mauritiana* Lamk.) trees in Haryana. *Harayana Journal of Horticultural Science*, 19(1-2): 106-111.

Ahmedov, V.A. and Halmatov, H.H. (1969) Pharmacognostic studies on *Ziziphus jujube* growing in Uzbekistan, [in Russian]. *Rastitel'nye Resury*, 5: 579-81.

Al-Akeely, M.A. (1985) Dictionary of Plants in Jizan Area. Dar Al-Bilad Press, Jeddah, Saudi Arabia.

Anonymous (1953) Propagation of fruit plants. *Agriculture and Animal Husbandry*, Uttar Pradesh, India. (1950), 1(3): 19-27.

Anonymous (1981) Project Report 1976-1980. All Indian Coordinated Fruit Improvement Project, Coordination Cell III, ICAR, India. Technical Document No. 1: 6-8, 35.

Anonymous (1987) Annual Progress Report. 1986-87. All India Coordinated Research Project on Postharvest Technology of Horticultural Crops, ICAR, New Delhi, India.

Anonymous (1989 a) Annual Progress Report. 1988-89. All Indian Coordinated Research Project on Post-harvest Technology of Horticultural Crops. ICAR, New Delhi, India.

Anonymous (1996) *Lakh utpadan ke Adhunik Tarike* [in Hindi]. Indian Lac Research Institute, ICAR, Namkun, Ranchi, India. p.11-12.

Awasthi, O.P., Pathak, R.K. and Pandey, S.D. (1994) Effect of sodicity and salinity levels on survival and total chlorophyll content of ber genotypes (*Ziziphus* spp.). *Horticultural Journal*, 7 (10: 51-54.

Babu, R. H. and Kumar, P.S. (1986) A note on the seed characters of certain cultivars of ber *(Ziziphus mauritiana)*. *Progressive Horticulture*, 18(3-4): 297-298.

Babu, R. H. and Kumar, P.S. (1988) Studies on the growth and flowering of certain cultivars of ber (*Ziziphus mauritiana* Lamk.) of Hyderabad. *Haryana Journal of Horticultural Science*, 17(1-2): 24-27.

Bagle, B.G. (1992) Incidence and control of fruit fly (*Carpomiya vesuviana* Costa) of ber (*Ziziphus mauritiana* Lamk.). *Indian Journal of Plant Protection*, 20(2): 205-207.

Bailey, L. H. (1947) *The Standard Cyclopaedia of Horticulture*. Macmillan & Co., New York.

Bajwa, G. S., Bal, J.S., Brar, S.S. and Minhas, P.S. (1993) Chemical weed control in ber orchards. In *Integrated Weed Management for Sustainable Agriculture*, Proceedings of the International Symposium Indian Society of Weed Science, III: 225-227.

Bajwa, G.S., Sandhu, H.S. and Bal, J.S. (1986) Effect of pruning severity on growth and bearing of ber. *Indian Journal of Horticulture*, 43(3/4): 203-206.

Bajwa, G.S., Sandhu, H.S. and Bal, J.S. (1987) Effect of different pruning intensities on growth, yield and quality of ber (*Ziziphus mauritiana* Lamk.). *Haryana Journal Horticulture Science*, 16(3-4): 209-213.

Bajwa, G.S., Sindhu, M.S. and Shergill, T.S. (1990) A note on the effect of various weed control treatments on nutrient and water losses by weeds in ber orchard. *Haryana Journal Hortiucltture Science*, 15(3-4): 224-227.

Bajwa, M.S., Singh, P. and Singh, R. (1972) Ber cultivation in Punjab. *Indian Horticulture*, 16(4): 7-12, 32.

Bakshi, J.C. and Singh, P. (1974) The ber-a good choice for semi-arid and marginal soils. *Indian Horticulture*, 19: 27-30.

Bal, J.S. (1981a) A note on sugars and aminoacids in ripe ber. *Progressive Horticulture*, 13(3/4): 41-42.

Bal, J.S. (1982) A study on biochemical changes during room and refrigerator storage of ber. *Progressive Horticulture*, 14(2/3): 158-161.

Bal, J.S. (1992) Identification of ber (*Zizyphus mauritiana* L.) cultivars through vegetative and fruit characters. *Acta Horticulturae*, 317: 245-253.

Bal, J.S., Jawanda, J.S. and Singh, S. N. (1979) Developmental physiology of ber var. Umran. 4. Changes in amino-acids and sugars (sucrose, glucose and fructose) at different stages of fruit ripening. *Indian Food Packer*, 33 (4): 33-35.

Bal, J.S., Singh, P. and Mann, S.S. (1978) Changes in total yellow pigments, protein and total carbohydrates during ripening of ber fruits. *Progressive Horticutlure* 19(1): 73-75.

Bal, J.S., Singh, S. and Gupta, M.R. (1988b) Variation in the concentration of biochemical constituents in different parts of ber fruit. *Punjab Horticultural Journal*, 28(1-2): 77-80.

Bankar, G. J. and Prasad, R. N. (1992) Relationship between stomatal distribution and growth of ber rootstocks. *Indian Journal of Horticulture*, 49(2): 169-171.

Baratov, K. B., Shipkova, L. V., Babaev, I. I., Massover, B. L. (1974) The contents of vitamin C and P and total sugars in some *Zizyphus jujube* forms cultivated in Tadznik SSR [in Russian]. In *Nauchnye Osnovy Pitaniya Zdorovogo I Bol nogo cheloveka* T.1.Alma Ata, Kazakh SSR Kazakhstan 158-160, from *Referativnyi Zhurnal* (1975), 4(55):810.

Basha, J.M.G. (1952) Experiments on the control of the fruit borers of jujube (*Ziziphus spp.*)-*Carpomyia vesuviana* Costa and *Maridarchis scyrodes* Meyrick in South India, *Indian Journal of Entomology*, 14(3):229-238.

Batra, H.N. (1953) Biology and control of *Dacus diversus* Conquillet and *Crpomyia vesuviana* Costa and important notes on other fruit flies in India. *Indian Journal of Agricultural Science*, 23:87-112.

Bhambhota, J.R. and Singh, A. (1971) A quick method of raising ber rootstock. *Punjab Horticultural Journal*, 11: 251-253.

Bhandari, C. R. (1969) *Vanaushadhi-Chandroday*. An encylopaedia of Indian Botany and Herbs. Part 7. Chaukhamba Sanskrit Series Office, Banaras, India. pp. 97-99.

Bhandari, M.M. (1978) *Flora of the Indian Desert*. Scientific Publishers, Jodhpur, India.

Bhansali, A. K. (1975) Monographic Study of the family Rhammanceae of India. Ph. D. thesis, University of Jodhpur, India.

Bhatia, S. K. and Gupta, O.P. (1985) Chemical changes during development and ripening of ber fruits. *Punjab Horticultural Journal*, 25: 62-66.

Bhatnagar, P. and Lakra, R. K. (1992) Biology and control of hairy caterpillar, *Euproctis fraterna* Moore (Lepidoptera : Lymontriidae) on jujube (*Ziziphus mauritiana* Lamk.). In *Bio-ecology and Control of Insects Pests:* Proceedings of the National Symposium on Growth, Development and Control Technology of Insect Pests, pp, 150-155.

Bisla, S.S., Chauhan, K.S. and Godara, N.P. (1980) Evaluation of late ripening germplasm of ber (*Zizyphus mauritiana* Lamk.) under semi-arid regions. *Haryana Journal of Horticultural Science*, 9(1/2): 12-16.

Bisla, S.S. Singhrot, R.S. and Chauhan, K.S. (1984) Effect of growing medium and urea application on seed germination and growth of ber (*Ziziphus mauritiana* Lamk.). *Harayana Journal of Horticultural Science*, 13(3/4): 118-122.

Bohra, H.C. and Ghosh, P.K. (1981) Nutrtive value of pala for ruminants in Boradi (*Ziziphus nummularia*), a Shrub of the Indian Arid Zone-Its Role in Silvipasture, eds, Mann, H.S. and Saxena, S.K. Central Arid Zone Research Instiytute, Jodhpur, India. pp. 42-47.

Bowden, W.M. (1945) A list of chromosome number in higher plants. *American Journal of Botany*, 32 : 191-201.

Brandis, D. (1906) Indian Trees. Fifth impression 1971. Bishen Singh, Mahendra Pal Singh, Dehra Dun, India, pp. 169-172.

Chadha, K.L., Gupta, M.R. and Bajwa, M.S. (1972) Performance of some grafted varieties for ber (*Ziziphus mauritiana* Lamk.) in Punjab. *Indian Journal of Horticulture*, 29(2): 137-50.

Chandra, V. (1964) Cultivated varieties of ber (*Ziziphus mauritiana* Lamk.) in India. *Bulletin of the National Botanical Gardens, Lucknow, India*. I. Karna Varanasi, p.95. II. Narma Varanasi, p.98. III Bekata Varanasi, p. 101.

Chattopadhyay, P.K. and Dey, S.S. (1992) Note on standarisation of some aspects of ber propagation. *Indian Journal of Horticulture*, 49(1): 47-49.

Chauhan, K.S. and Gupta, A.K. (1985) Effect of foliar application of urea on the fruit drop and physico-chemical composition of ber (*Ziziphus mauritiana* Lamk.) fruits under arid conditions. *Haryana Journal of Horticultural Science*, 14(1/2): 9-11.

Cheema, G. S., Bhatia, S.S. and Naik, K.C. (1954) Commercial Fruits of India, With Special Reference to Western India, Macmillan & Co., Calcultta, India.

Cherian, M.C. and Sunderam, C.V. (1941) Proceedings of the 20th Indian Scinece Congress, III: 191.

Chevalier, A (1947) Les jujubier on *Ziziphus* die I'ancien monde et I'utilisation de leurs fruits [in French]. *Revue Internationale de botanique appliqué et d'agriculture tropicale* (Paris), 27 (301/2): 470-483.

Chiarugi, A.N. (193) Parthenocarpia in *Ziziphus sativa* Gaertn. *Bot. Ital.*, 37: 287-312.

Chopra, D.P. and Srivastava, G. (1973) Ber varieties for western Rajasthan. *Indian Horticulture*, 17(4): 6-7, 30.

Chovatia, R.S., Patel, D.S., Patel, A.T. and Patel, G.V. (1992) Growth, yield and physico-chemical characters of certain varieties for ber (*Ziziphus mauritiana* Lamk.) under dryland conditions. *Gujarat Agriculture University Research Journal*, 17(2): 56-60.

Chovatia, R.S., Patel, D.S., and Patel, G.V. (1993) Performance of ber (*Ziziphus mauritiana* Lamk.) cultivars under arid conditions. *Annals of Arid Zone*, 32 (4): 215-217.

Chundawat, B.S. and Srivastava, H.C. (1978) *Haryana Men Ber Ki Kheti*, Directorate of Extension, Haryana Agricultural University, Hisar, India. p.38. [in hindi].

Clifford, S.C., Corlett, J.E., Joshi, S. and Jones, H.G. (1996) Photosynthetic studies relating to drought response of fruit trees for semi-arid areas. Plant Biology Abstracts, SEB Annual Meeting, 47(supplement), p.339.

Cooke, T (1902) *The Flora of the Presidency of Bombay*. Vol. 1, Part II. Secretaryt of State for India in Council, Dehradun, India.

Cyong, J.C. and Hunabusa, K. (1980) Cyclic adenosine monophosphate in fruits of *Ziziphus jujube*. *Phytochemistry*, 19(12): 2747-2748.

Cyong, J.C. and Takahashi, M. (1982) Identification of guanosine 3:5'-monophosphate in the fruit of *Ziziphus jujube*. *Phytochemistry*, 21(80): 1871-1874.

Dahiya, S.S., Ahlawat, V.P. and Khera, A.P. (1985) Soil versus foliar nutrition of ber trees with urea. *Internatinal Journal of Tropical Agriculture*, 3(3): 192-195.

Dahiya, S.S. and Dhankhar, O.P. (1982) Studies on salt tolerance of ber. *Haryana Journal of Horticutlural Science*, 11(1/2): 53-58.

Dahiya, S.S., Dhankar, O.P. and Khera, A.P. (1981) Studies on the effect of soil salinity levels on seed germination of ber *(Ziziphus rotundifolia)*. *Haryana Journal of Horticultural Science*, 10: 20-23.

Dalziel, J.M. (1937) The Useful Plants of West Tropical Africa. Crown Agents, London, 2nd reprint, 1955.

Darlington, C.D. and Wylie, A.P. (1955) *Chromosome Atlas of Flowering Plants.* Allen and Unwin, London.

Dastur, J.F. (1952) *Useful Plants of India and Pakistan. A Popular Handbook of Trees and Plants of Industrial, Economic and Commercial Utility,* D.B. Tarporewala Sons & Co. Ltd., Bombay, India.

De Candolle, A. (1886) *Origin of Cultivated Plants.* Bibliotheque Scientifique Internationale, 43, Paris.

Depommier, D. (1990) *Ziziphus mauritiana*: Cultivation and use in Kapsiki country, northern Cameroon, *Bois et Forests des Tropiques* No.218, 57-62, CTFT-CIRAD, ICRAF, Nairobi, Kenya.

Desai, U.T. and Patil, A.V. (1978) Studies on the floral biology of ber (*Ziziphus mauritiana* Lamk.) cultivars, Umran and Illaichi, *Udyanika*, 2(1/2): 26-32.

Dhaliwal, G.S. and Sandhu, I.P.S. (1984) Effect of pruning on vegetative growth, flowering and fruit set in ber (*Ziziphus mauritiana* Lamk.) cv. Umran. *Haryana Journal of Horticultural Science*, 11(3/4): 208-212.

Dhankar, O.P., Makhija, M. and Singhrot, R.S. (1980) Effect of salinity levels on germination of seeds and growth of transplanted seedlings of ber (*Ziziphus rotundifolia*). In Proceedings *of the International Symposium of the Arid Zone Research Development*, Jodhpur, India, ed, Mann, H.S.pp. 351-356.

Dhatt, A.S., Grewal, G.P.S. and Dhillon, W.S. (1993) Effect of N, P and K treatments on growth and quality of Umran ber (*Ziziphus mauritiana* Lamk.). *Punjab Horticultural Journal*, 33(1/4): 70-75.

Dhingra, R.P., Singh, J.P. and Chitkara, S.D. (1973) Varietal variations in physico-chemical characters of ber (*Ziziphus mauritiana* Lamk). *Haryana Journal of Horticultural Science*, 2 (3/4): 61-65.

Dinesh Kumar (2001). *Zizyphus mauritiana* - A promising multipurpose tree for sustained productivity of Alluvial degraded soils in Indo-Gangetic plants of Uttar Pradesh. Nation Conference as Resource Conservation and Water shed Management. CSWCRI, Detrradun. May, 23-25. pp. 73.

Don, G. (1831) *A General History of Dichlamydeous Plants*, London.

Dutta, S. (1954) Jujubes of Assam. *Indian Journal of Horticulture*, XI : 53-56.

Evreinoff, V.A. (1964) Notes sur le jujubier (*Ziziphus jujube* Gaertner) [in French], *Journal of Agriculture tropicale et de botanique appliquee*, 11: 177-187.

Funkhouser, S. (1998). Local natural resources management of musau fruit (*Ziziphus mauritiana*) in the mid-Zambezi valley. Abstr. *International Workshop on Ziziphus mauritiana*, Harare, Zimbabwe, 13-16 July, 1998.

Gangwar, V.S. (1983) Biology of *Phyllodiplosis jujube* Grover infesting leaves of *Ziziphus jujube* Lamk. (*Cecidomyiidae* : Diptera). *Cacidologia Internatinale*, 4(1/2/3): 57-73.

Ghosh, S. N. and Sen, S.K.(1988) Effect of seed treatment on germination, seedling growth and longevity of ber (*Ziziphus mauritiana* L.) seeds. *South Indian Horticulture*, 36(5): 260-261.

Godara, N. R., Chauhan, K. S. and Bisla, S. S. (1980) Evaluation of mid-season ripening ber (*Ziziphus mauritiana* Lamk.) germplasm. *Haryana Journal of Horticultural Science*. 9(3/4): 101-105.

Goncharova, N.P., Isamukhamedov, A.S.H and Glushenkova, A.I. (1990) Lipids of *Ziziphus jujube*. *Chemistry of Natural Compounds*. 26(1): 16-18.

Gopani, G.D. (1976a) *Arthik Drishi Se Bor ki Kheti*. Gujarat Agricultural University, Junagadh, India. [in Gujarati].

Gupta, I.C., Sharma, B.D. and Singh, P.M. (1989a) Effect of bicarbonate and chloride in irrigation water on germination and growth of ber and soil properties. *Journal of the Indian Society of Soil Science*, 37: 428-430.

Gupta, J.H., Ram Nath and Srivastava, V.P. (1977) Chemical control of powdery mildew of ber (*Ziziphus mauritiana* Lamk.). *Progressive Horticulture*, 9 (1): 81-83.

Gupta, M.R. (1977) Physico-chemical characters of some promising ber cultivars grown at Bahadurgarh (Patiala). *Punjab Horticultural Journal*, 17(3/4): 131-134.

Gupta, M.R. and Singh, S. (1979) Effect of pruning on the growth, yield and fruit quality in ber. *Punjab Horticultural Journal*, 17(1/2): 54-57.

Gupta, O.P. (1983) Delicious candy from ber fruits. *Indian Horticulture*, April-June, p. 25-27.

Gupta, O.P. and Mehta, N. (1987) Effect of post-harvest applications of fungicides, chemical and precooling treatments on the shelflife of Gola ber fruits. *Haryana Agricultural University Jouranl of Research*, 17: 146-152.

Gupta, O.P., Kainsa, R.L., Chauhan, K.S. and Dhawan, S.S. (1981b) Post-harvest studies in ber fruits (*Ziziphus mauritiana* Lamk.) III. Evaluation of cultivars for candy making. *Haryana Agricultural University Journal of Research*, XI: 480-492.

Gupta, O.P., Siddiqui, S. and Chauhan, K.S. (1987) Evaluation of various calcium compounds for increasing the shelf life of ber (*Ziziphus mauritiana* Lamk.). *Research and Development Reporter*, 6(1): 35-40.

Gupta, O.P., Siddiqui, S. and Gupta, A.K. (1989b) Effect of pre-harvest sprays of various chemicals on the storage of ber (*Ziziphus mauritiana* Lamk.). *Research and Development Reporter*, 6(1): 35-40.

Gupta, O.P., Singh, J.P. and Chauhan, K.S. (1981) Evaluation of various packings on the storage behaviour of ber cv. Umran. Research *Reports and Project Proposals, First National Workshop on Arid Zone Fruit Research, Hisar, India.* All India Coordinated Fruit Imprvement Project, ICAR, New Delhi, India.

Gupta, P.C. and Madaan, R.L. (1977a). A new leaf spot diseases of *Ziziphus mauritiana*. *Indian Phytopathology*, 30: 271-272.

Gupta, P.C. and Madaan, R.L. (1977b). Diseases of fruits from Haryana. A new leaf spot disease of *Ziziphus mauritiana* Lamk. *Current Science*, 46: 237-238.

Gupta, P.C., Madaan, R.L. and Grover, R.K. (1978) Occurrence of powdery mildew of ber in Haryana and its control. *Indian Phytopathology*, 31: 440-443.

Gupta, R.B. and Godara, N.R. (1989) Effect of time and severity of pruning on growth, quality and yield in ber (*Ziziphus mauritiana* Lamk.) cv. Umran. *Progressive Horticulture*, 21(1-2): 15-20.

Gupta, R.B., Godara, N.R. and Chauhan, K. S. (1990) Effect of time and severity of pruning on flowering and fruiting of ber (*Ziziphus mauritiana* Lamk.) cv. Umran. *Haryana Journal of Horticultual Science*, 19(1-2): 45-49.

Han, B.H., Park, M.H. and Wah, S.T. (1987) Strcuture of diaechualkaloid-A, a new pyroolidine alkaloid of novel skelton from *Ziziphus jujube* var. *inermis*. *Tetrahedrom Letters*, 28(34): 3957-3958.

Hooda, P.S., Sindhu, S.S., Mehta, P.K. and Ahlawat, V.P. (1990) Growth, yield and quality of ber (*Ziziphus mauritiana* Lamk.) as affected by soil salinity. *Journal of Horticultual Science*, 65(5): 589-593.

Hooker, J.D. (1875) *Flora of British India*. Part I. L. Reeve and Co., London.

Hore, J. K. and Sen, S.K. (1994) Role of pre-sowing seed treatments on germination, seedling growth and longevity of ber (*Ziziphus mauritiana* Lamk.) seeds. *Indian Journal of Agricultural Research*, 28(4): 285-289.

Hussian, A. K. M. and Khan, T.A. (1962) Lac cultivation in East Pakistan. *The Pakistan Journal of Forestry*, 12(3): 220-22.

Hussian, F., Ahmed, M., Shaheen, G. and Durani, M.J. and Shaheen, G. (1993) Phytosociology of the vanishing tropical dry deciduous forests in district Swabi, Pakistan. I. A. community analysis. *Pakistan Journal of Botany*, 25(1): 51-66.

Jain, R. K. and Chitkara, S.D. (1980) Effect of different packings in transportation on quality of ber (*Ziziphus mauritiana* Lamk.) fruit. *Annals of the Arid Zone*, 19(3): 289-291.

Jain, R.K., Chitkara, S.D. and Chauhan, K.S. (1981) Studies on physico-chemical characters of ber (*Ziziphus mauritiana* Lamk.) fruits cv. Umran in cold storage. *Haryana Journal of Horticultural Science*, 10(3/4): 141-146.

Jawanda, J.S. and Bal, J.S. (1978) The ber-highly paying and rich in food value. *Indian Horticulture*, 23(3): 19-21.

Jawanda, J.S. and Bal, J.S. (1980) A comparative study on growth and development of ZG-2 and Kaithli cultivars of ber. *Punjab Horticultural Jouranl*, 20(1/2): 41-46.

Jawanda, J.S., Bal, J.S., Josan, J.S. and Mann, S.S. (1980a) Studies on the storage of ber fruits. I. Room temperature. *Punjab Horticultural Journal*, 20(1/2): 56-61

Jawanda, J.S., Bal, J.S., Josan, J.S. and Mann, S.S. (1980b) Studies on the storage of ber fruits. II. Cool temperature. *Punjab Horticultural Journal*, 20 (3/4): 171-178.

Jawanda, J.S., Bal, J.S., Josan, J.S. and Mann, S.S. (1981) Ber cultivation in Punjab. *Punjab Horticultural Journal*, 21: 17-22.

Johnston, M.C. (1972) Rhamnaceae. In *Flora of Tropical East Africa*, eds. Milne-Redhead, E. and Polhill, R.M. Crown Agents, London.

Jones, H. G. (1992) *Plants and Microclimate : A Quantitiative Approach to Environmental Plant Physiology* (2nd edn.) Cambridge University Press.

Joon, M.S., Singh, R.R. and Daulta, B.S. (1984) Effect of foliar sprays of zinc and urea on yield and physico-chemical composition of ber fruits cv. Gola. *Haryana Jouranl of Horticutlrual Science*, 13(3/4): 110-112.

Joshi, N.J. (1960) Ber grafting in Banaskantha division. *Indian Forester*, 86: 169-171.

Jyotishi, R.P., Kashyap, R. and Maurya, K.N. (1967) Budding of ber, *Ziziphus jujube* for top working can be done throughout the year. *JNKVV Research Journal*, 1:91-92.

Kainsa, R.L., Singh, J.P. and Gupta, O.P. (1978) Surface microflora of freshly harvested ber (jujube) (*Ziziphus mauritiana* Lamk.) and their role in spoilage of fruits. *Haryana Agricultrual University Journal of Research*, 8(2): 86-89.

Kajal, R. S. and Singhrot, R. S. (1986) Vegetative propagation of ber (*Ziziphus mauritiana* Lamk.). IX. Studies on the effect of sowing depths and dates on seed germination and seedling growth. *Haryana Journal of Horticultural Scinece*, 15 (3/4): 156-161.

Kalikiti, F. (1998) *Ziziphus mauritiana* in Siavonga district, Zambia. *In International workshop on Ziziphus mauritiana*, Harare, Zimbabwe, 13-16 July, 1998.

Kanwar, J.S. and Singh, S.N. (1981) A note on the longevity of ber (*Ziziphus mauritina* Lamk.) budwood as affected by the wrapping material. *Science and Culture*, 47(6): 220-221.

Kaundal, G.S., Grewal, S.S. and Bal, J.S. (1984) Influence of meterological factors on bud take efficiency of ber (*Ziziphus mauritiana* Lamk.) cv. Umran. *Jouranl of Research, Punjab Agricutlural University*, 21(3): 372-374.

Khangura, J. S. and Sandhu, G. S. (1976) Efficacy of dilute and concentrate sprays of insecticides in the control of ber leaf-hopper, *Zyginidia pakistanica* (Ahmed) in Punjab, *Pesticides*, 10(12) : 47-49.

Khera, A.P., Bisla, S.S., Chauhan, K.S. and Daulta, B.S. (1981) Studies in root activity in ber cv. Umran (*Ziziphus mauritiana* Lamk.) trees by redioactivity using P^{32}. *Haryana Journal of Horticultural Science*, 10(3/4) : 171-173.

Khera, A.P. and Singh, J.P. (1976) Chemical composition of some ber cultivars (*Ziziphus mauritiana* Lamk.). *Haryana Journal of Horticultural Science*, 5 (1/2): 21-24.

Khokhar, I. (1978) Isoquinoline alkaloids from the fruits of *Ziziphus sativa*. *Pakistan Journal of Scinetific Research*, Lahore, Pakistan. From *Medicinal & Aromatic Plants Abstracts*, 1: 7902-0510.

Khoshoo, T. N. and Singh, N. (1963) Cytology of northwest Indian trees. *Ziziphus jujube* and Z. *rotundifolia*. *Silvae Genetica*, 12 Heft. Pp. 141-180.

Khurdiya, D.S. (1980) Studies on dehydration of ber (*Ziziphus mauritiana* Lamk.) fruits. *Journal of Food Science and Technology*, 17: 127-130.

Khurdiay, D.S. and Roy, S.K. (1986) Solar drying of fruits and vegetables, *Indian Food Packer*, 40(4): 28-39

Khurdiya, D.S. and Singh, R.M. (1975) Ber and its products. *Indian Horticulture*, 20: 5, 25.

Kim, Y.S., Hong, K.H. and Kim, W.S. (1981) Studies on the drying methods of jujube (*Ziziphus mauritiana* Mill.) [in Korean]. *The Research Reports of the Office of Rural Development, Horticulture and Sericulture* (Korea R.), 23: 34-38.

Kirtikar K.R. and Basu, B.D. (1975) *Indian Medicinal Plants*. Vol. I. Bishen Singh Mahendrapal Singh, Dehra Dun, pp. 588-596.

Krishna, B. M. and Kulasekaran, M. (1984) Studies on seed germination in wild ber (*Ziziphus rotundifolia*). *South Indian Horticulture*, 32(3): 153-154.

Krivencov, V. I., Karakhanova, S.V. and Savina, G. G. (1970) Distribution of vitamin C in fruits of *Ziziphus jujube* [in Russian]. *Byulleten Gosudars tovennogo Nikitskogo Botanicheskogo Sada*, 2(13): 57-59.

Kuliev, A. A. and Guseinova, N. K. (1974) The content of vitamin C, B_1, B_2 and E in some fruits. *Regerativnyi Zhurnal*, 2: 69-73.

Kulkarni, A. R., Bhat, N.R., Supe, V. S. and Kale, P. N. (1992) Influence of storage conditions and pre-sowing treatments on germination and seedling growth in ber. *Journal of the Maharashtra Agricultural Universities*, 17(1): 132-133.

Kumar, N., Abha, T. and Mehta, P.K. (1990) Effect of different ESP levels on micronutrient status of the leaves of ber. *Haryana Journal of Horticultural Science*, 19(1-2): 89-95.

Kumar, P.S. and Babu, R. H. (1987) Physico-chemical characters of some ber (*Ziziphus mauritiana* Lamk.) cultivars grown at Hyderabad. *Punjab Horticultural Journal*, 27(1-2): 17-21.

Kumawat, S.R. and Swaminathan, R. (1990) Relative preference of jujube (*Ziziphus mauritiana* Lamk.) cultivars by bark eating caterpillar (*Indarbela quadrinotata* Walker). *Indian Journal of Entomology*, 52(2): 336-338.

Kundi, A.H.K., Wazir, F.K., Abdul, G. and Wazir, Z.D.K. (1989a) Physico chemical characterstics and organoleptic evaluation of different ber (*Ziziphus jujube* Mill) cultivars. *Sarhad Journal of Agriculture;* 5(2): 149-155.

Kundi, A.H.K., Wazir, F.K., Abdul, G. and Wazi, Z.D.K. (1989b) Morphological chracterstics, yield and yield components of different cultivars of ber (*Ziziphus jujube* Mill). *Sarhad Journaal of Agriculture*, 5(1): 53-57.

Kundi, S. and Singhrot, R.S. (1990) Propagation studies in ber (*Ziziphus mauritiana* Lamk.). X. Effect of Waxol dipping on the transplanting success of budlings. *Haryana Journal of Horticultural Science*, 19(3-4): 225-231.

Kundu, S. S. (1983) Studies on effect of different rootstocks on seed germination, seedling growth and budding success in ber. M.Sc. thesis, Haryana Agricultural University, Hisar.

Kundu, S. S., Pareek, O.P. and Gupta, A. K. (1995) Effect of time and severity of pruning on physico-chemical characterstics and yield of ber (*Ziziphus mauritiana* Lamk.) cv. Umran. *Haryana Journal of Horticultural Science*, 24(1): 23-30.

Lenham, W.B. (1926) Jujubes in Texas. Texas Agricultural Experiment Station, Texa, USA, Circular No. 41. p. 28.

Li, J. and Hu, X. (1994) Rejuvenation and utilization of wild *Ziziphus mauritiana*. *Forest Research*, 7(2): 224-226.

Liu, M.J. and Cheng, C.Y. (1995) A taxonomic study of the genus *Ziziphus*. *Acta Horticulture*, 390: 161-165.

Madaan, R.L. and Chand, J.N. (1985) Epidemiological studies on Alternaria leaf spot and fruit rot disease of ber (*Ziziphus mauritiana* Lamk.). *Mycology and Plan Pathology*, p. 26-27.

Madalageri, B.B., Reddy, P.N. and Basawaraj, B. (1977) Ber cultivars HB2, KB1 and HB1 have commercial possibilities. *Current Research*, 6(7): 113-114.

Mann, G. S. and Bindra, O.S. (1976) Evaluation of different jujube (*Ziziphus mauritiana*) cultivars for resistant to the ber fruit fly. *Punjab Horticultural Journal*, 16(1/2): 64-47.

Mann, G. S. and Bindra, O. S. (1977) Field screening of different jujube cultivars against the bark-eating caterpillar. *Indian Journal of Horticulture*, 34 (3):313-315.

Maposa, M. and Chisuro, D. (1998) Importance of *Ziziphus mauritiana (masau)* in the Mukumbura area of Zimbabwe-From a farmer's and extensionist's point of view. *In International Workshop on Ziziphys mauritiana*. Harare, Zimbabwe, 13-16, July, 1998.

Mathur, N., Ramawat, K.G. and Sonie, K.C. (1993) Plantlet regeneration from seedling explants of *Ziziphus* and silver nitrate and nutrient requirement for callus and morphogenesis. *Gartenbavwissenschaft*, 58(6): 255-260.

Mathur, N. and Vyas, A. (1996) Physiological changes in *Ziziphus mauritiana* by different VAM fungi. *Indian Forester*, 122(6): 501-506.

Mawari, P.B. and Singh, S.P. (1992a) Effect of method and time of budding success in ber (*Ziziphus mauritiana*) cv. Gola. Part II. *Harticulture Journal*, 5(2): 79-83.

Mawari, P.B. and Singh, S.P. (1992b) Effect of method and time of budding success in ber (*Ziziphus mauritiana*) cv. Gola. Part I. Horticulture Journal, 5 (1): 31-35.

Meel, O.P., Chharia, A.S., Kumar, R. and Joon, M.S. (1991) Physico-chemical changes in the fruit of Sandhura Natural ber (*Ziziphus mauritiana* Lamk.) during growth and development. *Haryana Journal of Hortiucultural Scinece*, 20(1-2): 65-72.

Mehra, K.L. (1967) History of the use of jujube (*Ziziphus* species) in ancient India. *Indian Journal of Horticulture*, 24 (1&2) : 33-37.

Mehrotra, N.K., Singh, R. and Singh, S.N. (1987) Studies on removal of N, P and K in ber (*Ziziphus mauritiana* Lamk.) cv. Umran. *Punjab Horticultural Journal*, 27(3-4): 200-202.

Mehta, P.K. (1982) Studies on effect of soil sodicity and boron on seed germination, growth and mineral composition of young ber (*Ziziphus* spp.) plants. Thesis Abstracts, 8 (1): 47-48.

Minamide, T., Goto, M. and Iwata, T. (1986) Forms of calcium compounds and their changes after harvest in fruits and vegetables [in Japanese]. *Journal of the Japanese Society for Horticultural Science*, 54(4): 507-513.

Ming, W. and Sun, Y. (1986) Fruit trees and vegetables for arid and semi-arid areas in northest China. *Journal of the Arid Environments*, 11: 3-16.

Morinaga, T., Fukushima, E., Kano, T. and Yamasaki, Y. (1929) Chromosome number of cultivated plants II. *Botanical Magazine*, Tokyo, 43: 589-594.

Moti, Dhar, L. and Chaturvedi, O.P. (1976) Propagating some subtropical and tropical fruits by budding. *Punjab Horticultural Journal*, 16: 33-38.

Murthy, B.N.S. and Reddy, Y.N. (1989b). Improvement of seed germination in jujube (*Ziziphus mauritiana* Lamk.) with growth regulators and nitrogenous compounds. *Indian Journal of Agricutlural Science*, 59(4): 269-270.

Murthy, B.N.S. and Reddy, Y.N. (1990b) Temperature dependence of seed germination and seedlings growth in ber (*Ziziphus mauritiana* Lamk.) and their modification by pre-sowing treatments. *Seed Science and Technology*, 18(3): 621-627.

Nehra, N.S., Chitkara, S.D. and Singh, K. (1984) Studies of morphological chracters of some wild forms and cultivated varieties of ber. *Punjab Horticultural Journal*, 24(1/4): 49-59.

Nehra, N.S., Sareen, P.K. and Chitkara, S.D. (1983) Cytological studies in genus *Ziziphus*. *Cytologia*, 48: 103-107.

Nerd, A. and Mizrahi, Y. (1998) Ber experience in Israel. Abstract. *International Workshop on Ziziphus mauritiana*, Harare, Zimbabwe, 13-16 July, 1998.

Nijjar, G.S. (1975) Ber cultivation. *Punjab Horticultural Journal*, 15(1-2): 3-8.

Om Prakash and Dinesh Kumar (2001). Tree root management for higher producitivity in yamuna ravenous lands under semi-arid conditions. National conference on Resource Conservation and Watershed Management. CSWCRTI, Detrodun. May, 23-25, pp. 173.

Pandey, R.C., Pathak, R.A. and Pathak, R.K. (1990) Physico-chemical changes associated with growth and development of fruits in ber (*Ziziphus mauritiana* Lamk.). *Indian Journal of Horticulture*, 47(3): 286-270.

Pandey, S.D., Pathak, R.K. and Awasthi, O.P. (1993) Note on effect of salinity levels on nutrient status in ber. *Indian Journal of Horticulture*, 50(1): 46-48.

Pandey, S.D., Pathak, R.K. and Dwivedi, R. (1991) Effect of sodicity and salinity levels on chlorophyll, free proline and amino acid contents in ber leaves. *Horticulture Journal*, 4(1): 33-36.

Pandey, V. (1999a). Effect of NAA and GA3 on fruit retention, growth, yield and quality of ber cv. 'Banarasi Karaka'. *Orissa Journal of Horticulture*, 27(1): 69-73.

Pandey, V. (1999b). Effect of heading back on vegetative growth and flowering of ber (*Z. mauritiana* Lam.) cultivars. *Orissa Journal of Horticulture*, 27(2): 57-61.

Panwar, J.S. (1981) Post Harvest physiology and storage behaviour of ber fruit (*Ziziphus mauritiana* Lamk.) in relation to temperature and various treatments (cv. Umran and Kaithli). Thesis Abstracts. *Haryana Agricultural University, Hisar*, 7(1): 64-65.

Pareek, O.P. (1977) Arid horticulture. In *Desertification and its Control*, ed. Jaiswal, P.L., Indian Council of Agricultural Research, New Delhi, India. p. 256-262.

Pareek, O.P. (1978a) Grow ber for higher income in arid regions. CAZRI, *Jodhpur, India, Leaflet* No. 1, p.4.

Pareek, O.P. (1978b) Quicker way to raise ber orchards. *Indian Horticulture*, 23, (1): 5-6.

Pareek, O.P. (1983) *The Ber*. ICAR, New Delhi, India.

Pareek, O.P. (1998) Arid Zone fruit research in India. *Indian Journal of Aricultural Science*, 68(8), special issue: 508-514.

Pareek, O.P. (2001). Ber. ICUC, Southampton, UK.

Pareek, O.P. and Gupta, O.P. (1988) Packaging of ber, date palm and phalsa. In *Packaging of Fruits and Vegetables in India*, eds. Venkataratnam, L. Anwer, M.M., Satyanarayana, G., Chelvan, R.C. and Rameshwar, A. Agri-Horticultural Society, Hyderabad, India. pp. 91-103.

Pareek, O.P. and Sharma, S. (1993) Genetic resources of under-exploited fruits. In *Advances in Horticulture, Vol. I-Fruit Crops. Part I*, eds, Chadha, K.L. and Pareek, O.P. Malhotra Publishing House, New Delhi, India. pp. 189-225.

Pareek, O.P. and Vashishtha, B.B. (1983) Delicious ber varieties of Rajasthan. *Indian Horticulture*, 28(2): 13-15.

Pareek, O.P. and Vashishtha, B.B. (1986) Vegetative and fruit characterstics of ber (*Ziziphus mauritiana* Lamk.) cultivars of Rajasthan grown in the Thar Desert. *Annals of the Arid Zone*, 25(3): 214-221.

Pareek, O.P. and Vishal Nath (1996). Ber. In Coodinated Fruit Research in Indian Arid Zone-A two decades profile (1976-1995). *National Research Centre for Arid Horticuture*, Bikaner, India: 9-30.

Pareek, O.P., Vishal Nath and Vashishtha, B.B. (1999). *In situ* technique to establish ber orchards in arid region. *Extension Bulletin 8. National Research Centre for Arid Horticulture*, Bikaner, India.

Patel, B. and Patel, V. J. (1979) Impact of chemicals on ber (*Ziziphus mauritiana* Lamk.). *Pesticides*, 13(3): 28-30.

Parel, B.H., Upadhyay, V.R., Muralidharan, C.M. and Judal, G.S. (1989) Comparative efficacy of different insecticides against ber fruit fly (*Carpomyia vesuviana* Costa). *Indian Journal of Plant Protection*, 17(1): 39-45.

Patil, P.K., Patil, V.K. and More, R.M. (1981) Influence of various levels of soil ESP on the growth and nutritional status of ber (*Ziziphus mauritiana* Lamk.), variety Umran. *Haryana Jouranl of Horticultural Science*, 10(3/4): 166-170.

Popli, S., Kathpal, T.S. and Lakra, R.K. (1980) Note on dissipation of malathion residues from jujube fruits and leaves. *Indian Journal of Agricultural Science*, 50(2): 191-192.

Prasad, A. (1964). Studies on floral biology of jujube (*Ziziphus mauritiana* Lamk.) *Agra University Journal of Research* (*Science*), 13(2): 41-48.

Ragumoorthi, K.N. and Arumugam, R. (1992) Chemical control of ber fruitly (*Carpomyia vesuviana* Costa). *Indian Journal of Plant Protection*, 20(1): 32-36.

Rajput, C. B. S. and Singh, J. (1976) Effect of urea sprays on the chemical composition of ber fruits (*Ziziphus mauritiana* Lamk.). *Journal of Horticultural Scinece*, 52(2): 371-372.

Ram, A. (1940). Ber crop. *Punjab Fruit Journal*, 5: 447.

Randhawa, G.S. and Biswas, G.S. (1966) Studies on morphology and chemical composition of some jujube varieties. *Indian Journal of Horticulture*, 23: 101-107.

Rao, A.V. and Dass, H.C. (1989) Growth of fruit plants as influenced by nitrogen fixing bacteria. *Annals of Arid Zone*, 28(1-2) : 143-147.

Raturi, G.B. and Chadha, K.L. (1993) Horti-silvaie-pastoral system. In *Advances in Horticulture, Vol.II-Fruit Crops* : Eds. Chadha, K.L. and Pareek, O. P. Malhotra Publishing House, New Delhi, India.

Rawal, R.D. and Saxena, A.K. (1989) Evaluation of different fungicide against black leaf spot of ber. *Indian Journal of Horticulture*, 46(3): 413-414.

Reddy, B.G.M., Patil, D.R., Channabasappa, K.S., Hulamani, N.C. and Patil, S.G. (1995) Performance of some cultivars of ber (*Ziziphus mauritiana* Lamk.) under arid zones of Karnataka. *South Indian Horticulture*, 43(5/6): 146-148.

Reddy, Y.N. (1983) A rational approach to determine the pruning level in ber (*Ziziphus mauritiana* Lamk.) var. Umran. In *Research Papers* and *Reports, Second National Workshop on Arid Zone Fruit Research*, Udaipur, India : 242-243.

Reddy, Y.N. (1988) Potential of dryland fruit crops to make social forestry programme more productive. *National Seminar, Dryland Horticulture*, Hyderabad, India. Abstr. P.37.

Reddy, Y.N. and Chadha, K.L. (1993) Pruning in ber and phalsa. In *Advances in Horticulture, Vol. II-Fruit Crops* : Eds. Chadha, K.L. and Pareek, O.P., Malhotra Publishing House, New Delhi, India L 707-715.

Reddy, Y.N. and Murthy, B.N.S. (1990) Studies on germinability and seedling vigour at different intervals of seed storage in ber (*Ziziphus mauritiana* Lamk.). *Indian Journal of Horticulture*, 47(3): 314-317.

Rendle, A.B. (1959) The Classification of Flowering Plants. Vol. II. *Dicotyledons.* Cambridge University Press, Cambridge, UK.

Rendle, (1971) *The Classification of Flowering Plants. Vol. II. Dicotyledons. Cambridge* University Press, Cambridge, UK.

Ridley, H.N. (1922) *The Flora of the malay Peninsula. Vol. I Polypetalae.* L. Reeve & Co., London, UK.

Sadhu, M.K., Ghosh, S.K. and Bose, T.K. (1978) Mineral nutrition of fruit plants. II. Effect of different levels of nitrogen, phosphorus and potassium on growth, flowering, fruit set and tissue composition of jujube (*Ziziphus mauritiana* Lamk.). *Mysore Journal of Agricultural Science*, 12(1): 101-105.

Saen, T. (1986) Host plants of the fruitflies in Thailand. *King-Mongkut's Agricultural Journal*, 4(1): 1-15.

Samra, J.S. (1985) Comparative sodium accumulation and its toxicity in mango, guava and ber. *Indian Journal of Horticuture*, 42(3/4): 178-183.

Sandhbhor, D.R. and Desai, U.T. (1991) Influence of post-harvest treatments on the shelf life of ber (*Ziziphus mauritiana* Lamk.) cv. Umran. *Maharashtra Journal of Horticulture*, 5(2): 24-28.

Sanjeev Kumar, Ojha, C.M., Bhagwan, D., Awasthi, O.P. and Nainwal, N.C. (1992) Potentiality of ber (*Ziziphus mauritiana* Lamk.) cultivars for candy making. *Progressive Horticutlure*, 24(1-2): 74-78.

Sebastian, M.K. and Bhandari, M.M. (1990) Edible wild plants of the forest areas of Rajasthan (India). *Journal of Economic and Taxonomic Botany*, 14(3): 689-694.

Shanmugavelu, K.G. (1970) Effect of gibberellic acid on seed germination and development of seedlings of some tree plant species. *Madras Agricultural Journal*, 55: 311-314.

Sharma, A. (1992) Seasonal incidence of *Larvacarus transitans* on *Ziziphus* in Rajasthan. *Annals of Arid Zone*, 3(3) : 231-232.

Sharma, K.D., Pareek, O.P. and Singh, H.P. (1982) Effect of runoff concentration on growth and yield of jujube. *Agricultural Water Management* (Netherlands), 5(1): 73-84.

Sharma, K.D., Pareek, O.P. and Singh, H.P. (1986) Micro-catchment water havesting for raising jujube orchards in an arid climate. *Transactions American Societ of Agricultural Engineers*, 29(1): 112-118.

Sharma, K.K., Jawanda, J.S., Singh, M.P. and Bajwa, M.S. (1980) Effect of different spacings and pruning intensities on the growth and fruit yield of Umran ber. *Punjab Horticultural Journal,* 20(1/2): 47-51.

Siddiqui, S. and Gupta, O.P. (1989a) Determination of maturity standards of ber (*Ziziphus mauritiana* Lamk.) fruits. *Haryana Agricultural University Journal of Research,* 19(3): 260-262.

Siddiqui, S. and Gupta, O.P. (1989b) Effect of pre-harvest spray of calcium on shelf life of ber (*Ziziphus mauritiana* Lamk.) fruits. *Research and Development Reporter,* 6(1): 172-176.

Siddiqui, S. and Gupta, O.P. (1990) Evaluation of zero energy chamber for storage of ber (*Ziziphus mauritiana* Lamk.) fruits. *Haryana Agricultural University Journal of Research,* 20(3): 221-224.

Siddiqui, S. and Gupta, O.P. (1995) Effect of post-harvest application of some chemicals on the shelf life of ber (*Ziziphus mauritiana* Lamk.) fruits. *Haryana Journal of Horticultural Science,* 24(1): 19-23.

Siddiqui, S., Gupta, O.P. and Yamdagni, R. (1989) Effect of pre-harvest sprays of chemicals on the shelf life of ber (*Ziziphus mauritiana* Lamk.) fruits cv. Umran. *Haryana Journal of Horticultural Science,* 18(3-4): 177-183.

Singh, A.R., Shukla, P.K. and Singh, K. (1989a) Effect of boron, zinc and NAA on the chemical composition and metabolites of ber (*Ziziphus mauritiana* Lamk.) fruits. *Haryana Journal of Horticutlural Science,* 18(1-2): 23-28.

Singh, B.P., Singh, S.P. and Chauhan, K.S. (1981a) Certain chemical changes and rate of respiration in different cultivars of ber during ripening. *Haryana Agricultural University Journal of Research,* 11(1): 60-64.

Singh, D., Bakhshi, J.C. and Singh, K. (1970) Flowering and fruiting behaviour of ber, variety Banarasi Karaka (*Ziziphus mauitiana* Lamk.). *Punjab Horticultural Journal,* 10: 21-28.

Singh, D. and Jindal, P.C. (1980) Physico-chemical characters of some promising ber cultivars grown at Gurgaon (Haryana). *Haryana Journal of Horticultural Scinece,* 9(3/4): 114-117.

Singh, H.P. and Sindhu, S.S. (1985) Control of ber powdery mildew. *Indian Horticulture,* 29(4): 27.

Singh, J.P. and Gupta, O.P. (1983) Evaluation of various packings of ber fruit in relation to decay loss caused by various microbes. *Haryana Agricultural University Journal of Research,* 13(4): 593-595.

Singh, J.P. and Singh, I.S. (1973) Some promising varieties of ber. *Indian Horticulture,* 18(2): 3-4, 28.

Singh, K. (1979) Vegetative Propagation of Ber. M.Sc. thesis, *Haryana Agricultural University, Hisar*, India.

Singh, K. Singhrot, R.S. and Chauhan, K.S. (1981b) Effect of budding height and diameter of rootstock on budding success and budling growth of ber. *Haryana Journal of Horticultural Science*, 10(1/2): 69-71.

Singh, K., Singhrot, R.S. and Chauhan, K.S. (1982b) Vegetaive propagation of ber (*Ziziphus mauritiana* Lamk.). VI. Effect of various transplanting treatments on the growth of ber seedlings and success of budding. *Haryana Journal of Horticultural Science*, 11: 5-12.

Singh, K.K. (1964) *The Ber in India*. Farm Bulletin, ICAR, New Delhi, Delhi.

Singh, K.K., Chadha, K. L. and Gupta, M.R. (1973a) Ber cultivation in Punjab. Punjab Agricultural University, Ludhiana.

Singh, L. (1988) Propagational Studies in Ber (*Ziziphus mauritiana* Lamk.). Ph.D. thesis, *Haryana Agricutlural University, Hisar*, India.

Singh, L. B. (1952) Preliminary trial on top-working of ber (*Ziziphus mauritiana* Lamk.) in Uttar Pradesh. *Indian Journal of Horticulture*, 9(1): 16-19.

Singh, L.B. (1961) Raising of ber (*Ziziphus mauritiana* Lamk.) and aonla (*Phyllanhus emblica* L.) seedlings for budding in pot. *Annual Report - Fruit Research Station, Saharanpur*, India, 1954-56. pp.34-36.

Singh, M.P. (1984a) Field screening of some jujube cultivars against the attack of bark eating caterpillar, *(Indarbela quadrinotata* Walker). *Madras Agricutlrual Journal*, 71(6): 416-417.

Singh, M. P. (1984b) Studies on the field resistance of different jujube cultivars to the fruitfly *(Carpomyia vesuviana* Costa). *Madras Agricultural Journal*, 71(6): 413-415.

Singh, M.P. and Vishishtha, B.B. (1984) Field screening of some ber cultivars for resistance to ber fruit fly, *(Carpomyia vesuviana* Costa). *Indian Journal of Plant Protection*, 12(1): 55-56.

Singh, P., Bakhshi, J.C. and Singh, R. (1971) Identification of ber cultivars through vegetative characters. *Punjab Horticutlural Journal*, 11(3/4): 176-187.

Singh, P., Bakhshi, J.C. and Singh, R. (1972a) Identification of ber (*Ziziphus mauritiana* Lamk.) cultivars through fruit characters. *Punjab Horticultural Journal*, 12(2/3): 120-133.

Singh, R. and Gupta, O.P. (1987) Evaluation of various packages for ber (*Ziziphus mauritiana* Lamk.) under various storage condtions. *National Symposium on Role of Food Technology in Rural Developmnet*, Haryana Agricultural University, Hisar, India.

Singh, R. and Khanna, R.C. (1968) Some north Indian cultivars of ber. *Indian Horticulture,* 12(2): 23-26.

Singh, R. and Tomar, N.S. (1988) Performance of some jujube (*Ziziphus mauritiana* Lamk.) cultivars in semi-arid zone of Punjab. *Indian Journal of Agricultural Science,* 58(5): 382-383.

Singh, R.R., Chahuhan, K.S. and Singh, H.K. (1986) Effect of various doses of N, P and K on physio-chemical composition of ber fruit of cultivar Gola. *Progressive Horticulture,* 18(1/2): 35-38.

Singh, R.R., Jain, R.K. and Chauhan, K.S. (1983a) Floweirng and fruiting behaviour of ber (*Ziziphus mauritiana* Lamk.) under Gurgaon conditions. *Haryana Agricultutral University Journal of Research,* 13(1): 112-114.

Singh, R.S. (1997) Note on the effect of intercropping on growth and yield of ber (*Ziziphus mauritiana* Lamk.) in semi-arid regions. *Current Agriculture,* 21(1/2): 117-118.

Singh, R.S., Vashishtha, B.B. and Prasad, R.N. (1998) Micrometeorology of ber (*Ziziphus mauritiana* Lamk.) orchard grown under rainfed arid conditions. *Indian Journal of Horticulture,* 55(2): 97-107.

Singh, S. and Ahlawat, V.P. (1995) Physico-chemical attributes and mineral composition of ber leaves as affected by foliar application of urea and zinc sulphate. *Haryana Journal of Horticultural Science,* 24(2): 94-97.

Singh, S., Krishnamurthy, S. and Katyal, S.L. (1967) Fruit culture in India. ICAR, New Delhi, India.

Singh, S.B., Chhabra, R. and Abrol, I.P. (1983b) Effect of exchangeable sodium on the growth and mineral composition of jujube and guava. *Indian Journal of Agricultural Science,* 53(6): 446-450.

Singh, S.B., Maheswari, S.K. and Singh, P.N. (1995) Field evaluation of fungitoxicants against powdery mildew of ber (*Ziziphus mauritiana* Lamk.). *Annals Plant Protection Sciences,* 3(2): 168-169.

Singh, U.R., Singh, D.V. and Singh, A.P. (1973c) Study of some budded cultivars of ber (*Ziziphus mauritiana* Lamk.). *Progressive Horticulture,* 5(2): 61-66.

Singh, Z., and Sandhu, A.S. (1984) Effect of pruning time on productivity and physico-chemical characters of ber (*Ziziphus mauritiana* Lamk.) cv. Umran. *Journal of Research, Punjab Agricultural University,* 21(4): 521-524.

Singhrot, R.S., Bakhshi, J.C. and Singh, K. (1970) Vegetative propagation of ber (*Ziziphus mauritiana* Lamk.). I. Relative performance of seedlings raised in field and in alkathene bags. *Punjab Horticultural Journal,* 10(3/4): 181-186.

Singhrot, R.S. and Kajal, R.S. (1986) Studies on the vegetative propagation of ber (*Ziziphus mauritiana* Lamk.). I. Effect of after budding treatments on the budding success and budding growth. *Haryana Journal of Horticultural Science,* 15(3/4): 162-169.

Singhrot, R.S. and Makhija, M. (1979a) Vegetative propagation of ber (*Ziziphus mauritiana* Lamk.). II. The transplanting success of grafted ber plants at various stages of scion and shoot growth. *Haryana Journal of Horticultural Science*, 8(1/2): 4-8.

Singhrot, R.S. and Makhija, M. (1979b) Vegetative propagation of ber (*Ziziphus mauritiana* Lamk.). III. Effect of time of sowing and acid treatment on ber seed germination and seedling performance. *Haryana Journal of Horticutlural Science*, 8(3/4): 168-172.

Sonawane, B.R. and Dorge, S.K. (1971) On the biology and control of fruit borer on ber. *Poona Agriculture College Magazine*, 61 (1/2): 1-4.

Srinivasachar, D. (1940) Embryological studies of some members of Rahamnaceae. *Proceedings of the Indian Academy of Science*, 2: 107-116.

Srinivasa, V.K. (1952) Chromosome numbers in the genus *Ziziphus*. *Current Science*, 21: 224-225.

Sussenguth, K. (1953) Rhamnaceae [in German]. In *Die Naturlichen Pflanzenfamilien* Engler, A. and Prantle, K. Duncker & Humblot, Berlin, Germany: 121-132.

Syamal, M.M. and Rajput, C.B.S. (1989) Effect of pruning on growth, fruiting and fruit quality of ber (*Ziziphus mauritiana* Lamk.). *Indian Journal of Horticulture*, 46(3): 364-367.

Tagiev, T.M. (1992) Promising areas for cultivating jujube in Azerbaijan [in Russian]. *Sadovodsto i Vinogradstvo*, No. 1: 19-20.

Tanaka, Y. and Sanada, S. (1991) Studies on the constituents of *Ziziphus jujube* Miller [in Japanese]. *Shoyakugaku Zasshi*, 45 (2): 148-152.

Tasmotov, L.T. (1963) *Ziziphus jujuba* in industry [in Russian]. *Sadovodstov*, No.6: 30-31.

Teaotia, S.S. and Chauhan, R.S. (1963) Flowering, pollinati on, fruit set and fruit drop studies in ber (*Ziziphus mauritiana* Lamk.). I. Floral biology. *Punjab Horticultural Journal*, 3: 60-70.

Teaotia, S.S., Dube, P.S., Awasthi, R.K. and Upadhyay, N.P. (1974) Studies on physico-chemical characterstics of some important ber varieties (*Ziziphus mauritiana* Lamk.). *Progressive Horticulture*, 5(4): 81-88.

Toky, O.P. and Bisht, R.P. (1993) Above ground and below ground biomass allocation in important fuelwood trees from arid northwestern India. *Journal of Arid Envinroments*, 25(3): 315-320.

Tomar, N.S. and Singh, R. (1987) Performance of six promising ber (*Ziziphus mauritiana* Lamk.) cultivars grown at Bhatinda. *Haryana Journal of Agricultural Sciemce*, 16(1-2): 52-58.

Tomoda, M. Shimuju, N., Gonda, R. (1985) Pectic substances. II. The location of O-acetyle groups and the Smith degradation of *Ziziphus* Pectin A. *Chemical and Phamaceutical Bulletin*, 33(9): 4017-4020.

Trehan, K.N. (1956) Brief notes on crop pests and their control in the Punjab (India). *Journal of the Bombay Natural History Society*, 54(3): 581-626.

Trimen, H. (1893) *A Handbook to the Flora of Ceylon.* Part I. *Ranunculaceae-Anacardiaceae.* Dulau & Co., London, UK.

Troup, R.S. (1921) *The Silviculture of Indian Trees. Vol. I.* Clarendon Press, Oxford, UK.

Tuohy, J.M., Prior, J.A.B. and Stewart, G.R. (1991) Photosynthesis in relation to leaf nitrogen and phosphorus content in Zimbabwean trees. *Oecologia* (Heidelberg), 88(3): 378-382.

Vashishtha, B. B. (1983) *Biosystematics of Ber cultivars*, Ph.D. thesis, University of Jodhpur, Jodhpur, India.

Vashishtha, B.B. (1987a) Horticultural qualities in Boradi (*Ziziphus nummularia*), eds. Mann., H.S. and Saxena, S.K. : 31-34.

Vashishtha, B.B. (1997b) *Ziziphus* for drylands- a perennial crop solving perennial problems. *Agroforestry Today*, July-September, 1997: 10-12.

Vashishtha, B.B. and Pareek, O.P. (1979) Flower morphology, fruit set and fruit drop in some ber (*Ziziphus mauritiana* Lamk.) cultivars. *Annals of Arid Zone*, 18(3): 165-169.

Vashishtha, B.B., Pareek, O.P., Chandrasekharan, I. and Ghamin, A. (1989a) Chemical races identified by chemo-taxonomy in ber (*Ziziphus mauritiana* Lamk.) *Annals of Arid Zone*, 28(1-2): 133-135.

Vashishtha, B.B., Singh, M.P., Singh, G.K. and Pareek, R.A. (1999) Breeding in ber (*Ziziphus mauritiana* Lamk.) for the resistance to fruitfly. Abstr. *Fourth Agricultural Science Congress-Sustainable Agricultural Export*, Jaipur, India.

Vavilov, N.I. (1951) The Origin, Variation, Immunity and Breeding of Cultivated Plants. *Chronica Botanica.*, NY, USA.

Verma, A.N. and Singh, R. (1974) Note on the screening of jujube (*Ziziphus mauritiana* Lamk.) cultivars for incidence of *Indarbela quadrinotata* Walker. *Haryana Journal of Horticultural Science*, 3(1-2): 62-64.

Verma, A.N. and Singh, R. (1975) Chemical control of bark eating caterpillar (*Indarbela quardinotata* Walker) by painting the bark with insecticides. *Haryana Journal of Horticultural Science*, 4(3/4): 145-149.

Verma, A.N., Singh, R. and Khurana, A.D. (1972) Chemical control of hairy caterpillar. *Indian Journal of Agricultural Science*, 42(10: 928-931.

Verma, R.C. and Gujar, K.D. (1994) *Production and Marketing of Horticultural Crops in Rajasthan*-Final Report to ICAR, New Delhi, College of Agricutlure, Rajasthan Agricultural University, India.

Vishal Nath and Bhargava, R. (2000). Flowering scenario of commercial cultivars of jujube (*Z. mauritiana*) under Indian Arid ecosystem. *Journal of Applied Horticulture*. 2(2): 98-101.

Vishal Nath and Bhargva, R. (1998). Shelf life of Ber (*Ziziphus mauritiana* Lamk) as affected by post harvest treatments and storage environment. Progressive Horticulture, 30(3-4) 158-163.

Watt, G. (1893) A Dictionary of the Economic Products of India. Vol. VI, Part IV, Second Reprint 1972, Cosmos Publications, Delhi, India.

Wong, K.C., Chee, S.G. and Tan, C.H. (1996) Voltaile constituents of the fruit of *Ziziphus jujube* Mill. var. *inermis*. *Journal of Essential Oil Research*, 8(3): 323-326.

Yadav, G.R. and Singh, R.Y. (1985) Effect of different time intervals and fungicides for control of powdery mildew of ber. *Indian Journal of Mycology and Plant Pathology*, 15(3): 261-264.

Yadav, G.R., Nirman, R.S. and Prasad, B. (1980a) Powdery mildew of ber in Uttar Pradesh and its control. *Progressive Horticulture*, 12(1): 27-32.

Yadav, L.S. and Godara, N.R. (1987) Effect of planting distance and severity of pruning on physico-chemical chracters of ber (*Ziziphus mauritiana* Lamk.) cv. Umran. *Haryana Journal of Horticultural Science*, 16(1-2): 45-51.

Yadav, R.C., Sharma, K.D., Singh, H.P. and Pareek, O.P. (1980) Some hydro horticultural relations in a ber orchard (*Ziziphus mauritiana* Lamk.). *Annals of Arid Zone*, 19(3): 297-300.

Yamadagni, R., Gupta, A.K. and Ahlawat, V.P. (1985b) Performance of different cultivars of ber (*Ziziphus mauritiana* L.) under rainfed conditions. *Annals of Arid Zone*, 24(2): 175-177.

Yamadagni, R., Kumar, R. and Jindal, P.C. (1980) Effect of graded doses of nitrogen on fruit set, fruit drop and yield of ber (*Ziziphus mauritiana* Lamk.) cv. Kaithali. *Progressive Horticulture*, 12(2): 5-7.

Yang, L.Q. and Niu, Z.Y. (1992) Expeirmental research on drying characterstcis of jujube. In *Agricultural Engineering and Rural Development*. Proceedings of Conference held in Beijing, China, October, 12-14, 1992, Vol. I. International Academic Publisher, North Western Agricultural University, Yangling, China.

—x—

3

POMEGRANATE

Pomegranate (*Punica granatum* L.) is one of the important fruits of tropical and sub tropical regions of the country, which belongs to family Punicaceae. It is also cultivated in semi-arid and arid tracts of the country but requires irrigation water for better quality fruit production (Prasad *et al.*, 1996). Versatile adaptability, hardy nature, less cost in orchard management, high yield potential, fine table and therapeutical values and excellent keeping quality has made this fruit more lucrative and remunerative. Pomegranate is considered as very nutritive fruit and rich in vitamin B, A and potassium. In India, it is largely used as dessert fruit and for fresh juice (Pareek *et al.*, 2002). Pomegranate fruit is also used for making delicious juice and health drinks. The edible portion of pomegranate fruit is aril which is nearly 68 per cent of the total fruit containing 78 per cent moisture, 0.7 per cent mineral matter, 1.6 per cent protein, 0.1 per cent fat, 14.5 per cent carbohydrate and 5.1 per cent fiber. The fruits are rich in vitamins (0.06 mg thiamine, 0.1 mg riboflavin, 0.3 mg niacin and 16.0 mg vitamin C per 100g pulp) and minerals (10 mg calcium, 12 mg magnesium, 70 mg phosphorus and 0.3 mg iron per 100 g pulp). Khodade *et al.*, (1990) observed chemical changes in pomegranate fruits during growth and development. The pomegranate is important as a special diet for sick and aged persons. Ellagic acid which is particularly plentiful in pomegranates, prevents carcinogen oxidation of cellular membranes (Annon, 2004). It is highly profitable fruit crop under marginal agro climate of arid and semi-arid regions. In India, it is cultivated for domestic consumption as well as export purpose.

Pomegranate is commercially grown for its sweet acidic fruits, which seem to have been used form the prehistoric period for dessert purpose. Fresh grains are used to garnish desserts and salads. These may be rolled in small cream cheese balls and served as appetizers. In Iran, gentle pressing to release a small hole made in the rind loosens the whole fruit. Juice of wild pomegranate is used for the preparation of citric acid and sodium citrate for medicinal purpose. Stem bark and root contain several alkaloids of the pyridine group, which have medicinal value. The juice is also used for the treatment of dysentery and diarrhoea. The fruit peel relieves bronchitis. The peel extract kills the worms in human stomach. Fruit peel extract is also used for tanning clothes to impart natural colour. Pomegranate seed contains about 15 per cent oil having refractive index, iodine value and very low melting point. The oil has potential for industrial use. Ornamental type pomegranate bears beautiful flowers and evergreen foliage and such genotypes are used for beautification purposes.

In global market, India plays a significant role in export of pomegranate fruits. Recently, exports of pomegranate from India have increased and extended to several countries. The export of pomegranate fruits from India to Middle East, Germany, Netherland, Holland, U.K. and Behrin is of Rs. 70 million per year. For export purpose, cultivars Ganesh, Phule Arakta, G-137, Bhagava and Mridula are most suitable (Anon, 2003). During the year 2000-2001, the export of pomegranate fruits to Europe has diversified as 140 tons to Italy, 80 tons to Netherlands and 31 tons to Belgium. It shows that there is wide acceptability for Indian pomegranate in world market and this need to be expanded by tapping more markets (Anon, 2001). Among Far East countries, pomegranate from India was exported to Philippine (35 tons in 1998-1999), Singapore (22 tons in 1999-2000) and Hong Kong (15 tons in 1998-99). There is a little export to African countries except South Africa, where 47 tons of fresh pomegranates were exported during 2000-2001 (Anon, 2001).

Origin and Taxonomy

Pomegranate is a native of Mediterranean region. *Punica granatum* L. belongs to family Punicaceae. Genus *Punica* has two species i.e. *Punica protopunica* and *Punica granatum*. *Punica protopunica* is found wild in the Socotra Island and *Punica granatum* is cultivated in tropical and subtropical part of the world. It is mentioned in ancient literature that pomegranat has great antiquity and is known to have been cultivated in the Middle East for more than 5,000 years. Pomegranate variability of do exist in India, Afghanistan, Pakistan, Egypt, Russia, Spain, Europe and Iran, California and sub tropical parts of USA where it is being grown from time immemorable. In Afghanistan, Kabul and Kandhar area are most suitable for pomegranate cultivation.

Pomegranate is a bushy plant, which attains 5-7 m height with multiple branch system. The leaves are elliptic to lanceolate, dark green in colour measuring 7.5 cm long. The inflorescence in pomegranate is cymes. Male and hermaphrodite flowers are borne in one cluster. Flower bud of spring flush are borne on mature wood of one year old shoot whereas July-August flush, comes on current season's growth. Orange red flowers are borne in clusters mainly in the axis towards the end of branch lets. Terminal flowering is also found in pomegranate. The calyx is tubular, persistent, five to seven lobbed, the petioles lance shaped inserted between the calyx lobes. The ovary is embedded in the calyx tube and contains several compartments in two series, one above the other. It is a diploides with the basic chromosome number n = 16 or 18. Time of dehiscence of anthers in pomegranate varies with variety and existing environmental conditions particularly temperature and humidity. Although, male and hermaphrodite flowers bear fertile pollens, the fertility rate in hermaphrodite flower is quite high. The stigma attains maturity one day prior to anthesis and receptivity remains at peak up to two days of anthesis. Both self and cross-pollination have been recorded in pomegranate however, higher percentage of fruit set has been observed by hand pollination and open pollination (Bose and Mitra, 1984).

There are many cultivars of pomegranate which are grown in different countries. For Indian arid zone, the main selection criteria aims at identification of genotypes having i) high productivity, ii) large fruit size (over 750 g), iii) small and mellow (1-4 kg/cm^2) seeds, iv) dark red and bold aril, v) high TSS (above 17^0 brix), low acidity (below 1.9%) and tannin (below 17^0 brix) and vii) thin and red fruit colour besides resistance to extremes of temperatures, pests and diseases. Some well-known commercial cultivars from other countries have been also been introduced and adapted in arid and semi-arid parts of the country (Bose and Mitra, 1984). The fruit is large, orange to red in colour, obscurely six sided, with a smooth leathery skin. Botanically, the fruit is "blasta" which is divided into several chambers containing many thin, transparent, vesicles of reddish juicy pulp, each surrounding at angular elongated seed (Plate 3.1).

Plate 3.1. Red aril of mature Pomegranate fruits

Area and Distribution

Pomegranate is commercially cultivated in Iran, Afghanistan, Russia, Israel, North and Latin American countries, Africa and India. where as in India, it is cultivated in Maharashtra, Karanataka, Gujarat, Rajasthan and Andhra Pradesh.

India is world's leading producing country of pomegranate. It is almost grown in all parts of the country but commercial cultivation exists only in Maharashtra, Karnataka, Gujarat, Rajasthan, U.P, Andhra Pradesh and Tamil Nadu. More than 1.00 lakh hectare area is cultivated under pomegranate, producing 5,09480 MT of fruits. Maharashtra state accounts for 78% of total production of the country. Sholapur, Nasik, Sangli, Ahmednagar, Pune, Satrara are the prominent districts of Maharashtra where its production is concentrated. Among different states Maharashtra is the main pomegranate producing state where the area under pomegranate cultivation is about 77,176 ha with production of 4.76 lakh tonnes. Next to Maharashtra maximum pomegranate cultivation takes place in Karnataka. In Karnataka, its production is mainly concentrated in Bijapur, Chitradurga, Belgaon and Bangalore districts covering an area of 10,689 ha with annual production of 1,15616 tonnes. In Rajasthan, the area under pomegranate is 427 ha with production of 4,511 tonnes.

Soil and Climate

The loam soils with medium texture, pH 7.6-8.2 having good moisture holding capacity is preferred. Sandy loam to deep loamy soils with good drainage is ideal for its cultivation where the fruit attains good colour and quality. Although, it is not specific to soil requirement, it is sensitive to fluctuations in soil moisture particularly during fruit bearing stage. It can tolerate salinity and can be grown on slightly alkaline soils, however growth is adversely affected at 25.5 ESP. Soils having better fertility and drainage is good for quality fruit production in pomegranate.

Pomegranate performs well under subtropical climate. It grows well under semi-arid and arid agro climate (Singh *et al*, 1989). For flowering and fruit set, in general 30 to 35^{0}C temperature is appropriate. It requires dry and sunny weather during flowering and fruit development period for better quality. Warm days and cool nights help in the development of good color and sweetness in the aril. High vapor pressure deficit owing to high temperature and low humidity coupled with high temperature makes it susceptible to different biotic agents. Sudden fluctuations in atmospheric humidity caused cracking of fruits and reduce their quality. It can withstand frost but leaves are damaged by very low temperature. The plants of pomegranate also withstand high temperature (45^{0}C) during summer in arid region.

Plant Genetic Resource Management and Varieties

The cultivar with thin rind and thick rind grown in Iran are Bala-Myursal, Kara-Bala-Myusral, respectively where as Bashkalinski, Ali Shirin Anar, Kazzale Anar, etc. are promising Russian cultivars. The American variety Ruby Red, Early Red and Wonderful, the Egyptian cultivar Banati and Manfaloti and Tunisian cultivar Gabesi are also commercially grown. In India, the popular cultivars of pomegranate are Ganesh, Muskat, Jalore Seedless, Dholka, Kabul Red, Vellodu, Bassein Seedless and Jodhpur Red. Some promising clonal selections have been made from these varieties. These are G137, P23, P26, Jyothi, Yercaud 1 and Co1. Some promising hybrids such as Mridula, Ruby, etc. have also been developed. For *anardana*, seedling selections having high acidity have been made at IIHR, Bangalore. Vishal Nath and Singh (1998) mentioned the important cultivars for different states of India. The major characteristics of some of the important cultivars are as follows.

Ganesh

It is a leading variety of pomegranate and developed through seedling selection. It is an evergreen cultivar having spreading growth habit and produce yellow fruit with red tinge on the surface having deep pink aril and soft seed. The average fruit weight is 229g and keeping quality is good. This variety is suitable for domestic consumption as well as for export (Plate 3.2).

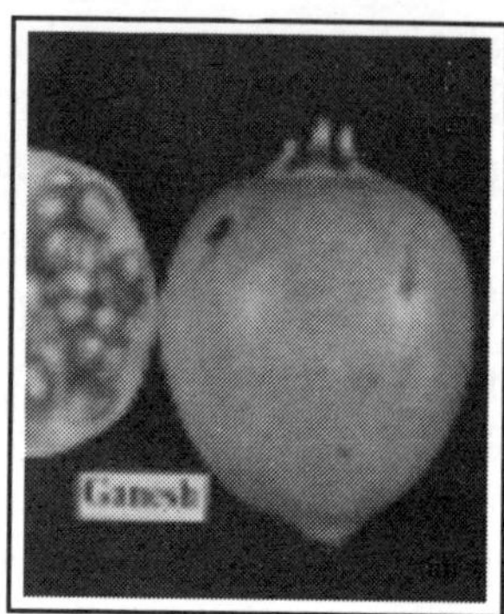

Plate 3.2. Pomegramate Cultivar Gamesh (i) Plant in full bearing (ii) bunch of fruits suitable for export (iii) mature fruit and aril.

Jalore Seedless

This is a cultivar suitable for semi-arid and arid region. It is commercially grown in Rajasthan state. The fruits are large, reddish green with sweet (15.2^0 Brix TSS), pink aril and soft seed (Plate 3.3).

Plate 3.3. Pomegramate cv. Jalore Seedless (i) mature fruits (ii) high quality fruit with aril arrangements.

Bassein Seedless

It has spreading growth habit, bears red fruits with light pink aril, soft seeds and 16.2^0 Brix TSS. Jyothi, a clonal selection from this cultivar produces fruits having deep red aril colour. It is suitable for cultivation in Karnataka.

Muskat

Muskat is an evergreen cultivar. It produces yellowish fruits with pink aril and medium hard seeds. Two clonal selections, P23 and P26 having higher fruit weight, light pink, aril and soft seeds have been made from this cultivar.

Mridula

Mridula is a hybrid between Ganesh and a Russian cultivar, Gulsha Rose Pink. It has dark red fruit and blood red aril, soft seeds and high TSS. Mridula is also a leading variety in Maharashtra and is being exported to Middle East (Plate 3.4).

Plate 3.4. Pomegramate cultivar Mridula (i) Plant in bumper fruiting (ii) T.S. of the fruit showing blood red aril

Ruby

This is a hybrid cultivar developed at IIHR, Bangalore as a result of complex crosses between F_2 of Ganesh x (Kabul x Yercaud) and Ganesh x Gulsha Rose Pink (Pareek and Vishal Nath, 1996). The red coloured fruits are of about 250g having dark red aril and soft seeds (Plate 3.5).

Plate 3.5. Pomegranate cv. Ruby developed at IIHR, Bangalore

Phule Arakta

This variety is also a selection from open pollinated population of a cross between Ganesh and a Russian variety. This variety has been released by MPKV, Rahuri. The fruit of this variety has dark red aril, soft seeds, attractive fruit colour and high TSS. Fruits mature within 120- 135 days and gives yield of 30-35 kg /tree. It fetches high prices in the market. This variety is being grown commercially in Maharashtra and Karnataka and there is heavy demand for export to UK and Gulf countries (Anon, 2002).

Bhagava

This variety is recommended for commercial cultivation in Maharashtra. Fruits are very attractive saffron colour and high demand in UK, Germany, European and Gulf countries. Fruits have better keeping quality than other varieties. It is a high yielding late maturing (170-190 days) variety. It produces yield of 30-40 kg /tree. There is less fruit drop and cracking in this variety. It is suitable for export to Middle East countries (Anon, 2002).

G-137

This variety has been released from MPKV, Rahuri and is suitable for cultivation under Maharashtra conditions. It is a high yielding and medium size variety. This variety is also exported to Gulf countries (Plate 3.6).

Plate 3.6. Pomegranate cv. G-137

GKVK-1

This variety is grown in Maharashtra and Karnataka state. It is a medium yielding variety. Fruit of this variety is used for fresh consumption as well for juice making.

Amlidana

For Anardana purpose, this variety has been released from IIHR, Bangalore. Its aril is sour in taste and the tree is high yielding. Now a days, there is a heavy demand of sour types pomegramate for garnishing the dishes and making candy.

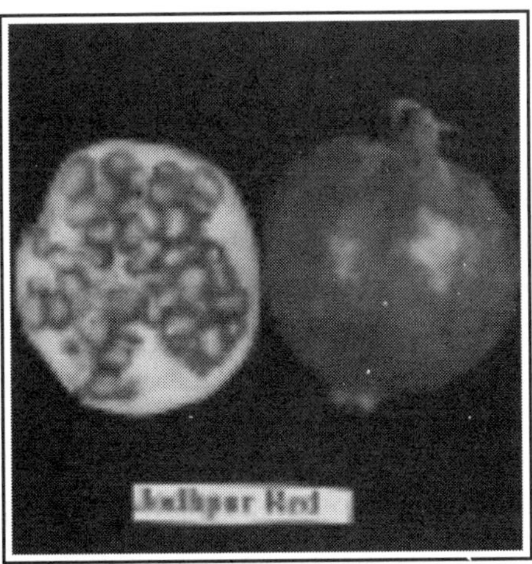

Plate 3.7. Pomegranate cv. Jodhpur Red

The physico-chemical characters and yield of some important cultivars and hybrids of pomegramate (Pareek and Vishal Nath, 1996) have been given in Table 3.1.

Table 3.1 : Physico-chemical composition of pomegranate cultivars, selections and hybrids.

Variety	Fruit weight (g)	Fruit size (LxB) (cm)	Aril colour	Seed mellow-ness	100 grain wt. (g) (kg/cm^2)	Juice (%)	TSS (°B)	Acidity (%)
Ganesh	229	8.1x7.3	Light pink	Soft (3)	16.4	77.4	14.6	0.3
Muskat	327	8.5x8.2	Whitish pink	Medium hard (5.0)	19.2	70.4	15.6	0.49
Bassein Seedless	284	8.4X8.4	Light pink	Soft (2.9)	18.1	74.3	16.2	0.37
Jalore Seedless	301	8.5x7.9	Pink	Soft (1.8)	-	-	15.2	0.42
Dholka	292	7.8x8.1	Whitish pink	Soft (3.5)	18.1	71.0	14.0	0.41
Jodhpur Red	200	7.8x7.1	Whitish pink (Plate 3.7)	Hard (10.0)	14.7	57.8	15.2	0.61
Jyoti	219	7.9x7.5	Deep red	Soft (2.7)	26.3	83.0	15.0	0.58
IIHR Selection	195	6.6x7.6	Light pink	Soft (3.6)	18.8	71.5	13.7	0.55
G-137	269	7.9x7.5	Light pink	Soft (1.3)	28.6	65.2	15.5	0.36
P-23	385	7.8x7.5	Light pink	Medium hard (4.7)	32.1	84.0	16.0	0.42
P-26	379	7.6x7.5	Light pink	Soft (4.3)	32.5	84.0	15.5	0.41
Ruby	249	7.2x7.8	Dark red	Soft	40.0	78.0	17.2	0.64
Sel-5	414	9.2x8.9	Pink	Soft (1.0)	-	-	16.3	0.36
Mridula	244	7.5x7.7	Blood red	Soft (1.0)	-	-	17.9	0.47
Sel- 242	221	7.6x7.4	Red	Soft (1.6)	-	-	19.1	0.55

Source : Pareek and Vishal Nath (1996)

Besides some exotic varieties of pomegranate are also being maintained in India mainly for crop improvement programme. These are:

Wonderful

The fruit is a large, deep purple red with glossy appearance. The kernels and juice are deep crimson with good flavour. The seeds are small, tender and the rind is of medium thickness. It is the only recognized variety grown in California for the market.

Ruby Red

The fruit is about the same size as Wonderful. In this cultivar, crimson- purple colour develops early. However, juice tests show that both varieties mature at the same time. All of the Ruby Red fruits mature at the same time, allowing one picking which reduces the splitting problem. It does not store as well, as Wonderful.

Spanish Sweet of Paper shell

It is large fruited variety with pale yellow rind washed with pink blush. The kernels are pale pink with a sweet flavour.

Shirin Anar

It is a highly seeded Russian variety mainly used as male parent to impart sweetness.

Gulsa Rose Pink

It is also an exotic variety with bright fruit and aril colour used mainly for improving aril colour in crop improvement programme.

Plant Propagation and Nursery Management

Pomegranate can be propagated through seed as well as through vegetative means. Seed propagation leads to low yield and poor fruit quality however, it is employed in developing breeding lines. Fresh seeds germinate easily without any treatment but trees are not grown commercially from seed because the seedling does not come true to type. Such seedlings produce variable plants fruits of different shape and size. Therefore, the researchers looking for specific traits raise the new population through seed.

Pomegranate is commercially propagated through semi, hard and hard wood cuttings. Treatment of cutting with 1000 ppm IBA as basal quick dip or by dipping in 250 ppm IBA for 1 minute improves rooting (Singh, 1994). The cuttings are planted in polythene tubes filled with a mixture of soil, organic manure and sand in equal proportion. These can also be planted in open nursery at 15 –20 cm spacing in rows with about one- third (5-8 cm) of top exposed. Cutting taken from the mature, 6-12 mm thick branches emerging from the base of main stem give better rooting. About 15-20 cm long hardwood stem cutting are taken from 1-2 year old wood. Cutting can be planted in the open nursery during spring season. However, these can also be planted during rainy season under poly house. For fast multiplication of limited stock, smaller size cutting (3-4 buds) taken from semi hard wood or hard wood or soft wood cutting can be rooted in small pots filled with perlite- peat mixture under intermittent mist with bottom heat.

Air layering (Gooti) is also a successful method of pomegramate propagation. Application of 10,000 ppm IBA with lanolin paste at the upper side of the ring and covered with moist moss grass and tie with white polythene resulted better rooting in air layers. It is most successful during rainy season due to optimum moisture in the atmosphere. The layers can be separated after about 30-90 days and are set in the nursery for proper root development and proliferation before planting in the field. This method is commercially used in Maharashtra for large scale multiplication of true-to-type plants.

Planting

Commercial orchard of pomegranate is planted at 5 x 5m or 6 x 4m spacing in square or rectangular system. High density plantation at 5 x 2 m spacing can also be done in pomegranate. The square or rectangular system of planting facilitates inter culture operations in the commercial orchard. Pits of 0.6 x 0.6 x 0.6 cm size, dug about one month before planting, are filled with top soil, pond slit and organic manure mixture in 1:1:1 proportion. In areas having termite problem, 50g Methyl parathion dust (5%) or Chlorpyriphos dust (10%) is also mixed in each pit. A basal dose of 15- 20 kg well rotten manure is applied in each pit. Sharma *et al.,* (1999) recommended that under hot arid climatic conditions having sandy soils, the pomegranate pits should be filled with a mixture of top soil and pond silt in equal proportion along with 10 kg well rotten goat manure for better survival and plant growth. Rainy season is the best time of planting in warmer regions (Pareek *et al,* 2000) to utilize moisture. In cooler regions, plants undergo dormancy during winter and planting can be done towards the end of winter before the onset of spring growth. High density planting has been reported suitable in pomegranate for high yield when planting is done at 5x2m spacing (Table 3.2) in cv. Ganesh. For high density plantation (5 x 2 m spacing) under semi-arid agro climatic conditions receiving 800 mm rainfall, higher water and fertilizer input (22 irrigations per year and 625 gN:250 g P:250g K per plant) has resulted nearly 2.5 times more yield than the normal spacing. However, owing to enhanced diseases and pests problem in close spacing, it is only recommended under high management practices (Vishal Nath and Singh, 2000). In some parts of Maharashtra, Pomegranate is planted at a spacing of 4.5x 3.0 m for better yield (Anon., 2002).

Table 3.2 : Effect of high density planting on fruit yield of pomegranate cv. Ganesh

Sl. No.	Spacing (m)	Yield (q/ha)
1.	5 x 2	85.7
2.	5 x 3	62.0
3.	5 x 4	38.4
4.	5 x 5	36.6
	C.D. at 5%	25.3

Source: Pareek and Vishal Nath (1996).

Training and Pruning

Pomegranate plant is bushy in growth habit and thus produces considerable number of shoots from the base, which create crowding and may lead to infestation by insects such as shoot borers. When trained on single stem, the fruiting area goes too high which creates difficulty to spray on the trees, harvest fruits and protect from the bird damage. Retention of 3 - 4 stems from ground level is, therefore advisable. Since pomegranate has a tendency to give out suckers, extra suckers from ground level should be removed regularly. The main stem should be topped at a height of about 70 cm to induce branching. The tree is given a desired shape during the first 3 years by proper selection of secondary and tertiary branches. Care should be taken to remove downward growing and criss-crossing branches. The branches must be well-spaced and only one branch should be allowed from a node to avoid weak crotches.

Pomegranate bears on terminal and auxiliary short spurs arising from the mature shoots. Hence, it does not require regular annual pruning except the removal of water suckers, diseased and pest affected or dried branches. After about 10 years, those old main stems which become less productive should be renewed. For this, the old stem is cut back from near the ground level and new shoot emerging from ground level is trained to prepare a renewal stem. In semi arid conditions, pomegranate can be trained in single stem also.

Orchard Management

Pomegranate is an important arid fruit crop. It has high export as well as domestic demand. To produce the quality fruits of pomegranate, appropriate orchard management practices needs to be followed.

Nutrient Management

A tissue nutrient status of less than 0.99% N, 0.136% P, 0.417% K, 1.081% Ca, 0.405% Mg and 0.109% S in the 60 days old 8th leaf pair from the growing tip, respectively indicate deficiency of these nutrients in pomegranate (Pareek and Vishal Nath, 1996). 10 kg organic manure along with 125g nitrogen, 50g phosphorus and 50g potash is applied to one-year-old tree. The dose is raised every year by the same rate so that after five years a yearly dose of 625 g nitrogen, 250 g phosphorous and 250 g potash is applied to each tree besides the basal dose of 50 kg organic manure. Time and method of application of nutrients play significant role in production of quality pomegranate. Therefore, nitrogen is applied in two spilt doses, one at the time of first irrigation in spring and the second after 3 weeks of the first application whereas full dose of phosphorus and potash are applied once along with the first dose of nitrogen. Organic manure should be applied 20-30 days before the first irrigation.

Micronutrient is also important in pomegranate. The deficiencies of micronutrients such as zinc, iron and boron can be overcome by two foliar sprays of zinc sulphate (0.06%), ferrous sulphate (0.4%) and borax (0.2%), respectively. These can be applied separately or in combination during flowering and fruit set.

Water Management

Although pomegranate is a drought hardy fruit plant, irrigation is essential to obtain better growth, fruit yield and quality. To achieve better survival of plants, light irrigation is necessary just after planting. Regular irrigation must be applied to the orchard from onset of spring growth until the onset of monsoon. In case of extended dry spells, irrigation is required throughout the fruit development period. Irrigation may be given for one hour on alternate days through 2 drippers per tree each of 4 liter / hr flow rate capacity. The most critical stage for irrigation is during 60-120 days after fruit set. Pomegranate can be grown with saline irrigation water in semi arid region (Singh *et al.,* 1989).

Drip irrigation considerably saves water and improves water use efficiency. Drip irrigation on the basis of 20 per cent wetted area on alternate days proved to be the best in semi-arid regions. Fruit cracking is a problem under arid agro-climatic conditions, which directly influenced by atmospheric temperature and moisture. Irrigation at 75% CPE reduces fruit cracking in such regions. Under rain fed conditions, *in-situ* water harvesting by micro-catchments increase plant growth and fruit production in low rainfall areas. Size and slope of the micro-catchments would differ according to soil and other ecological parameters. Practices such as weed control and mulching by the available organic waste materials help to conserve soil moisture by checking evaporation and weed growth. Water losses through plant surface can be minimized by planting windbreaks and shelterbelts and by use of anti -transpirants like Kaolin (10%), PMA (10^{-5}M) and Power oil (1.5%), etc.

Orchard Floor Management

In pomegranate, normal spacing of 5 x 5 m or 6 x 4 m gets fully occupied only after 5 years. Therefore, to make use of the interspaces during the initial years, intercropping with leguminous crops like cowpea, cluster bean, gram and vegetables like chillies, cabbage, cauliflower, peas, tomato, carrot, radish, eggplant, etc. not only gives additional income but also improves the soil physical properties. The selected intercrops in pomegranate orchard should be of short duration. When plants attain full growth and start commercial yield, intercropping may be discontinued (Pareek *et al,* 2000). In Jalore district of Rajasthan, gram, barley, spices are grown as intercrops.

Weeds in pomegranate orchard create shelter of many pests and diseases during rainy season and compete for water and nutrients. Mechanical weeding by inter cultural operations or chemical weed control by application of herbicides such as glyphosate, minimizes the weed population (Pareek and Vishal Nath, 1996). The basin of plants should be kept clean by removal of weeds and grasses to reduce disease and pest pressure and enhance fruit yield and quality in pomegrante.

Disease Management

Pomegranate is infected by number of diseases in field as well as during post harvest handling which cause considerable loss to the pomegranate growers and traders.

Fungal leaf and fruit spot spoils the appearance and market value of fruits. Four sprays with fungicide like copper oxychloride (0.2%) or Carbendezim (0.1%) or Mancozeb (0.3%) at an interval of 15-20 days starting from the initiation of disease have given effective control.

Bacterial leaf and fruit spot is caused by *Xanthomonas campestris* var. *punicae*. It occurs in comparatively humid areas. Affected fruits become unmarketable due to their poor appearance. Three sprays with 500 ppm Paushamycin + 0.2% Copper oxychloride at 15 days interval starting from the initiation of the disease significantly reduces the incidence.

Pomegranate wilt/die-back is a complex problem and caused by fungus, pests and nematodes. This problem is serious in some parts of Maharashtra and Karnataka. It can be managed by adopting proper orchard management practices like proper distance of planting, irrigation, control of stem borers, etc. Soil application of neem cake (2.0 t/ha) at the time of bahar treatment can minimize the problem. Experiments conducted at MPKV, Rahuri have shown that this complex disease is controlled by application of Phorate 10 G (4kg/ha) and Carboxyl 50 WP (0.25 %) + Bordeaux mixture 1% and use of *Trichoderma viridi* up to a certain extent (Anon, 2002).

Pest Management

Plant protection measures are vital for successful fruit production in pomegranate. The major pests of pomegranate observed in India are termites, pomegranate butterfly and bark-eating caterpillar.

Termites are the major problem during establishment of plants particularly in arid eco-system. At the time of pits preparation, addition of chloropyriphos or methyl parathion dust (50g/pit) in filling mixture has been found effective. To control termite infestation in growing plants, application of chloropyriphos @ 5 ml per tree at the time of irrigation is effective.

Pomegranate butterfly (*Virachola isocrates* Fab.) also known as Anar butter fly causes severe damage to the fruits. Female butterfly lays eggs on calyx and the developing fruits during July. The emerging caterpillars feed on the seeds causing rotting and premature drop of fruits. The holes made by the caterpillars can be clearly observed on the fruit. Bagging of the fruits with butter paper reduces the infestation considerably. Two sprays, one each with 0.002% Deltamethrin and 0.2% Carbaryl 50 WDP at 21 days interval in rotation after fruit set has been found effective to control the pest.

Bark eating caterpillar (*Indarbela quadrinotata*) generally infests old and neglected plants. Damage by this pest can be reduced by proper orchard sanitation and training of trees. Removing insect webs from holes and affected parts and then, insertion of cotton swab soaked in carbon disulphide or petrol or kerosene in each hole followed by sealing of holes kills the cater pillar. The spraying of the affected parts by insecticides such as dichlorvos (0.08%), fenvalerate (0.04%), carbaryl (1%) and quinalphos (0.08%) effectively reduces the larval population.

Physiological Disorder and its Management

Cracking of fruits is a serious disorder in pomegranate. It can be observed at any stage of fruit development. The malady is mainly due to boron deficiency in young fruits while in developed fruits it is caused due to variation in soil moisture content and atmospheric humidity. At the time of fruit ripening, if the soil becomes too dry and then irrigated heavily or there is unexpected rains, cracking may occur. Some cultivars, like Gule Shah, Khog and Kazaki Anar are reported to be resistant but they do not have better yield and quality. In commercial cultivars, the cracking problem exists at varying intensity and therefore, its effective management schedule should be followed for commercial production. Regular irrigation to maintain soil moisture at desired level, spraying of calcium compounds or 0.2% Borax or 120 ppm GA_3 on young fruits minimizes fruit cracking (Chadha and Pareek, 1993).

Blackening of arils has been a serious threat to the pomegranate industry. Apparently, the mature fruit looks good and intact, but after opening some arils in a particular segment or in whole of the fruit is found brown or black in colour and giving foul smell in advance stage. This physiological disorder is also known as "internal Break down" of pomegranate. In the affected fruit, the pulp bearing seeds (arils) do not develop and attain the typical red colour and are somewhat flattened than plump. Flavor of the arils become abnormal and many of them have a streaked appearance due to fine white lines radiating from the seeds.

Initially, there is loss of pigment from aril of affected fruit and at later stage the arils become soft and shrivelled and turn brown which finally turns black. From the external appearance it is difficult to locate the fruit that is affected with disorder. A lot of research efforts are being made but till date the exact cause has not been pinpointed. There is reduction in TSS, sugars (total and reducing), ascorbic acid and calcium while there is increase of non-reducing sugars, tannins, boron and polyphenol oxidase and peroxidase activities in affected arils. But these are results and not the cause (Pareek and Vishal Nath, 1996).

Probably, and it is a complex caused by genotypes, season of fruiting, soil fertility, imbalance of nutrients and stage of maturity. Further, role of pathogen can also not be ruled out. Research in this direction needs to be taken up. Temporarly, the problem can be kept under control by spraying with Bordeaux mixture, Bavistin (0.1%) and Rovral (0.1%) after pruning (Anon., 2002).

Flowering, Fruiting and Yield

Flowering in any tree is governed by the internal physiological conditions. Krauss and Krabyl have indicated the role of C: N ratio in flowering process. C: N (ratio 60:40) induces better fruit bud differentiation and gives flowering under favorable environmental conditions. Accumulation of carbohydrates takes place when its rate of utilization of CHO is reduced due to reduced rate of growth during resting stage of plants. Temperature, soil moisture, nutrition are some other factors reduce or accelerate the rate of growth of plants. Normally, the deciduous trees go to rest due to very low winter temperatures. In the North India where winters are severe, pomegranate plant goes to rest and its vegetative growth is ceases during November to January. The trees under such conditions differentiate flower bud in the meristamatic tissues and flower naturally with the onset of spring (Plate 3.8). In Southern India (Maharastra, Gujarat, Karnataka, Andhra Pradesh, part of Madhya Pradesh), the summer and winter seasons are not distinct and hence the plants do not go to rest naturally during winter and grow continuously. Pomegranate under Southern Indian conditions put forth new growth continuously with three distinct flowering seasons i.e. *Ambia bahar* (January-February), *Mrig bahar* (June-July) and *Hasta bahar* (September-October). Thus, pomegranate under North Indian conditions naturally flowers profusely in February-March. In areas where it grows continuously, if pomegranate is allowed to flower naturally, they will produce little, indiscriminate

Plate 3.8. Pomegranate plant in full bloom

crop of inferior quality all through out the year posing market problems. Maintenance of such small crop also becomes difficult and uneconomical. It is, therefore, become necessary to regulate the growth and flowering in such a way that tree produces better crop of good quality in any one of the season by forcing the tree to go in rest in one of the desired season by controlling moisture supply and nutrition. The process of manipulating the crop performance is known as "Crop regulation" or "Bahar treatment".

The pomegranate starts fruiting about 4 years after planting (Plate 3.9). Based on seasonal changes, precisely, there are three flowering seasons known as Bahar in pomegranate. In Mrig Bahar, the flowering starts in June and July-August fruiting coincides with on set of monsoon, therefore, this bahar is not recommended for commercial fruit production in areas having clear monsoon phase. However, in areas like Bangalore and deccan plateau where water is scarce during hot weather, Mrig bahar crop gives good yield. Ambe Bahar flowering takes place during February-March and fruiting coincides with hot and dry summer in north India. Therefore, this crop is also not much economical in north India. In southern India, however, the areas having assured irrigation, Ambe bahar crop produces good yield. Hast Bahar flowering occurs during September-October and fruit development takes place from November to February. This is the main crop of north India. Although, pomegranate plant may be induced to bear fruit in any of the season, ordinarily, only one

Plate 3.9. Fruiting in pomegranate during initial pears

bahar should be taken from a tree. That too, the crop coinciding with long fruit development phase having hot day time and cool nights. Therefore, crop regulation is essential in order to obtain fruits from the desired season (Plate 3.10). This can be done by cultural treatments such as withholding irrigation or foliar spray of urea (10%) or even

Plate 3.10. Bumper fruiting in pomegranate after bahar treatment

root exposure during the undesired flowering seasons followed by application of manures and fertilizers and irrigation at regular interval. The period of withholding of irrigation would depend upon the type of soil. In light rocky soils stress of 30 days would be sufficient, in medium soil it may extent to 40-50 days while in heavy/deep soils some times it require stress of 60 days or even more. The operation helps in more vegetative growth and flowering and fruiting. Normally, a pomegranate tree bears 125-150 fruits. If all the fruits are retained, hardly 20 to 25 per cent would be of A- grade (over 750 g). Therefore, flower thinning may be done either manually or by spraying 500ppm ethephon ensuring that 50-60 fruits are retained on one plant.

Harvesting and Post Harvest Management

Pomegranate is a non-climacteric fruit and, therefore, only tree-ripe fruits are harvested. Fruits generally ripen in 130-150 days after fruit set. However, cultivar Bhagava matures within 170-190 days. Besides biochemical parameters, change in rind color from light green to yellowish pink or even red with waxy shining surface, closure of calyx on distal end of the fruit and presence of cracking sound of grains on pressing the fruit are some common indices of fruit maturity. Fruits are individually picked in the morning or late in the evening by separating with a sharp knife retaining about 1 cm stalk attached. In warmer season, when harvested in the evening, the fruits are wiped and collected in crates. A full-grown pomegranate bush normally produces 40-50 fruits of good size, which, under intensive management, can be as high as 100 fruits.

To fetch good price in the market, the pomegranate fruits must be graded. Before grading, sorting is done to discard malformed, immature, diseased and cracked fruits. Then the fruits are graded into the categories as:

A--------------over 750 g, spotless

B---------------500-750 g, spotless

C--------------400-500 g, spotless

D--------------300-400 g

E--------------250-300 g

Before packaging, the fruits are cleaned by muslin cloth or by dipping in ethylolate to remove dust and to give good appearance. A and B grade fruits are individually wrapped in tissue paper or plastic liner. Packaging (Plate 3.11) is done in corrugated fiber board cartons with cushioning material in 3.0kg and 5.0kg capacity packages for marketing and export (Anon, 2003).The average weight of fruit of cv. Ganesh is preferred for export to Middle East is (300-450g) while 250-300g spotless fruits have better demand in UK, Netherlands and Germany. Therefore, the grading and packing should be based on market demand.

Plate 3.11. Grading and packing of pomegranate for export market (i) Uniform graded fruits (ii) Handy packing of fruits for safe transport.

Pomegranates can be stored up to 2 months at 5^{0}C. For long periods, storage could be done at 10^{0}C to avoid chilling injury. Storage could be done for 6 months at 6^{0}C in 3% oxygen + 6% carbon dioxide atmosphere. Waxing of the fruits reduces water loss and their shriveling during storage however, it is not preferred in many countries because of the stickiness and smell of the emulsified wax.

Processing and Value Addition

Apart from fresh consumption, the pomegranate fruits can be processed into various value added products. Juicy pomegranate arils (sweet type) are crushed to prepare sparkling juice of high quality, which can be used for the preparation of syrup, squash, jelly and *anar* rub. The juice can be used as a refreshing punch base. Citrus fruit sections and slices of apples or pears or banana or other fruits marinated in pomegranate juice are attractive in salad and fruit cocktail. The juice can also be used in making gelatin desserts, icing for cakes and pudding sauces. The juice can be concentrated to prepare a thick, viscous product similar to tomato sauce or ketchup. Grenadine is made from its juice, which is used as flavoring in mixed drinks, as a topping for ice cream and desserts. In India, fruit grains of the acidic varieties are dried to prepare *anardana,* which is used for garnishing various household preparations such as vegetable curries and for the preparation of chutneys. Pomegranate juice can be easily fermented for the production of wine and good quality vinegar (Adsule *et al.*,1992).The methods to prepare different processed products from pomegranate are as follows:

Pomegranate Juice

Juice can be extracted either by (i) Pressing fruit quarters in a rack and cloth hydraulic press. This yields 42 per cent juice or (ii) Extracting in a spiral type screw press or (iii) Extracting by pressing arils in a basket press. This method provides 70 per cent juice with low tannin (0.12%). Excess tannin from the juice can be precipitated by gelatin and it is clarified by pectinase enzyme or by heating in a flash pasteurizes at 79-82 ^{0}C followed by cooling, seting up for 24 hrs, racking up and filtering. Sodium benzoate (600ppm) is added as a preservative. The methods are:

Pomegranate rub

Equal parts of juice and sugar are mixed and the syrup is concentrated on slow fire to a thick consistency (70-75^0 Brix).

Pomegranate Jelly

Mix 500 ml of pomegranate juice, 800g sugar and 2-4 table spoon of lime juice. Boil the mixture. Add 40-50 g pectin, stir the contents constantly. Bring to a rolling boil and boil hard for about half a minute. Remove from heat, skim and pour hot in jars and seal.

Pomegranate wine

Press whole fruit without crushing. Add sugar in the juice to make up 22-23^0 Brix TSS. Add potassium meta-bisulphite. Ferment the juice by adding wine yeast starter. The wine contains 11.9% alcohol, 0.85% total acidity, 0.07% volatile acidity and 11.5 ^{0}Brix TSS.

Anardana

Dehydrate arils of sour type pomegranate in a solar drier or a cross flow drier. The dried product has 7.8 –15.4% acidity. It is used to garnish different dishes.

Grenadine

Mix equal parts of pomegranate juice and sugar, allow standing for three days and then bringing to a boil and simmering for 3 minutes Strain the supernetant into sterilized jars and seal.

Pomegranate rind powder

Cut the rind into small pieces and dry at 60^0 C for 48 hrs. Grind the dried peel and sieve it with 60-mesh sieve and pack it. The rind powder yield is about 13 per cent. The powder is rich in carotene (12.52%), polyphenols (16.5%), minerals (5.35%) mainly potassium, phosphorous, magnesium, sodium, calcium, iron, zinc, etc.

Future Research Needs

To make the pomegranate production more profitable in the traditional areas and an alternative crop in the potential areas, research efforts needs to focused on following thrust areas:

1. Development of drought hardy varieties with quality fruits soft seed, attractive aril colour and high yield.
2. Development of varieties with longer storage life.
3. Emphasis should be made for identification and development of suitable varieties for cold arid region.
4. Development of varieties and sustainable management practices against the major pest and disease.
5. Development of varieties free from physiological disorder particularly the fruit cracking and aril blackening.
6. Development of plant ediotypes with high bearing capacity.
7. Acertaining the causes and development of management practices to control wilting of plants and internal break down of aril.

REFERENCES

Adsule, R.N, Kotecha, P.M and Kadam, S.S (1992) Preparation of wine from pomegranate. *Beverage and Food World*, 19 (4): 13-14.

Anonymous (2004) Frutis in National Security. Technical Bulletin-1B, Indian Institute of Horticultural Research, Bangalore. P : 1-41.

Anonymous (2003) Dalib Vrit (in Marathi), Maharashtra Pomegranate Growers Research Association, Pune.

Anonymous (2002) Biennial Report, AICRP on Arid Zone Fruits. CIAH, Bikaner

Anonymous (2001) Agro Export Statistics. APEDA. New Delhi.

Anonymous (2001) Statistical Report. National Horticulture Board (1999-2000), Ministry of Agriculture, Govt of India, Gurgaon (Haryana).

.Bose, T.K. and Mitra, S.K. (1984) Fruits in India, Naya Prakash, Kolkata.

Chadha, K.L. and Pareek, O.P. (1993) Advances in Horticulture, Vol. I-IV. M.P.H., New Delhi.

Khodade, M.S., Wahval, K.N. and Kale P.N. (1990) Physico-chemical changes during growth and development of pomegranate fruits. *Indian Journal of Horticulture* 47: 21-27.

Pareek, O.P. and Vishal Nath (1996) Pomegranate. In : Coordinated fruit research in Indian Arid zone - A two decades profile. NRCAH, Bikaner, pp.105.

Pareek, O.P., Vishal Nath and Vashishtha, B.B. (2002) Cultivation of arid and desert fruits. Krishi Mantralaya, Govt of India, New Delhi.

Pareek, O.P., Vashishtha, B.B. Vishal Nath and Singh R.S. (2000) *Shuska keshtra mein- Anar utpadan ki unnat Taknik*, Extension Bulletin-2, NRCAH, Bikaner, p16.

Pareek, O. P. and Vishal Nath (2006) Arid and minor fruits. In: Text book of Horticulture (Ed. K. L. Chadha) MPH, New Delhi.

Prasad, R N., Bankar G.J. and Vashishtha B.B. (1996) Problems and prospects of pomegranate cultivation in arid regions. Sym. on Recent Advances on management of Arid Ecosystem held at CAZRI, Jodhpur.

Sharma, B. D., Singh R. S., Vishal Nath and Pareek O. P. (1999) Effect of pit size and filling mixture on plant establishment of pomegranate (in hindi). Abstract: National seminar on Horticultural development in India in 21st centuary 14-15 February, 1999. Janta Collage, Bakewar, Etawah (UP).

Singh, R.S. (1994) Effect of growth substances on rooting of pomegranate cuttings. *Current Agriculture* 18:87-89.

Singh, R.S., Vishwa nath and Kumar A. (1989) *Sehat aur amdani ke liye Anar. Phal Phool.*12 (3):3-4.

Vishal Nath and Singh R. S. (2000) Pomegranate: For Health and Wealth. *Intensive Agriculture.* 38 (3-4): 10-15.

Vishal Nath and Singh R.S. (1998) *Shushka kshetra main anar ki paidawar. Mati Mudir.* 1 (3): 11-16.

—x—

4

AONLA

Aonla (*Emblica officinalis* Gaertn.) is an ancient, indigenous fruit of India. It belongs to the family *Euphorbiaceae*. It is a hardy tree and can be successfully grown in variable agro-climatic and soil conditions. Aonla or Indian gooseberry is cultivated in India since time immemorial and has become the important fruit crop of waste lands particularly in the arid and semi-arid regions of the country. By nature, the plants of aonla are bestowed with some specific characters like intensive and deeper root system, winter dormancy (under north Indian condition) in plants and summer dormancy in fertilized zygote, reduced leaf area, coincidence of fruit growth and development with moisture availability period (monsoon period) and selective absorption of cations and anions, which make it suitable to grow in varied agro-climatic and soil conditions of the fragile agro-ecosystem. At present, aonla is being cultivated in an area of about 50,000 ha with an annual production of 1.75 lakh tones in the country. The high production potential with low maintenance cost and enormous therapeutic and nutritive value of the crop has recently attracted the growers for aonla cultivation. The rainfed conditions with low and erratic rainfall pattern prevailing in these regions, however restricts the growth and development of plant and production from the orchards. Under severe drought conditions, the plants are prone to water stress. In spite of no assurance of irrigation in these regions, the moisture conservation technique is generally not being followed in aonla orchard, which aggravates the problem many folds. Soil moisture and soil temperature are the two most significant factors, which influence the evapo- transpiration, growth, development and biological activity of the plant. In arid and semi-arid areas, thermal stress and erratic rainfall create moisture stress in the plant particularly during summer.

Aonla is regarded as sacred tree and has great mythological significance. It finds mention in Vedas, Ramayan, Charak Samhita, Sushrut Samhita, Kadambari and other ancient literature. According to Hindu mythology, one day meal is arranged beneath a aonla tree during Kartik (October) month when trees are laden with fruits. Aonla is a nutritious fruit proving the proverb, "***take an aonla everyday keeps the doctor away***". It contains 500-1500 mg vitamin C, 0.2 mg nicotinic acid, 1.2 mg iron, 0.5g protein, 0.1g fat, 0.7g minerals, 2.0 - 3.4g fibre, 14-21.8 g carbohydrates, 0.02g phosphorous and 0.05g calcium in 100g pulp. Fruits are commercially used for preparation of *Chayanprash* and *triphala*, the well known Ayurvedic medicine. Being a rich source of vitamin C, *aonla* is helpful in curing scurvy, problems of teeth, gums, eye and stomach. Fresh fruits are also used for the preparation of products such as preserve (*murabha*), pickle, *chutney*, shreds, etc.

The ascorbic acid and other constituents are well retained in dried aonla fruits. The stability of ascorbic acid in dried aonla fruits and their astringency could be due to presence of poly-phenols in the fruits. Gallic acid present in the fruit has marked antioxidant property. Aonla fruit is highly valued among indigenous medicines. Dry aonla powder is said to be superior to synthetic vitamin C in treating diarrhoea, dysentery, jaundice, anemia, bronchitis and cough. The fruits are also processed into a number of products.

Origin and Taxonomy

Aonla (*Emblica officinalis* Gaertn), an important member of family Euphorbiaceae is native of Tropical South-Eastern Asia particularly Central and Southern India (Morton, 1960). It is also reported to be the native of Ceylon, Malaysia and China where it's considerable variability exits. Seedling plants of aonla are commonly observed in Vindhyan hills, North eastern hill regions and also in the north western Himalayan region up to 1350 m MSL. Genus *Emblica* comprises about 350-500 species (Balley, 1917), mostly shrubs, few herbs and trees. Chromosome numbers in aonla is X=7, 2n=28 (Perry, 1943) and Polyploidy with 2n=98 to 104 (Ammal and Bhagwan, 1957). Aonla is much branched tree and ranges 9-12 m in height. Seedling tree sometimes grow beyond 20 meters. The tree is observed to be deciduous under north Indian conditions. However, it is considered as evergreen in the tropic. Aonla tree bears two types of shoots. On the basis of growth characteristics, these have been categorized as long or indeterminate and short or determinate (Bajpai, 1965 and Ram, 1971). The indeterminate shoots are longer and continue to put new growth in the season. These shoots neither fall from the tree nor bear flowers irrespective of period of their emergence. On the other hands, determinate shoots appear on the nodes of indeterminate shoots and their number at each node may vary from 3-5 in different cultivars. These determinate shoots bear small sized (10-13 mm length, 2-3mm width) leaves arranged so closely that apparently the shoot appears to be a pinnately compound leaf (Plate 4.1).

Plate 4.1. Sketch Showing Shoot, flowers and fruits of aonla

The internodes are much shorter in the determinate shoots. These shoots are barren or floriferous with embricate leaves. First few proximal nodes of the determinate shoots are barren (without leaves) which are reduced to dark brown scarious cataphyllus. Succeeding nodes are with green but reduced leaves. Subtending cymules of male flowers are followed by nodes each with cymule of one central female flower (rarely two) and several lateral males. Distal half floriferous determinate shoots are normally sterile with typical leaves (Plate 4.2).

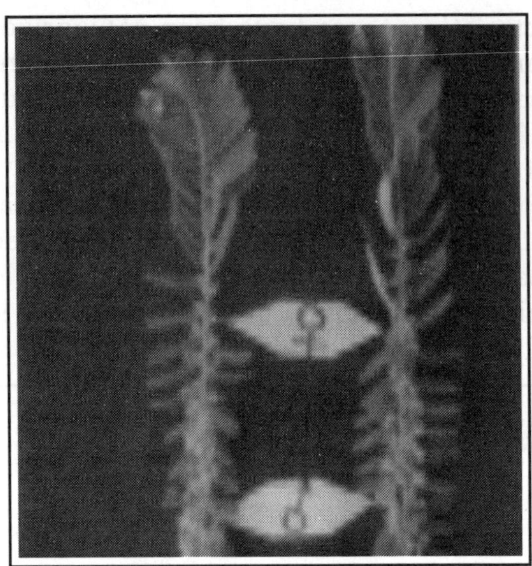

Plate 4.2. Determinate flowring shoot of aonla

Inflorescence in aonla is recemose type, flowers are minute, unisexual with short pedicel. Male flowers appear first in cluster, perianth 6, yellowish green to deep pink in colour with valvate aestivation. Androecium consists of three stamens, each profusely branched, filament attachment is basifixed or innate type, short and cohesion of anthers is syngenacious type. Female flowers have tiny green perianth and number of segments varies from 5 to 7 but commonly 6 are observed. Ovary hypogynous, Carpels 3-4, 3 chambered, placentation axile, two ovules per locule, margin straight to crecent shaped, ovarian chamber shallow to deep. Fruit is capsular (drupaceous), oval round to flattened round or triangular in shape with fleshly exocarp, colour whitish green to yellowish green, skin smooth to rough, semi translucent, segments 6-8, surface smooth to raised. The endocarp is rather thick and massive, eventually separating from a well developed collumella which is trigonous. Stone hard, round to triangular in shape; seed 6-8, light brown to deep brown in colour. Seed coat is hard, dark chestnut brown in colour with 2-3 locules containing 2 seeds in each locule.

Area and Distribution

The total area under aonla cultivation is more than 1 lakh hectare. The area under aonla plantation is increasing day by day. The plantations are mostly being taken on degraded land, wastelands and up lands in the watershed areas. In India, Uttar Pradesh is the major growing State of aonla where its commercial plantations are available especially in Pratapgarh, Allahabad, Sultanpur, Faizabad, Mirzapur, Kanpur, Etawah and many eastern districts. Aonla is also commercially cultivated in the States of Rajsthan, Maharashtra, Karnataka, Tamil Nadu, Gujrat, Haryana, Punjab and Himanchal Pradesh. It can be cultivated successfully in marginal soil and various kind of wasteland situations such as sodic and saline soil, ravines, dry and semi-dry regions including plateau area of Bundelkhand, Jharkhand, Orissa, Chhatishgarh and southern India. The plantations of aonla are also being taken up in the isoclimatic regions across the world particularly in Israel, African and Asian countries.

Soil and Climate

Aonla can be grown in a wide range of soils, however well drained, deep, fertile and sandy loam soil having good water holding capacity is considered to be the best. It can be successfully grown in heavy as well as light soil except very sandy soil. In sandy soils, aonla plants can be successfully grown with the provision of irrigation at frequent intervals (Pareek *et al.*, 1998). Aonla is also identified as one of the most potential fruit crop to be successfully grown in sodic soil/ wasteland, ravine lands and foothill of forest area. In saline and sodic wastelands, aonla plantation can be successful if some amendments are used. Calcarious soils are not suitable for its growth. Aonla can be successfully grown in hot arid climate also.

Aonla, being very hardy, survives in sodic soils but its growth and productivity are adversely affected. Seed germination is delayed and seedling growth is inhibited. The salts also cause interveinal chlorosis, leaf margin and tip burning, shoot dieback and in acute cases at about 30 ESP, even death of the seedlings. Application of 25kg FYM at the time of filling a pit of 1x1x1m size has been recommended as a general practice, however in highly sodic soils, amendments should be applied. Growth of aonla plants, in sodic soils having pH 10.3 to 10.7, ESP 69 to 73 and ECe 0.66 to 0.84 dsm^{-1} at Faizabad, was improved by application of amendments such as gypsum, pyrite, usar tod khad, press mud and pond soil. The best results were obtained by application of pond soil @ 120kg/pit. Since seedling plants have shown more tolerance to sodicity and salinity than budded plants, planting of six month old seedlings and then budding them *in situ* has been advised to establish aonla plantation in the problematic soils.

Effect of saline and sodic soils on survival of aonla plants and nutrient uptake by the plants under these conditions has been studied in some commercial cultivars of aonla. The studies have shown cultivar specificity in this regard. Cultivar Francis and Kanchan were more tolerant than Banarasi and Krishna because the former were able to take up more nutrients and less sodium under saline and sodic soil conditions (Table 4.1).

Table 4.1: Effect of salt stress on survival of plants and nutrient content (dry wt. basis) in leaves

Cultivar	Soil	Survival (%)	Nitrogen (%)	Phosphorus (%)	Potasium (%)	Calcium (%)	Magnesium (%)	Sodium (%)
Francis	Normal	100.0	2.63	0.158	0.900	1.366	0.479	0.162
	Saline	100.0	2.34	0.120	0.808	1.422	0.520	0.238
	Sodic	100.0	2.52	0.126	0.825	1.259	0.418	0.210
Kanchan	Normal	100.0	2.54	0.152	0.880	1.344	0.485	0.153
	Saline	100.0	2.47	0.123	0.838	1.403	0.495	0.195
	Sodic	100.0	2.40	0.126	0.820	1.240	0.416	0.233
Krishna	Normal	100.0	2.50	0.145	0.897	1.348	0.489	0.160
	Saline	87.5	1.88	0.108	0.780	1.445	0.534	0.248
	Sodic	87.5	2.00	0.110	0.751	1.054	0.410	0.278
Banarasi	Normal	100.0	2.45	0.149	0.865	1.357	0.480	0.159
	Saline	87.5	1.85	0.096	0.701	1.509	0.560	0.281
	Sodic	75.0	1.90	0.098	0.690	1.044	0.370	0.315
CD at 5%		21.47	0.03	0.008	0.031	0.040	0.012	0.010

Source: Pareek and Vishal Nath, 1996.

Aonla is a subtropical fruit but its cultivation in tropical climate is quite successful. The grownup tree is not much affected either by hot wind or frost. Trees shed their leaves and become dormant during winters to sustain cold weather. In young plantations, however, frost causes severe damage (Plate 4.3) and therefore, special cares such as thatching and light irrigation is advised in frost susceptible areas. The young plant should also be protected from hot wind during May-June at least up to the age of 3-4 years under North Indian conditions. The mature tree can some what tolerate freezing temperature as low as -2 to -3°C as well as high temperature as high as 46°C (Pathak *et al.*, 2003). The warm season appears conducive for the initiation of floral buds.

Plant Genetic Resource Management and Varieties

The genetic base in aonla is very narrow because of its commercial cultivation is restricted in some selected pockets. The seedling populations of aonla are commonly observed in Vindhyan Hills, North

Plate 4.3. Damage in aonla due to freezing temperature (-7.8ºC)
(i) View of 8 years old orchard (ii) Damaged young shoot

Eastern Hill regions and also in the North Himalayan region up to 1350 m MSL. Being indigenous fruit to India, great variability in *aonla* exists among the seedling populations particularly in Vindhayan and Arawali region. From existing genetic stock in the country, efforts have been made to collect promising genotypes of aonla at various research Institutes / Stations. Pareek and Vishal Nath, (1996) mentioned that although duplicate but 127 genotypes of aonla are being maintained in the country. First systematic work on genetic resource management in aonla was initiated during eighties at NDUAT, Faizabad where more than 20 variable types of aonla were collected from different variability regions. Other collections might have been made from these original collections with some ecotype variations at different research centres. Later, GAU, Gujarat collected some promising coloured types from Gujarat and Madhaya Pradesh. In Tamil Nadu and Jharkhand, collection of aonla germplasms were initiated during nineties. Rai *et al* (2001) reported 10 cultivars of aonla being maintained at HARP, Ranchi mainly from Faizabad and Pratapgarh of U P and Betala forests of Jharkhand. Some popular cultivars like Chakaiya, Banarasi and Francis are already under cultivation. Shy bearing in Banarasi and predominance of internal necrosis in Francis are serious demerits of these two cultivars and thus these are not recommended. Now, some promising selections have been made at NDUAT, Faziabad (NA-6, NA-7, NA-10), GAU, Anand (Anand-1 and Anand-2) and CISH, Lucknow (Laxmi-52) which are recommended for commercial cultivation. Growth, flowering and fruit characters of some of the aonla genotypes have been given in table 4.2 and 4.3 where as the detailed characteristics of some of the cultivars are given below:

Table 4.2 : Growth, flowering and fruit characters of some aonla genotypes

Genotypes	Growth habit	Inflorescence	Fruit shape	Fruit skin	Segments
Banarasi	Semi tall, spreading tree, deep red branchlets	Deep pink, medium in length	Triangular	Straw yellow, smooth	Paired and covex, prominent suture
Krishna	Semi tall, spreading tree, brownish red branchlets	Deep pink, short	Triangular	White green, smooth	Paired and leveled, prominent suture
NA 9	Tall, semi spreading tree, pinkish green branchlets	Deep pink, short	Flattened round	Light green, smooth	Unpaired and convex, prominent suture
Balwant	Semitall, semi spreading tree, brownish red	Deep pink, long	Flattened round	Yellow green with pink tinge, rough	Unpaired and leveled, suture less prominent
Francis	Tall, erect tree, yellow green drooping branchlets	Yellow green, medium in length	Flattened oblong	Light green, smooth	Paired and convex, suture less prominent
NA 7	Tall, semi spreading tree, greenish pink branchlets	Pink-green, long	Flattened oval	Yellow green, smooth	Paired and convex, suture less prominent
Chakaiya	Tall, upright tree, pinkish green branchlets	Deep pink, long	Flattened round	Light green, smooth	Paired and leveled, prominent suture
Kanchan	Tall, spreading tree, brownish red branchlets	Pink-green medium in length	Flattened oblong	Light green, smooth	Paired and convex, suture less prominent
NA 6	Tall, Semi spreading tree, pinkish green branchlets	Pink-green, medium in length	Oval round	Light green, smooth	Paired and leveled suture less prominent
NA 8	Tall, upright tree, light green branchlets	Pink-green, short	Flattened round	Light green, slightly rough	Unpaired and leveled, suture less prominent

Source: Pareek and Vishal Nath, 1996.

Table 4.3 : Growth, flowering and fruiting in ten year old plantation of aonla genotypes

Genotypes	Tree Height (m)	Tree Spread (m)	Trunk Girth (cm)	Female Flowers/ branchlet	Sex Ratio (f : m)	Fruit Size (LXD)	Fruit Wt. (g)	Yield/ tree (kg)	TSS (^{0}Brix)	Ascorbic acid (mg/ 100g)
Banarasi	3.21	4.03	37.2	2.3	1:254	3.6 x 4.4	42.0	19.0	12.5	428
Krishna	2.91	3.24	23.3	2.0	1:356	3.6 x 4.2	38.0	20.0	11.0	579
NA 9	2.90	2.51	30.8	2.8	1:241	3.8 x 4.5	50.0	39.8	12.0	500
Balwant	2.56	2.74	21.2	4.6	1:90	3.4 x 4.2	41.0	4.5	9.8	558
Francis	2.6	3.00	27.9	4.8	1:160	3.9 x 4.3	44.0	18.0	9.3	653
NA 7	3.06	3.01	23.9	9.7	1:61	3.8 x 4.0	35.3	60.0	7.7	482
Chakaiya	3.57	3.05	30.6	7.9	1:64	3.4 x 4.1	33.7	29.1	12.7	673
Kanchan	3.30	3.92	35.3	7.8	1:67	3.3 x 3.5	24.0	62.3	11.0	508
NA 6	3.17	3.65	31.3	10.8	1:29	3.6 x 4.3	34.1	49.5	10.0	439
NA 8	2.97	2.65	24.0	3.3	1:76	2.9 x 3.5	16.5	16.5	8.0	374

Source: Pareek and Vishal Nath, 1996.

Chakaiya

The tree has upright growth habit and bears profusely. This is a late maturing cultivar with small medium fruit size (30-35g/fruit). Flattened fruit apex, deep stem end cavity and levelled styler end are some typical characters of the fruit. Some fibre in the fruit pulp favours longer shelf life of the fruit. The fruit have strong attachment hence premature dropping is not a problem in this cultivar.

Krishna

This is a chance seedling of Banarasi cultivar which bears fruit moderately. The fruit is large (40-50g), conical at apex, skin smooth, yellow in colour with pink blush with less fibrous pulp. Fruits can be used for preparation of preserve, candy and juices.

Kanchan

This is a late maturing cultivar selected from Chakaiya seedlings. The tree has spreading growth habit and it bears profusely (Plate 4.4). Fruit are small size (30.2g), flattened shape and yellowish green in colour. Owing to more number of female flowers, it produces higher fruit yield. Fruits are suitable for making pickle.

Plate 4.4. Aonla cultivar Kanchan developed at NDUAT, Faizabad (i) A fruiting twig (ii) Close view of the fruits

Narendra Aonla 6

This is another selection from Chakaiya. Tree has spreading growth habit and bears heavy fruit. Fruit is small to medium in size and oval round in shape (Plate 4.5) and less fibrous and semi-hard flesh suitable for making preserve, candy and jam. The genotype has bright future for commercial cultivation.

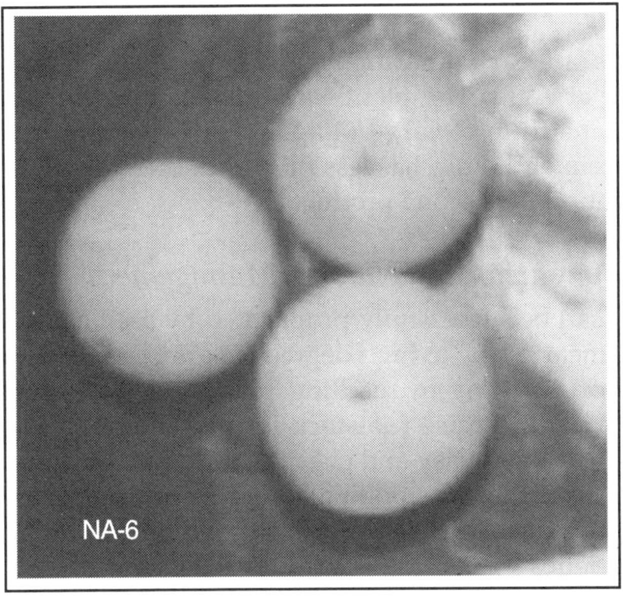

Plate 4.5. Aonla cv. NA-6

Narendra Aonla 7

This is a mid season, necrosis-free cultivar selected from seedling population of Francis. The plant is dwarf and has heavy bearing capacity. The fruits are large (40-50g/fruit), round, smooth with light yellow pulp (Plate 4.6) generally free from fibre. The cultivar is suitable for making preserve and other processed products.

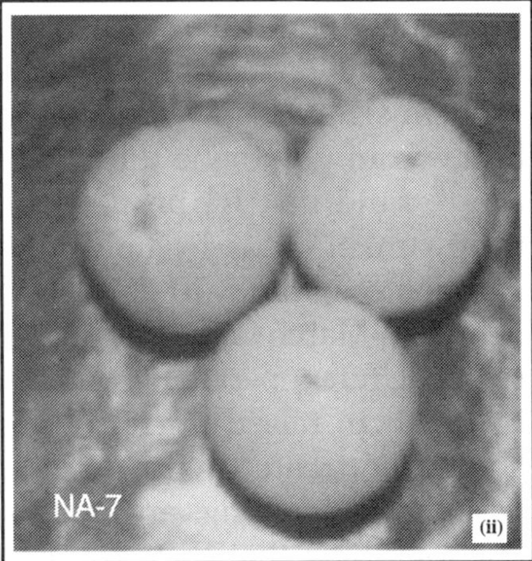

Plate 4.6. Aonla cultivar NA-7 developed as chance seedling of Francis (i) Plant in profuse bearing (ii) Close view of mature fruit

Balwant Bold

This is an early maturing cultivar selected from Banarasi population at Agra. The plants are tolerant to sodic soil. Fruits are attractive, medium to large (40-45g/fruit) with round shape and flattened stem end. Pulp has less fibre and, therefore, it is suitable for the preparation of various products.

Plant Propagation and Nursery Management

Aonla can be successfully propagated by seed as well as through vegetative means. Due to high degree of variability and long gestation period before fruiting in seedling plants, seed propagation is not advisable except to raise rootstock for vegetative propagation. Seed propagation, though easy and cheaper, it does not ensure true to type plant because of the cross pollinated nature of the species. In order to overcome the disadvantage of seed propagation, multiplication of suitable varieties by vegetative method is desirable.

For seed propagation or for raising rootstock for vegetative propagation, seeds should be extracted from mature fruits by drying in sun. After breaking the stone, seed kernels are obtained. For seed propagation however, embryo less seeds which float on water must be discarded. Pareek and Vishal Nath (1997) mentioned that aonla does not have seed dormancy and freshly collected water sinked seeds gave nearly 100 per cent germination. Dipping of seeds in 50ppm GA_3 for 24 hours improves seed germination. Newly emerging seedlings in the seed bed should be shifted within one month to the main nursery bed or polythene tube or bag. Pareek and Vishal Nath (1997) mentioned that about 50 per cent seedlings shifted in polythene bags or tubes (30 x 12 cm size) attain buddable size within 4-5 months. Seeds can also be sown in polythene tubes (40x15cm) filled with mixture of FYM + clay + sand in 1:1:1 proportion. These tubes are arranged in 40 cm deep sunken beds by holding them in upright position. The seeds germinate within 8-10 days. Aonla seeds remain viable for 3-4 years. February-March is the best time for sowing seed in nursery in the arid region. Under open field condition, sowing is done during July-August in 15-20 cm raised nursery beds of 3x1m size. Before sowing, seeds should be treated with fungicide (Bavistine @2g/kg seed). Newly germinated seedlings are shifted to flat beds at 20x20cm distance after one month so that they will be ready for budding during next July. Spray of 0.3 per cent Diathane M-45 has been found to protect the plants from diseases in the nursery. At the time of budding, the rootstocks should be of pencil thickness. Saroj *et al.*, (2000) reported that directly sowing of seeds into perforated poly container of 40 x 15 cm size gave higher percentage of buddable stock with more vigour and budding success (patch budding) than poly tubes of 25 x 10cm size. The technique saves 6 month period with higher (more than 90%) success.

Aonla can be commercially propagated through budding and soft wood grafting. Seedlings of pencil thickness are used for budding in aonla. Patch and modified ring budding are the most convenient and successful methods for aonla propagation. Pathak *et al.*, (1991) reported that patch budding is the most successful method of aonla propagation on one year old rootstock. July-September found to be the best period for budding under North Indian condition. In another study more than 80% success were achieved during May to August (Pathak *et al*, 1999) with highest success during July. Table 4.4 explains the success in aonla at two locations with different techniques of budding (Pareek and Vishal Nath, 1996).

Table 4.4 : Per cent success of budding in aonla at different locations

Method	**Time**	**Success (%)**	
		Rahuri	**Faizabad**
Patch budding	Mid May	-	93
	Mid June	85.0	93
	Mid July	91.6	100
	Mid August	81.6	93
T-budding	Mid May	-	73
	Mid June	80.0	53
	Mid July	81.6	53
	Mid August	68.3	33

Budding period can be extended up to September-October in north India but in that case only bud union takes place and the sprouting commence during February-March. In arid regions / areas with high sodicity and salinity, *in situ* budding by planting 6 months old seedling rootstock directly in the field have been recommended (Pareek, 1997 and Pareek *et al.*, 1998), Vishal Nath *et al.*, (2002) reported 80-90% success in aonla through *in situ* patch budding during June-September which further reduced during October. During February-March, the success rate was only 60-70%. Old seedling trees can also be converted into improved cultivars by top working. Recently, soft wood grafting through cleft method has been found most promising with more than 90% success on 8-10 month old seedling rootstock.

Planting

Appropriate planting technique plays an important role in successful establishment of aonla plantation. It is generally planted in square system of planting at a spacing of 8x8m. Pits of 1x1x1m size are prepared during May-June which is filled 15 days after digging with FYM and top soil mixture in 1:1 ratio. To protect the plants from termite, 50g methyl parathion dust (5%) is mixed in each pit. Plantation of aonla saplings is done in the centre of pit in July-August in arid region. However, with assured irrigation and protection facilities, plants can also be planted during February and October. Aonla has self incompatibility and therefore, ten per cent plants (15/ha) should be planted as pollinizers to obtain good fruiting. NA-7 is a good pollinizer for NA-6. In any case, plants of more than two cultivars in an orchard would solve the problem. Plantation of seedling trees on the borders serves as wind break and also provides seed stock for raising rootstock plants.

Provision of wind break has been considered essential for aonla plantation in arid region (Pareek *et al.*, 1998). For shelter belt, plantation of 2-3 staggered rows of fast growing tall tree species should be done. *Sheesham, neem, gonda, ardu,* etc. are good species for this purpose in arid region. Since these species would take time to make full shape, temporary shelter belts should be provided by planting fast growing bushy plant such as castor , *dhaincha,* etc. around the aonla plantations (Pareek *et al.,* 1998).

Training and Pruning

Training is very essential at initial stage to develop better frame work in aonla plant. It is to be trained to modified leader system. The frame work to be developed by encouraging the growth of 4-6 well spaced branches on the trunk at a height of 75-90 cm above the ground level. Aonla does not require regular pruning. Unwanted, criss-cross, diseased and dried branches are to be removed regularly. In a young plantation, staking of the plant may be useful in extremely windy location. Rejuvenation pruning to restore the canopy and enhance fruiting in old and senile aonla orchard has also been found successful at CISH, Lucknow and HARP, Ranchi.

Orchard Management

Orchard management is very much crucial for sustainable production in aonla. In current scenario, judicious use of nutrient and water and appropriate management of aonla orchard not only enhances the tree yield but also improves the orchard efficiency and thereby farm income.

Nutrient Management

For judicious use of manures and fertilizers and balance nutrition to aonla plants, nutritional diagnosis is of great importance. The aonla tree bears two types of shoots, ie. Short or determinate and Long or indeterminate. The indeterminate shoots continue to grow through the season. These shoots, irrespective of their emergence, neither are caduceus nor floriferous. The determinate shoots having short internodes emerge from the nodes of these shoots in clusters of 3-5. For the diagnosis of nutrient status in the plant, sampling should be done from the middle portion of indeterminate shoots. The status of major essential nutrients has been found to vary with age of shoots particularly in the newly emerging during April and become more or less stabilized when the shoots are 3 months old during July. Only these shoots should be sampled for nutritional diagnosis.

The characteristic deficiency symptoms of major nutrients have been developed in Chakaiya cultivar of aonla in sand culture at Faizabad. The characteristic symptom of nitrogen was general yellowing of leaves and retarded growth. In initial stages, tip and basal portions remained green, the middle portion turning yellow with green dots. The characteristic symptom of phosphorus deficiency was tip and marginal yellowing of leaves with middle portion remaining green even in acute cases. In initial stages, tip of leaf started yellowing and progressed towards the margin. The apical portion of P-deficient plant showed bushy appearance. The main symptom of potassium deficiency in aonla was yellowing and scorching of leaf tips along with slight portion of margins and remaining portions appearing dark green. The deficient plants were stunted. Under calcium deficiency plant showed yellowing symptom followed by burning and scorching of leaf tips which in severe cases extended towards the middle of leaves. The younger leaves were affected first in calcium deficient plants which were entirely different from NPK deficient plants. Owing to abnormal apical growth and increase in number of small branchlets, the plant showed malformed and bushy appearance under calcium deficiency.

Judicious application of manures and fertilizers increases fruit production in aonla. The quantity of manures and fertilizers depends on the fertility of the soil. These should be decided after proper soil and tissue analysis. Based on the leaf sample status in the middle portion of 3-4 months old intermediate shoots, the doses of nutrients at a particular place should be decided. Under normal soil conditions of arid region, application of 10 kg FYM, 100g N, 50 g P_2O_5 and 75-100 g K_2O per plant per year has been considered sufficient (Pareek *et al.*, 1998 and Singh *et al.*, 1993). The doses should be increased each year up to 10 years and then stabilized. Thus, 10 years or more plantations should received 100 kg FYM, 1 kg nitrogen, 0.5 kg phosphorus and 0.75 kg potassium. The best time for manure and fertilizer application is January-February and June-July when the plants are in floriferous and fruit development stage, respectively. Full dose of FYM and phosphorus and half nitrogen and potash is applied during spring season while the remaining half dose is applied during the rainy season (Singh *et al.*, 1993).

Deficiency of micronutrients causes poor growth in aonla. Plants having 0.85%N, 0.07%P, 0.21%K, 0.44% Ca and 0.21% Mg in leaves show deficiency symptom of respective nutrient. Singh *et al.*, (1993) described the nutrient deficiency symptom in aonla cv. Chakaiya and suggested that timely application of balance nutrient should be ensured for optimum yield. Application of 250-500g zinc sulphate/plant in sandy soils of arid region is found beneficial. Three foliar sprays with 0.6 per cent borax at 10-15 days interval have been found to eliminate the problem of necrosis (Singh *et al.*, 1993).

Water Management

Moisture stress affects growth and development of aonla tree during vegetative and reproductive phase and ultimately the yield of plant. Generally, plants experience thermal stress during sunny period under arid and semi-arid ecosystem, which leads to rapid drying of soil, which adversely affect the plant growth. Moisture conservation techniques such as mulching have been found to increase the availability of water and its retention in soil by decreasing the losses through evaporation and alleviation from the harmful effect of moisture stress in arid and semi arid regions (Singh *et al.*, 2006, Pareek *et al.,* 1998). In addition, mulches also impart manifold beneficial effects like suppression of extreme fluctuation of soil temperature, maintenance of soil fertility and suppression of weed growth, which ultimately leads to better yield (Singh *et al.,* 2006). The conservation of soil moisture by application of various mulches in aonla showed beneficial effects on growth and development of fruits under semi-arid ecosystem. Synthetic as well as organic mulches have been found beneficial in aonla orchard (Pathak *et al.,* 2003, Singh *et al.,* 2006). Pathak *et al.,* (2003) reported that paddy straw, banana leaves, sugarcane thrash and FYM have shown promising result in aonla orchard. Mulching with these organic wastes over a number of year have also been found helpful in improving the organic matter contents of soil, infiltration rate and restricting the upward movement of soluble salts. Organic mulches also encourage the increase in earthworm population and other beneficial microbes in the rhizosphere under dry land condition. Mulching with paddy straw, sugarcane trash and FYM over a number of years have been found to be helpful in improving the organic matter content, moisture holding capacity and physico-chemical properties of the orchard soil (Pathak *et al.,* 2003, Singh *et al.,* 2006).

Frequent irrigations at short intervals (3-4 days) for about one month are essential to ensure initial establishment of aonla in light soils of arid region. Afterwards, irrigation interval can be increased to 25-30 days. Flowering, fertilization and development of embryo in aonla takes place during the spring (March -April) and, therefore, for proper development of embryo, in sandy soils of arid lands, light irrigation during March-July at 10-15 days interval is most beneficial (Pareek *et al.,* 1998). Drip irrigation on 20% wetted area basis on alternate days (Shukla *et al.,* 2000) has been found to improve growth and yield of aonla. Ojha *et al.,* (2000) reported that drip irrigation at frequent intervals showed better growth in aonla as compared to prolonged intervals. In a normal monsoon year, irrigation should be avoided during September-October. During the winters, light irrigation particularly before the suspected frost occurrence is essential to save the young plants from damage (Pareek *et al.,* 1998) however in grown up orchard winter irrigation should be need based as it may inhance the uptake of nitrogen from the soil and put the plant in growth and partial flowering. As a matter of fact, the aonla trees should go under dormancy during winter to have better flowering and fruiting in the main season particularly under north Indian conditions.

Orchard Floor Management

Vegetatively propagated aonla plant starts fruit production after 3-4 years of planting and the commercial yield is achieved after 8-10 years. During the initial years, interspaces between rows and plants can be utilized for growing intercrops which not only provide additional income but also improve the physico-chemical properties of the soil and check weed population. Drought hardy leguminous, solanaceous and cucurbitaceous vegetables have proved to be the best intercrops in rainfed or partially irrigated aonla orchards. Vegetable *guar* (Durga Bahar) produces more than 50 q/ha green pods when grown in the interspaces of aonla trees. Chillies, brinjal and some leafy vegetables can be grown successfully in aonla orchards. Pathak *et al.,* (2003) recommended aonla based cropping system models for different conditions. Aonla + Ber / Guava (Two tier), Aonla + Ber + Phalsa (Three tier) models have been suggested by Singh *et al.,* (1993) for sodic waste land conditions. Vishal Nath *et al.,* (2006) developed aonla based multistoried cropping system models for water sheds of eastern plateau. Aonla as main crop at 10m spacing in square system with guava as filler plant at 5m within and between the aonla rows along with various intercrops (upland paddy, rangi, ground nut, black gram and deenanath grass) at lower storey proves more remunerative than sole cropping. The system improved the soil organic matter content and pH also.

Weeds compete for water and nutrients and, therefore, timely weeding of tree basin is essential during initial stage of orchard establishment. Hoeing of the basin and interculture operation around the tree trunk also breaks the soil crust and thereby ensure good root development. Mulching with black polythene (100 gauge) or paddy straw is beneficial to minimize the water loss and weed population.

Disease Management

Aonla is comparatively free from diseases, however some foliar and post harvest diseases and physiological disorder have been found to cause economical loss to aonla plants and fruits. Notable diseases are aonla rust or ring rust and sooty mould on trees and blue mould on stored fruits.

Aonla Rust or Ring rust

Aonla rust or ring rust has been found to cause considerable losses in the humid regions of the country. It is a serious disease of aonla infecting the leaves and fruits. The disease is reported to be caused by fungus *Ravenelia emblicae* var. *fructoidae*. Pareek and Vishal Nath (1996) mentioned that the incidence of ring rust is uncommon in Pratapgarh orchards (dry climate) whereas it is a serious problem in Udaipur and Jobner in Rajasthan (humid climate) where it cause great losses to the farmers. The disease breaks out from July and continues up to March with maximum incidence between August to December in orchards near Jaipur. The time of start of the disease and its rate of spread depends on extent of rainfall during the monsoon period. The incidence of the disease reached a peak in December-January when the maximum temperature ranged between 25.4^0 and 34.5^0C, minimum temperature between 5.2 and 24.6^0C, morning RH between 70 and 84% and evening RH between 26 and 53%. At Aruppukottai in Tamil Nadu, the disease occurred during July-February when the temperature decreased and rainfall incidences were maximum. Studies conducted at Udaipur during 1977-80 revealed that as much as 84 to 88 per cent disease can be controlled by 3 sprays of 0.1 per cent triademafon or benzimidazole at 15 days interval. The PDI was nearly 77 per cent in the control plant. A study on farmer' fields at Chomu (Jaipur) revealed that 4 sprays of 0.2 per cent chlorothalonil (Kavach) at 15 days interval starting just after initiation of symptoms gave as much as 75.24 per cent disease control resulting in increased fruit yield (Table 4.5). The disease can also be effectively controlled by 3-4 sprays of Blitox-50 or Diathane M-45 at 15 days interval.

Table 4.5: Effect of foliar sprays of different fungicides on aonla rust at Jobner

Treatment	Per cent disease control	Fruit yield (kg/tree)
Chlorothalonil 0.2%	75.24	87.47
Copper oxychloride 0.3%	34.95	76.00
Bioloxazol 0.1%	43.07	68.50
Mancozeb 0.3%	59.82	81.00
Wettable sulphur 0.2%	47.40	72.67
Thiophenate methyl 0.1%	47.42	77.83
Tridemafon 0.1%	69.78	83.50
Triadeomorph 0.1%	39.88	69.33
Control	-	67.20
CD at 5%	1.359	3.63

Source: Pareek and Vishal Nath, 1996

Physiological disorder

A physiological disorder known as necrosis has been observed in aonla fruits in Uttar Pradesh. The aonla variety Francis is prone to this disorder where nearly 60% fruits were found to be affected from the necrosis. The other varieties were comparatively free from the disorder (Singh *et al.,* 1993). Disorder symptom starts with browning of mesocarp which extend towards the epicarp resulting into brownish black area on the fruit surface in October. This disorder associated with boron deficiency. It can be controlled with three spray of borax at 0.6 percent at fortnightly interval, commencing from early September. Combined sprays of Zinc sulphate (0.4%) + copper sulphate (0.4%) and Borax (0.4-0.6%) during September-October have been suggested by Pathak *et al.,* (2003) against aonla necrosis.

Off season flowering

Off season flowering in aonla is now becoming a matter of serious concern for aonla growers in North and Eastern Inida. Rai *et al.,* (2002) reported that off-season flowering in aonla is cultivar specified. During the survey in the aonla growing areas of Uttar Pradesh and there in spot discussion with aonla growers and Development Officials in Pratapgarh, it was found that cultivar Chakaiya and NA-7 have high intensity of off season flowering whereas Kanchan and Krishna along with some local genotypes have less intensity. The problem causes 30-40% loss in yield due to disturbance of the tree physiology and commencement of more male flowers. Further, it was observed that prevailing low temperature during off-season flowering (December-February) do not allow the female flowers (although very few) to set or even to develop the fruits. The disorder seems to exhaust the tree before the main flowering season. The main reason of off-season flowering would be due to interruption in tree physiology by externally governed internal mechanisms. The impact of global warming and green house effect in this regard cannot be ignored. Managing the field and plants in a manner so that plant can take rest during winter is the only possible way to overcome the off-season flowering. Inter-culture operation during winter should also be avoided in bearing aonla orchard.

Pest Management

For good growth and fruit production, aonla trees should be protected from pest attacks. The aonla tree is affected by bark eating caterpillar, shoot gall maker, aphids, mealy bug, scale insects and anar butterfly at different locations. The important pests and their management techniques are as follows:

Bark eating caterpillar (Inderbela sp.)

Insect feed on stem and branches and makes tunnel in the main trunk or branches. The pest can be controlled by injecting kerosene oil in the hole and plugging with mud.

Shoot Gall maker (Betousa stylophora)

The young caterpillar bore into shoots and reach the pith during August-September. The damage tissue develops into gall like structure. Pruning of the affected parts and spraying of 0.05% monocrotophos during rainy season minimizes the shoot gall incidence.

Termite

It is a serious problem in light soils of arid region. Use of methyl parathion dust @ 50 g/plant at the time of pit filling followed by application of 5-10 ml/plant chlorpyriphos with irrigation water give effective control. Painting of the trunk up to 30cm with slaked lime + chlorpyiphos mixture checks the crawling of termites to the upper portions of the plant.

Flowering, Fruiting and Yield

In north India, aonla tree begins to shed its determinate shoots from February onwards, resulting in to the indeterminate shoots devoid of foliage by the end of February to mid March. The new determinate shoots on the previous season's growth begins to appear by the end of February and continue to appear till the first week of April. Blossom buds appear on the newly developed determinate shoots but shoots which emerge after mid April do not bear flower and remain vegetative. New flush in aonla comes after winter dormancy during March-April when vegetative flush combines with reproductive flush. The small pinkish green flowers are borne in the axils of leaves on the branchlet. Although the male and female flowers are borne on the same branchlet, the fruiting in aonla depends upon provision of sufficient pollinizers. Provision of 10% pollinizer has been found optimum in aonla. It has been observed that NA 7 is a good pollinizer for NA 6. Planting of two to three varieties in an orchard solves the problem of pollinizer. After pollination and fertilization in April, the newly developed embryo goes into dormancy during summer. With the commencement of rains during monsoon, development of fruit resumes which matures by November-December depending upon the cultivar. In south India, however, flowering takes place during June-July. Sometimes due to lack of fertilization, premature fruit drop occurs during September-October.

Aonla is a cross pollinated plant in which wind, honey bees and gravity play important role for effective pollination (Reddi and Bai, 1976). Common visiting hours of honey bees are late evening and mornings (Bajpai, 1957). Bajpai (1968) reported that there is no self incompatibility in aonla and cause of poor fruit set may be due to high percentage of male flowers. However, Pathak and Srivastava (1994) observed self incompatibility in initial studies in flowering and fruit set in Indian gosseberry. They also noted considerable variations with respect to sex ratio which varied from 28.7:1 to 355.5:1 in different cultivars. Number of female flowers varied from 2 to 8. NA-6 had the lowest sex ratio followed by NA-7, Kanchan, Chakaiya while NA-9, Banarasi, Krishna showed higher sex ratio. Pollen viability test with acetocarmin showed 82.3 to 96.4 per cent pollen viability, while only 16.7 to 36.9 per cent pollen germination was observed in the sucrose solution (Srivastava and Pathak, 1993).

Open pollination showed maximum fruit set and retention, followed by geitonogamy and minimum with bagging. Fertilization takes place within 36 hours after pollination. The zygote rests for 120-130 days and endosperm nucleus for 70-80 days after fertilization (Bajpai, 1968 and Ram, 1971). Growth studies have revealed that the fruit remains dormant while extension growth of the shoots continue at a rapid rate. With cessation of vegetative growth, initiation of fruit growth takes place. Because of resting condition of uninucleate zygote and endosperm, the fertilized ovary of aonla does not show any symptom of fruit growth from March to July. Fertilized ovaries (young fruitlet) remain encased in a cup like structure. Subsequent by after the break of the rest period, fruits start growing and continue to gain in size and weight. Stone weight also increased simultaneously as the fruit gain weight and size. Aonla genotypes viz., Chakaiya x Francis, NA 7 x Krishna, Banarasi x NA-6 and Kanchan x NA-6 showed better fruit retention and are being recommended for commercial exploitation (Pathak and Srivastava, 1994). However, there is need to have detailed information on this aspect.

Vegetatively propagated plant starts bearing after 2-3 years of planting while the seedling plant takes more than 8 years to start flowering and fruiting. In vegetatively propagated plants commercial yield is expected after 8-10 years which may be 150-200 kg fruit/tree depending upon the variety and orchard management. Cultivar NA-7 and Kanchan have been found to be the highest yielding varieties under north and eastern Indian conditions.

Harvesting and Post harvest Management

Maturity of fruit is indicated by the presence of shining green colour with reduced acridity and increased vitamin C content. Harvesting of fruits at appropriate maturity has great impact on quality, storage life and marketability (Singh *et al.*, 1993). Therefore, maturity indices for fruits have been suggested. Singh *et al.*, (1993) mentioned that Banarasi, Krishna and NA-10 are the early maturing varieties where as NA-7, NA-6, NA-9 are the mid season and Chakaiya and Kanchan are the late maturity varieties. Number of factors have been found associated with maturity of fruits. Important among are, the genetical, climatic and production factors. Harvest indices of aonla fruits under Faizabad condition (Table 4.6) has been suggested by Singh *et al.*, (1993). Harvesting in aonla is done through traditional techniques like individual fruit picking or shaking of branches followed by collection of fallen fruits. At any cost damage to fruits should be avoided during the harvesting. Fallen fruits are collected and packed into gunny bags. The care should be taken for its harvesting and post harvest management. Conveniently fruits are harvested in small bags by riding on the trees with the help of bamboo ladders.

Table 4.6: Harvest indices of aonla varieties

Criteria	Varieties				
	Chakaiya	**Banarasi**	**Krishna**	**Kanchan**	**Francis**
Surface colour	Whitish green	Whitish yellow	Straw yellow*	Yellowish green	Greenish white
Av. Fruit weight	32.0	47.0	42.0	27.0	45.0
Pulp/stone ratio	22.3	20.9	17.6	12.7	15.7
Specific gravity	1.07	1.08	7.10	7.10	1.08
TSS/acid ratio	5.5	5.7	7.1	5.1	6.0
Weeks from fruit set to maturity	23	17	18	20	20
Time	January	October end	November	December	Mid Nov.-Dec.

*with red tinged

Grading is another important operation in aonla fruits. It implies sorting of fruits into different groups based on their shape, size and defects. Singh *et al.*, (1993) described 'A', 'B' and 'C' grades in aonla fruits (Table 4.7). After harvest, fruits are graded into 'A' 'B' and 'C' grades. 'A" grade fruits are used for preparation of preserve (*murraba*), candy and pickle while 'B' grade fruits are used for the preparation of jam, syrup, squash, etc. 'C' grade fruits are small in size and uneven in shape and are used for making products like shreads and dried segments. Prematurely dropped fruits are also used for the preparation of mouth freshening and digestive products in combination with *harar, bahera,* asophoetida and spices.

Table 4.7: Description of Grades in aonla fruits

Grade	Description
A	Large sized fruit according to the variety. Banarasi-diameter 4.5 cm and above Chakaiya-diameter 4 cm and above, free from blemishes
B	Small sized fruit having diameter less than 4 cm and free from blemishes
C	Defective fruits i.e. blemished scaring and necrotic fruits

Stage of maturity at which fruits are harvested has greatest effect on quality, storage life and marketability. Fruit should be harvested at such a stage that it should not spoil between harvesting and till it reaches to consumer places. Apart from production of high quality aonla fruits, its ultimate delivery to consumer is utmost important. Therefore, proper backing and transportation of graded fruits should be ensured. Proper packaging of fresh fruits reduces the wastage during transport. The commercial basket and gunny bag have poor dimensional stability as well as stacking strength, yet they are the prime packaging containers for aonla fruits. .Baskets of pigeon pea stem and gunny bag of 40-50 kg capacity with paper liners are used for packing of aonla fruits.

Large sized sound fruits are mainly used for preserve, candy and pickles while small fruits are used for processing into Chavanprash, pulp, beverage and damaged or blemished fruits are used for drying and powder making. Packing of fruits according to grade is done for efficient marketing. Singh *et al.*, (1993) reported 17.2% loss in aonla fruits due to poor handling. They mentioned that although conventional baskets and gunny bags have poor dimensional stability and stacking strength, yet used as prime packaging container. Singh *et al.*, (1993) suggested that wooden crates with polythene liner are the most suitable packing material for long distance transportation of aonla fruits.

Storage of fruits for extended availability and minimization of market glut is very much important in aonla. Efforts have been made to extend the shelf life of fruits with different storage techniques (Vishal Nath, 1990; Kumar and Vishal Nath, 1992; Kumar *et al.*, 1992; Vishal Nath *et al.*, 1992a, b and c and Kumar and Vishal Nath, 1993). Fruits of cultivar Chakaiya and Francis can be stored up to 9 days where as Banarasi can be stored for 6 days at ambient temperature (Kumar and Vishal Nath, 1992 and Vishal Nath *et al.*, 1992 a and b). Under Zero-energy-cool chamber however, fruits can be stored up to 12 days (Kumar and Vishal Nath, 1993). Singh *et al.*, (1993) reported that fruits of different aonla cultivars can be stored up to 18 days under Zero-energy cool chamber (Table 4.8). Kumar *et al.*, (1992) and

Table 4.8: Losses in aonla fruits during storage in zero energy cool chamber

Varieties	Days after storage								
	12			15			18		
	Wt. Loss (%)	Patho-logical (%)	Vit. C. Loss (%)	Wt. Loss (%)	Patho-logical (%)	Vit. C. Loss (%)	Wt. Loss (%)	Patho-logical (%)	Vit.C Loss (%)
Chakaiya	0.8	6.5	4.7	1.1	7.2	9.3	1.7	8.5	15.5
Kanchan	0.8	7.8	5.3	1.1	8.2	12.8	2.7	10.6	18.4
Francis	0.9	4.5	5.7	1.1	6.3	11.5	2.3	8.1	17.1
Banarasi	1.5	8.5	6.2	1.9	10.8	12.3	3.2	13.1	18.0
Krishna	0.9	8.1	10.0	1.2	11.0	18.8	2.8	12.8	27.7

Singh *et al.*, (1993) reported that fruits of cultivar Chakaiya can be stored up to 45 and 75 days in respectively, 10 and 15 per cent salt solution (Table 4.9) without decay loss. Vishal Nath *et al.*, (1992 c) reported that aonla fruits of cultivar Chakaiya can be stored for 9-10 days at ambient conditions with post harvest treatment of 4% borax and 1% calcium nitrate. It has been observed that physiological loss in weight, pathological loss and reduction in vitamin 'C' content are the major losses during storage. Weight loss in fruits are due to the loss of water from the fruit surface by respiration and transpiration of the living cells (Vishal Nath, 1990) whereas the pathological loss is mainly due to growth of *Penicillium islandicum* (Setty, 1958, Vishal Nath, 1990) and *Aspergillus niger* (Singh *et al.*, 1993). Loss of vitamin 'C' during the storage of fruits are due to oxidation of ascorbic acid into dehydro-ascrobic acid by the enzyme ascorbic acid oxidase in presence of light and heat (Vishal Nath, 1990, Pathak, 1988). Therefore, to enhance the shelf life of aonla fruits and fruits an integrated approach right from maturity standard harvesting to post harvest treatment and storage conditions should be taken into consideration.

Table 4.9: Physical appearance and pathological loss during storage of aonla fruits under salt solution

Days after storage	Salt solution (%)			
	10		15	
	Physical appearance	Pathological loss (%)	Physical appearance	Pathological loss (%)
0	Sound fruits with greenish colour		Sound fruits with greenish colour	-
15	Sound fruits with greenish colour	Nil	Sound fruits with greenish colour	Nil
30	Sound fruits with pale yellow colour	Nil with pale yellow	Sound fruits with pale yellow colour	Nil
45	Sound fruits with pale yellow colour	Nil	Sound fruits with pale yellow colour	Nil
60	Slightly bruished fruit with straw yellow colour	Nil	Sound fruits with pale yellow colour	Nil
75	Fruits become very soft and mashed and lossed its appearance		Sound fruit and Nil surface become slightly brown	

Processing and Value Addition

Aonla fruits are not consumed as fresh because of its astringent taste. The excellent nutritive and therapeutic value of fruits offer great potential for processing into various value added products (Vishal Nath,

1993; Bhagwan Deen, 1992; Singh *et al.*, 1993; Vishal Nath, 1996; Vishal Nath and Sharma, 1998, Pathak, 1988; Vishal Nath, 1999, Vishal Nath, 2004). Singh *et al.*, (1993) reported that NA-6 and NA-9 have low fibre content and phenol along with average amount of vitamin C and minerals and therefore suitable for processing into preserve, candy jam and pickle. Vishal Nath and Sharma (1998) reported that Chakaiya is suitable variety for Jam and pulp based products (nectar, squash, syrup) whereas Banarasi found suitable for candy and pickle. Due to retention of high vitamin 'C' content even after drying the aonla fruits, Gupta and Pareek (1993) suggested that aonla fruits can be sun dried after destoning and powdered. Sethi (1986) reported that blanching of fruits before drying retains appreciable quantity of ascorbic acid. Aonla fruits can be processed into ready-to-serve beverage, nectar, syrup, squash and various other value added products at small scale and at industrial level (Vishal Nath, 1996, 1999) and these products can be stored for appreciable period at room temperature with acceptable quality and nutritive value (Vishal Nath and Pandey, 1996). Techniques for preparation of various products have been standardized at NDUAT, Faizabad (Singh *et al.*, 1993). Details of some of the processed products from aonla has been described by Singh *et al.*, 1993; Vishal Nath, 1993; Vishal Nath and Sharma, 1998 ; Srivastava and Kumar, 1993.

Read-to-serve (RTS)

This can be prepared by taking 10% pulp and 12% total soluble solids with 0.3% acidity. Prepared beverage is bottled (200ml capacity) and crown corked. Then bottles are pasteurized in boiling water for 20 minutes, air cooled and stored for use. RTS blend can also be prepared by adding 2% ginger juice in aonla pulp. Flow sheet for preparation of ready-to-serve beverage is given in Fig. 4.1.

Squash

Squash can be prepared by taking 40-45% pulp and 50% total soluble solids with 1% acidity and 350 ppm SO_2. Prepared squash is bottled (750 ml capacity), capped and stored for use. Flow sheet for preparation of aonla squash is given in Fig. 4.2.

Syrup

The ideal recipe for preparation of syrup is 40-45 per cent pulp, 68% total soluble solids and 1.2 per cent acidity. Prepared syrup is filled into clean and sterilized bottles (750 ml) and kept for storage. The technique for preparation of syrup is similar to that of squash.

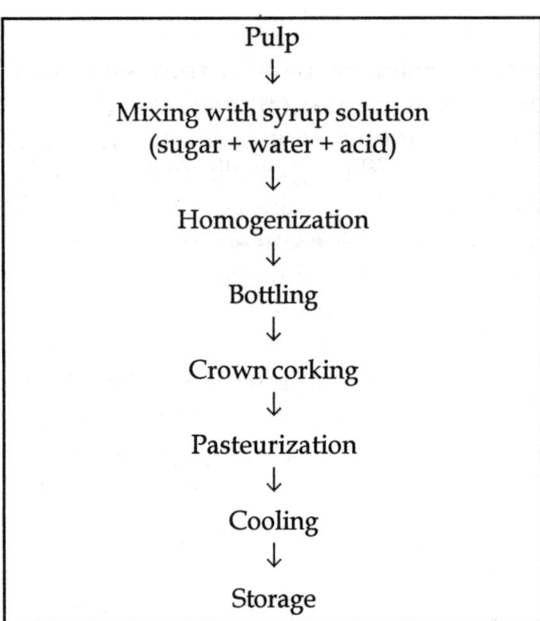

Fig. 4.1. Flow sheet for preparation of ready -to -serve

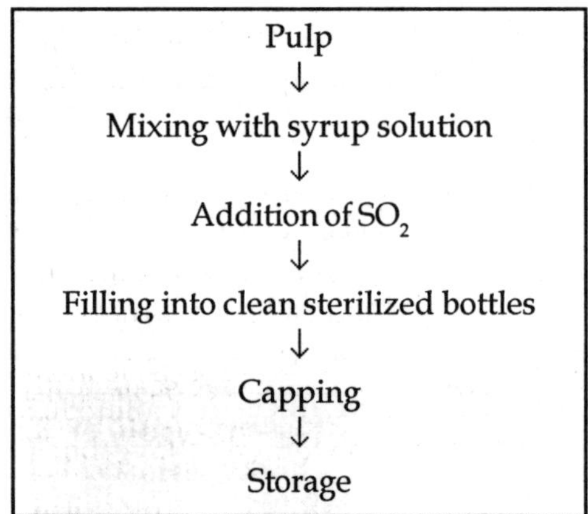

Fig. 4.2. Flow sheet for preparation of squash

Jam

Varieties which have low content of fibre is most suitable for jam making. Aonla fruit jam of the composition 45% pulp with 68% total soluble solids and 0.5% acidity produced a highly acceptable product. Fruit pulp are mixed with sugar and citric acid and cooked to desired consistency. Prepared jam is filled into clean and sterilized bottle (500 g capacity) and these are sealed and stored for use. The technique used for preparation of aonla jam is given in Fig. 4.3.

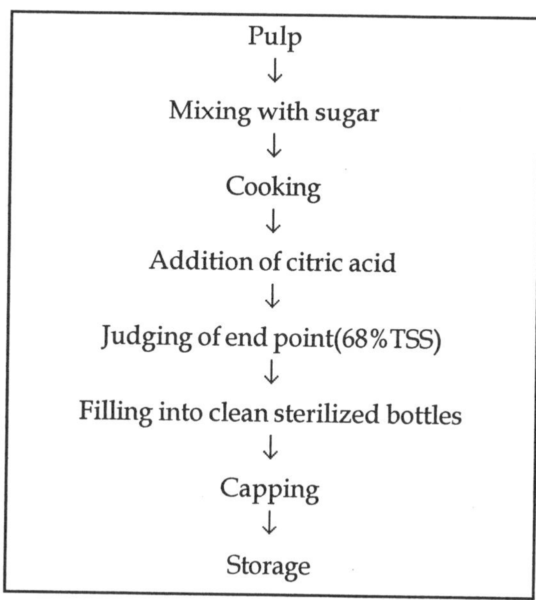

Fig. 4.3. Flow sheet for preparation of aonla jam

Shred

Aonla shred with 4 per cent dry salting has been found better. Product is very rich in vitamin C content even in dry form. The stability of ascorbic acid in dried product of aonla is assigned to the presence of leuco-anthocyanins or poly phenols. The technique used for preparation of shred is given in Fig. 4.4.

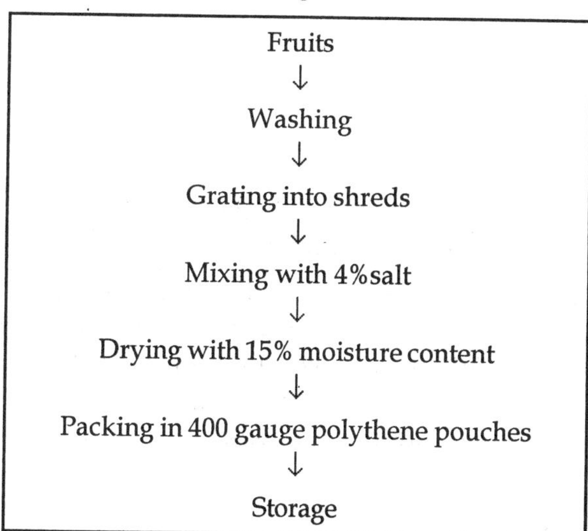

Fig. 4.4. Flow sheet for preparation of shreds from aonla

Candy

Fruits of Narendra Aonla-6 and Narendra Aonla- 9 have been found most suitable for candy making. Excellent quality of candy can be prepared from aonla fruits with 75% total soluble solids.The flow sheet for preparation of candy is given in Fig.4.5.

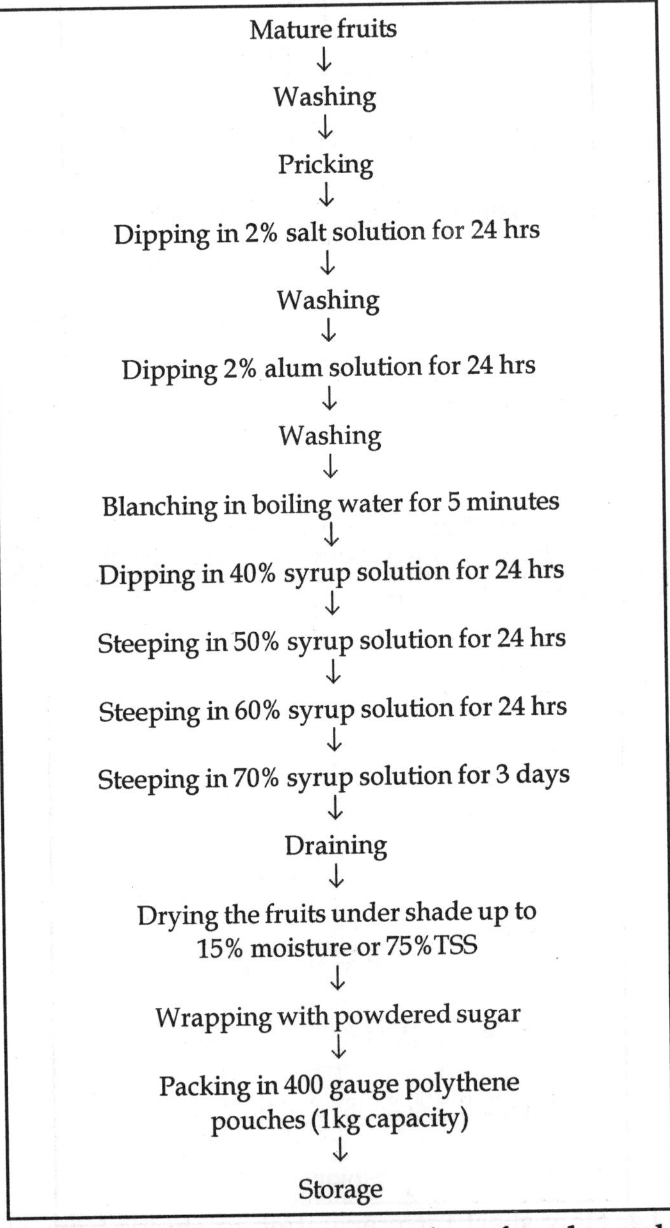

Fig. 4.5. Flow sheet for preparation of aonla candy

Pickle

Recipe containing 1 kg aonla segments, 150g salt, 10g each of turmeric and red chilies powder and nigella seeds, 30 g fenugreek powder and 300 ml cooking oil considered to be ideal. Flow sheet for preparation of aonla pickle is given in Fig.4.6.

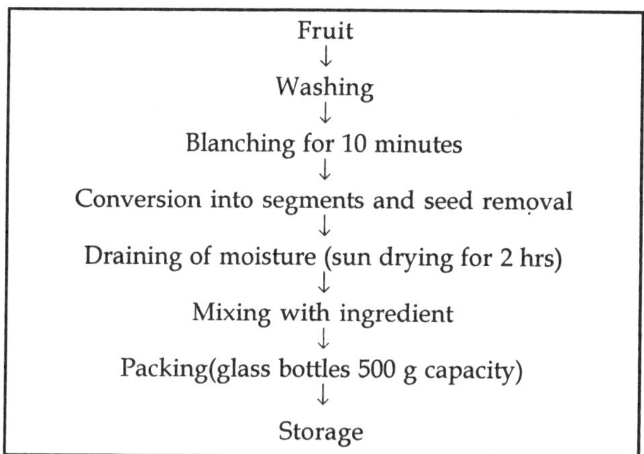
Fruit
↓
Washing
↓
Blanching for 10 minutes
↓
Conversion into segments and seed removal
↓
Draining of moisture (sun drying for 2 hrs)
↓
Mixing with ingredient
↓
Packing(glass bottles 500 g capacity)
↓
Storage

Fig. 4.6. Flow sheet for preparation of aonla pickle

Chutney

Both hot and sweet chutney can be prepared from aonla fruit. Procedure for making of sweet chutney is as under. Recipe containing aonla fruit 1.25 kg, sugar 1 kg, salt 50 g , ginger 15 g ,hot spices 25 g, red chillies 10 g and 10 ml glacial acetic acid has been found ideal for making chutney. The flow sheet for preparation of chutney is given in Fig. 4.7.

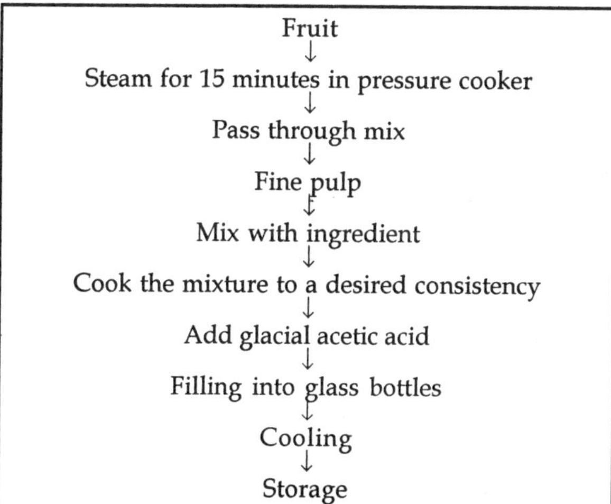
Fruit
↓
Steam for 15 minutes in pressure cooker
↓
Pass through mix
↓
Fine pulp
↓
Mix with ingredient
↓
Cook the mixture to a desired consistency
↓
Add glacial acetic acid
↓
Filling into glass bottles
↓
Cooling
↓
Storage

Fig. 4.7. Flow sheet for preparation of chutney from aonla

Future Research Needs

Though a lot of work on varietals improvement, propagation, nutrient and water management, IPM, organic farming, processing and value addition is in progress in aonla at different research institution, yet there is urgent need to address some of the issues through basic and strategic research for the benefit of aonla growers in the country.

1. The complete physiology of the crop has not been explained so far. As a result the problems like off-season flowering, heavy fruit drop, fruit abnormalities, immature leaf fall, etc are emerging as serious problems. The actual cause of these problems need to be established.
2. Aonla is a medicinal plant and therefore complete technologies needs to be developed for organic farming of aonla without reduction in crop yield.
3. Availability of quality planting material for large scale plantation is still a problem. Faster multiplication techniques using micro-propagation needs to be standardized and popularized in this crop.
4. Aonla fruits are highly perishable in nature. Although storage techniques for fruits have been developed, however, for bulk storage to a considerable period, cost effective cool storage models needs to be designed and tested for aonla particularly for the aonla growing regions of the country.
5. Only few commercial products of aonla are available in the market. Possibilities for product diversification, its quality and consumer acceptability needs to be explored and new range of products needs to be developed for availability to the common man.

REFERENCES

Ammal, E. K. J. and Bhaghawan, R. S. (1957) Proceedings of Indian Academy of Sciences B-46 : 312-314.

Bajpai, P. N. (1957) Bloosm biology and fruit set in *Phylanthes emblica* L. *Indian Journal of Horticulture*. 14 : 99-102.

Bajpai, P. N. (1968) Studies on flowering and fruit development in aonla (*Emblica officinalis* Gaertn.). Horticultural. Advancement. VII : 38-67.

Bhagwan Deen (1992) Studies on screening of aonla (*Emblica officinalis* G.) genotypes for processing. M.Sc. (Ag) thesis, NDUAT, Faizabad.

Chundawat, B. S. (1990) Arid Fruit Culture. Oxford and IBH Publishing Co. Pvt. Ltd., New Delhi : 110-118.

Gupta, O. P. and Pareek, O. P. (1993) Dehydration of fruits. In advances in Horticulture Vol. 4 (Eds K L Chadha and O P Pareek) M P H, New Delhi : 1931-1956.

Kumar, S., Vishal Nath, Singh, I. S. and Mainwal, M. C. (1982) Preliminary studies on storage of aonla fruits in salt solution. *Progressive Horticulture* 24(1-2) : 83-86.

Kumar, S. and Vishal Nath (1992) Storage stability of aonla fruits at ambient temperature. *New Agriculturist* 3(1) : 79-82.

Kumar, S. and Vishal Nath (1993) Storage stability of aonla fruits – A comparative study of zero-energy cool chamber versus room temperature. *Journal Food Science and Technology*. 30(3) : 202-203.

Morton, J. F. (1960) *Economic Botany* 14 : 119-127.

Ojha, C. M., Singh, H. K. and Pathak, R. K. (2000) Effect of drip irrigation on growth of aonla and guava plants under sodic soil.*Progressive Horticulture* 32(2):138-142.

Pareek, O. P. (1987) All India Co-ordinated Research Project on Arid fruits. In : 50 years of Horticultural Research, ICAR, New Delhi : 150-158.

Pareek, O. P. and Vishal Nath (1996) Aonla. In Coordinated fruit research in Indian Arid Zone - Two decades profile. NRCAH, Bikaner : 1-105.

Pareek, O. P. and Vishal Nath (1997) Propagation of ber and aonla. *Agri-Horticultural Society News letter* 3(3) : 4-7.

Pareek, O. P., Vishal Nath and Singh, R. S. (1998) *Shushka Kshetra main aonla Utpadan ki Unnat takneek*. NRC for Arid Horticulture. Extn. Bull.3 : 1-10.

Pathak, S. (1988) Post Harvest Technology of aonla (*Emblica officinalis* G.) fruits. Doctoral thesis, NDUAT, Faizabad.

Pathak, R. K., Ojha, C. M., Dwivedi, R and Hari Om (1991) Studies on effect of method and duration of budding in aonla. *Indian Journal of Horticulture*. 47 : 207-212.

Pathak, R. K. and Srivastava, A. K. (994) Flowering and fruit set studies in Indian goose berry. 24th International Horticulture Congress. Japan.

Pathak, R. K., Srivastava, A. K., Dwivedi, R. and Singh, H. K. (1993) Aonla Descriptor. NDUAT, Faizabad.

Pathak, R. K., Pandey, D. and Vishal Nath (2003) Aonla : Fruits for degraded land. *Chotanagpur Horticulture* 29(1-4) : CII and 6-7.

Pathak, R. K., Pandey, D., Misra, A. K., Haseeb, M. and Tandon, D. K. (2003) The aonla. CISH Extension Bulletin 15 : 1-32.

Perry, A. P. (1943) *American Journal of Botany*. 30 : 527-41.

Rai, M., Pan, R. S., Krishna Prasad, V. S. R., Vishal Nath, Singh, A. K. and Kumar, S. (2001) Plant genetic resource management of horticultural crops in Eastern India. *Indian Journal of Plant Genetic Resources*. 14 (3) : 331-336.

Rai, M., Vishal Nath, Singh, H. S., Dwivedi, R. and Gangapadhaya, K. K. (2002) Overcoming off-season flowering in aonla. *Indian Horticulture*. 47 (3) : 12-13.

Ram, S. (1971) Studies on physiology of fruit growth in aonla (*Emblica officinalis* Gaertn.). Ph.D. Thesis, Kanpur.

Reddi, C. S. and Janki Bai (1979) A study on the pollination in *Emblica officinalis* Gaertn., *New Botany*. 4 : 155-160.

Saroj, P. L., Vishal Nath and Vashishtha, B. B. (2000) Effect of polycontainers on germination, seedling vigour, root characters and budding success in aonla *Indian Journal of Horticulture*. 57(4) : 300-304.

Setty, K. G. H. (1958) Blue mould of aonla. *Current Science*. 28(5) : 20p.

Sethi, V. (1986) Effect of blanching on drying of aonla. *Indian Food Packer*, 40 (4) 7-10

Shukla, A. K., Pathak, R. K., Tiwari, R. P. and Vishal Nath (2000) Influence of irrigation and mulching on plant growth and leaf nutrient status of aonla (*Emblica officinalis* G.) under sodic soil. *Journal of Applied Horticultural* 2(1) : 37-38.

Singh, I. S., Pathak, R. K., Dwivedi, R. and Singh, H. K. (1993) Aonla Production and Post Harvest Technology. NDUAT, Technical. Bulletin. 1-34.

Singh, S. K., Singh, S., Apparao, V. V., Meshran, D. T. and Bagle, B. G. (2006) Mulching for sustainable production of aonla under rainfed condition *Agricultural Update*. 1(3) : 6-8.

Srivastava, A. K. and Pathak, R. K. (1993) Studies on sex ratio, pollen viability and suitable pollinizers in aonla (*Emblica officinalis* Gaestn.). Golden Jublee Symposium at IIHR, Bangalore.

Srivastava, R. P. and Kumar, S. (1993) Preservation of fruits and vegetables Principles and Practices : IBH Publication, Lucknow.

Vishal Nath (1990) Studies on storage of aonla (*Emblica officinalis* G.) fruits M.Sc.(Ag.) thesis, NDUAT, Faizabad.

Vishal Nath, Singh, I. S. and Kumar, S. (1992a) Evaluation of aonla cultivars for their shelf life at ambient temperature. Narendra Dev Journal Agricultural Res. 7(1) : 117-120.

Vishal Nath, Singh, I. S. and Kumar, S. (1992b) Studies on storage of aonla fruits. *Narendra Dev Journal Agricultural Research* 7(1) : 121-126.

Vishal Nath, Singh, I. S., Kumar, S. and Pandey, A. K. (1992c) Effect of post harvest treatments on shelf life of aonla fruits. *Progressive Horticulture.* 24(1-2) : 79-82.

Vishal Nath (1993) Studies on processing of aonla (*Emblica offinalis* G.) fruits. Ph.D. Thesis, CSJM Kanpur University, Kanpur.

Vishal Nath and Pandey, A. K. (1996) Storage stability of aonla products at ambient temperature. *New Agriculturist.* 7(1): 73-78

Vishal Nath (1996) Processing technology of aonla for beverage industry. *New Agriculturist.* 7(2): 143-146.

Vishal Nath and Sharma, R. K. (1998) Screening of aonla (*Emblica officinalis* G.) cultivars for processing. *Progressive Horticulture.* 30(1-2) : 76-77.

Vishal Nath (1999) Delicacies of aonla. *Indian Horticulture.* 44 (3) : 15-16

Vishal Nath, Rai, M. and Das, B. (2002) *In situ* propagation for ensured orchard establishment. *Chotanagpur Horticulture.* 15-19 (1-4) Special issue : 5-6 & 36.

Vishal Nath (2004) Value addition in arid Horticultural crops. Advances in Arid Horticulture, Vol. I (Eds. P.L. Saroj *et al.*) : 483-504.

Vishal Nath, Das, B., Yadav, M. S., Sikka, A. K. and Singh, R. V. (2006) Fruit Crop based Agro-forestry system for watersheds of Eastern India. *Journal of Research* (BAU), 18 (1) : 59-68.

Vishal Nath, Pathak, R. K., Pandey, D. and Das, B. (2004) Aonla Utpadan ki Aadhunik takneeken. In: *Parvatiya Bagwani* (Ed. H P Singh *et al*) : 185-213.

—x—

5

DATE PALM

Date palm (*Phoenix dactylifera* L.) is one of the ancient fruit crops. It is believed to be indigenous to countries around Persian Gulf. Date cultivation has historical records dating back to 6000 BC along the Tigris and the Euphrates rivers in Iraq and to 2000 BC at Mohanjodaro, along the river Indus. Several thousand years before its cultivation, pre-historic man used the fruit from wild palms and carried the seeds over the wide area from India though the middle-east and later to north Africa. The date palm (*Phoenix dactylifera* L.) was known to ancient people as the "Tree of Life" and is probably one of the first fruit plant to be cultivated. Representations of the date palm, dating back the 3000 B.C. appear on Sumerian temple walls in present day in Iraq. In Egyptian hieroglyphics, the date palm is used as the symbol for a year and its frond is the symbol for a month. Date palm logs were used to roof royal tombs as early as 2700 B.C.

Date palm fruits are eaten as raw dates (fresh fruits), dry dates (*chhuhara*) and soft dates (*pind khajoor*). These are used in religious ceremonies and are considered auspicious. Different products viz., sugar, starch, vinegar, juice, toffies, wine, chutney, jam, pickles etc. are prepared from date fruits. The date palm fruits have been recognized for their highly nutritive and calorific food values. Date fruits provide abundant quantities of iron, potassium, calcium, nicotinic acid and small amount of protein, copper, magnesium, chlorine, sulphur, vitamin A, B_6 and B_{12}. Date pulp contains moisture (20%), sugar (60-65%), fibre (2.5%), protein (2%) and less than 2 per cent fat, mineral matter and pectic substances. It's fresh fruits can help supplement the dietary needs of the desert people and provides about 3550 calories per kilogram fruits.

Leaves of date palm trees are utilized for the preparation of baskets, brooms, ropes, building material, fuel and paper. The date palm trees are very valuable as a cover crop for ground crops in hot and dry regions, because the trees have single unbranched and upright trunk. The trees like pomegranate, citrus, papaya and phalsa can be planted as inter crop in the date palm orchard (Chundawat, 1990). Similarly, vegetable, flowers, fodder crops can also be raised in date palm orchard. The plantation is also useful for the sand dune stabilization and maintenance of ecological balance in the deserts and arid regions.

Origin and Taxonomy

Date palm scientifically known as *Phoenix dactylifera* L. belongs to the family-Palmaceae. The date palm is a monocotyledonous plant with strong straight unbranched stem growing to the height more than 30 ft. It is characterised by pinnate leaf palms by the upward and lengthwise folding of the pinnae and peculiarly furrowed seeds. *Phoenix dactylifera* L., the commercially grown date palm species produces superior quality fruits. This species has characteristic feature of producing axillary offshoots / suckers which may arise near the base of stems unlike the two other species viz. *Phoenix sylvestris*(L.) Roxb. and *Phoenix canariensis* Chaubaud (Canary Island palm). Canary Island palm is mainly planted for ornamental purposes. *Phoenix sylvestris*(L.) Roxb is found growing wild in almost all parts of India and produces fruits of very inferior quality. This palm is used for production of crude sugar (*khajoor ka gur*) and a drink called "*neera*".

Date palm being dioecious in nature, bearing male and female flowers on separate trees, has chromosomes 2n=36. Intra sepecific crosses of *Phoenix dactylifera* (L) with *P. sylvestris* (L) Roxb. and *P. canariensis* Chabaud are successful and have shown metaxenic effect. Staminate (male) and pistillate (female) flowers are produced on separate palms in the axils of leaves of the previous years growth in all the species of date palm. The date palm does not possess a well-developed tap root system. It has a large number of secondary roots more or less equal in thickness, which arise, in the form of a dense cluster from the base of the stem. The secondary roots out side the relatively well marked central portion (stele) have visibily a great number of large air passages in peripheral (extra stelar) region resembling to the roots of plants adopted for growing in soils which contains excessive amount of water such as rice (*Oryza sativa*). The large air passage in the date root structure indicates its requirement for plentiful air supply.

Area and Distribution

Commercial cultivation of dates in the world lies between 20° and 35° N and S latitude from the equator. Beyond it, commercial cultivation is rare owing to occurrence of high rainfall during fruit growth and ripening period. Iraq, Saudi Arabia, Iran, United Arab Republic and Algeria are the main date palm producing countries in the world. Beside these countries Pakistan, Morocco, Libya, Tunisia, Sudan, Muscat, Oman, The Aden, United State of America and Bahrain also produce dates in substantial quantities (Vishal Nath and Singh, 2003). Dates are also produced in small quantity in Spain, Chad, Mexico, Yemen, Israel, and Somalia. These countries have a population of 60 million palms, out of which Iraq alone has 20 million palms. Apart from the above, a few date plantations exist in Peru, Brazil, Argentina, South Africa, Australia, Cyprus, Ethopia and India. Before partition of our country in 1947, the major date palm growing areas were Multan, Muzaffargarh, Jhang, Deragazi khan, Bhawalpur, Kherpur which are now in Pakistan (Chandra *et al.*, 1992).

In India, it is believed that date palm has been introduced by the soldiers of Alexander in the 4th century B.C in the Indus valley. Babar has mentioned in his memoirs that the date was grown in India in the early sixteenth century. Dates were also grown in India during the regime of Akbar (AD 1555-1605) (Randhawa, 1980). In Rajasthan, date cultivation was first introduced by the then ruler Ganga Singh of the erstwhile Bikaner state through the plantation of few suckers of Halawy, Khadrawy and Zahidi at Ganganagar, a part of Thar descrt, after development of Gang-canal in the state. Grooves of seedling date palm are found on the coastal belt from Anjar to Mandavi in Kutch district of Gujarat. Presently, over 12 lakhs date palm trees are in Kutch area. In north-western Rajasthan, scattered plantation of few thousand dates tree are confined mainly in the districts of Ganganagar, Bikaner, Jodhpur and Jaisalmer. Its cultivation has great potential in western districts of Rajasthan viz., Jaisalmer, Barmer, Bikaner and Jodhpur; besides, Nagour, Churu, Sikar and Ganganagar districts of Rajasthan. Visualizing the performance of some improved cultivars in Indian Arid Zone, new thrust on date palm cultivation has been given in the recent years (Vishal Nath and Singh, 2003). As a result, new areas are coming under improved dates varieties in Bikaner and Jodhpur in Rajasthan, Abohar in Punjab, Hisar in Haryana and Mundra in run of Kachchh.

Soil and Climate

Date palm is very specific to soil and climatic requirement. It thrives well in deep sandy loam soil having adequate facilities for sub soil drainage for optimum productivity. The soil should be fertile, having good water holding capacity. The soil pH in the date palm growing areas generally ranges between 8-10 in Rajasthan. At least in the upper 2 m surface of the soil, there should not be hard concentration of gravels or calcium carbonate layer. Date palm can tolerate more alkali or salt (mainly sulphates of sodium and magnesium and chlorides of these metals) as compared to other fruit plants.

Climatic coherence at a particular place with floriferous phase of the date palm decides the fate of its production. Successful date production requires moderate temperature and dry weather during winter and prolonged hot dry summers and almost rain free period at the time of fruit ripening (July-August). As per an Arabic saying "Date palm should be grown with its feet in running water and its head in the fire". Date palm grows well in sand but it is not arenaceous. It has air spaces in its roots and may grow well where soil water is close to the surface, but it is not aquatic. It grows well in saline conditions, but is not a true halophyte and does better in quality soil and water. Its leaves are adapted to hot, dry conditions, but it is not a xerophyte and requires abundant water. The date palm is adapted to areas with long, very hot summers with little rain and low humidity, but with abundant underground water. The mean temperature required at the time of flowering and fruit ripening is 25 °C and 40 °C, respectively. The pattern of maximum temperatures at the time of fruit maturity determines the rate of maturity and fruit quality.

It needs practically dry rainy season for attaining *pind* stage. The excess rain during fruit ripening period may even rot off the fruits. The fruits have to be harvested at whatever stage of maturity it may be, before the onset of monsoon. Late onset of monsoon, lesser total rainfall and number of rainy days are the primary requirements of the date palm cultivation as these conditions enable good quality and quantity of fully ripen soft dates (*pind khajoor*). Heavy rainfall followed by clear weather is less harmful as compared to a light shower followed by prolonged drizzle or cloudy weather. Similarly, heavy rain during early stage of ripening is less harmful than light rains at later stage of ripening. In India, on the basis of base temperature as 10° C, the requirement of the heat summation units for date palm varies in the range of 1950 to 3650 depending upon cultivars. However, the rate of maturity of the fruits and the development of its quality depend on the pattern of daily maximum temperature and heat units during the fruit ripening period. The situation is beneficial in the Indian Thar desert due to early ripening; the fruits escape spoilage from the ensuing rains. Vishal Nath and Singh (2003) mentioned that since at Jodhpur and Bikaner, the accumulation of heat units during March to May is more than other places, spathe opening (Last week of February to last week of March) and *doka* formation (First week of February to first week of July) take place earlier than Abohar (near Ganganagar) and Hisar. In Jaisalmer and Bhuj (Kachchh), the initial rate of accumulation of heat unit is slower than Jodhpur and Bikaner (Pareek, 1984). Four districts of Rajasthan namely Jaisalmer, Barmer, Bikaner and Jodhpur located at western parts of Rajasthan are idéally suited for date palm cultivation and have vast potential. The annual average rainfall of Jaisalmer, Barmer, Bikaner and Jodhpur is 160mm, 310mm, 305mm and 370mm, respectively. In these areas, rainfall distribution is erratic. The average maximum temperature during summer months generally remain above 40 ^{0}C and vary often reach around 45-48 ^{0}C. The accumulation of heat summation between flowering and fruit ripening period at Jaisalmer and Bikaner is sufficient enough for the date fruits to attain *dang* stage on the tree itself. Chandra *et al.,* (1990) reported the heat summation unit requirements of different date palm varieties to reach colour turning stage (early *doka*) and *pind* stage at Bikaner (Table. 5.1).

Table 5.1. Heat unit required for promising date palm varieties at Bikaner (from spathe initiation to full *dang* stage)

Varieties	Heat summation units (base 10 °C)	
	Early *doka*	Fully *dang*
Halawy	1951	3101
Medjool	2648	3650
Zahidi	2322	3479
Khalas	2357	3281
Barhee	2323	Not attained
Sewi	2244	3281
Zagloul	2227	3631
Hayani	2460	3168
Shamran	2411	3567
Khadrawy	2213	3342

Plant Genetic Resource Management and Varieties

In India, wide rang of variability of wild date palm (*P. sylvestris*) yielding small size acrid fruit exist in many parts. Large fruited cultivars which are showing promise in arid parts of the country are mainly introduction from Gulf countries, USA and Pakistan (earlier undivided India). Recently, some promising date palm cultivars (Chip chap, Sakkarai, Birham) have been introduced at CIAH, Bikaner from Iraq. However, variability of large fruiting type does exist in Run of Kachchh (Plate 5.1). About 34 cultivars imported from different countries have been successfully grown at Date palm Research Station, RAU, Bikaner but only Halawy, Barhee, Khuneizi, Medjool, Zahidi, Khalas and Sewi have shown promise. Considering the Rajasthan agro-climatic conditions, the cultivar should have (a) non-astringent and early maturing fruits for fresh eating (b) early maturing high sugar and rain tolerant fruits for processing particularly for preparation of *chhuhara* and *pind khajoor*, (c) tolerance to *Graphiola* disease besides high yield potential and dwarf growth habit. At present, the country has more than 25 improved cultivars of date palm. Characteristics of some of the cultivars acclimatized in Indian arid zone have been enumerated.

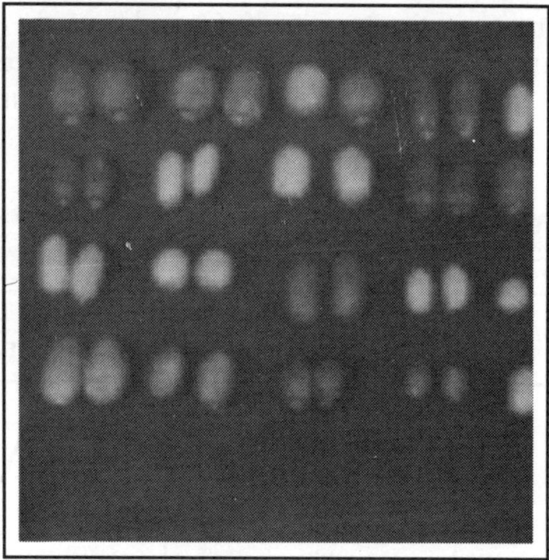

Plate 5.1. Variability in Date palm fruits

Halawy

It is one of the most suitable varieties for commercial date palm cultivation. The trees of this variety are tall and vigorous in growth. The fruit colour is light orange with yellow shade (Plate 5.2). The fruit is oblong elliptical with obtuse apex and almost horizontal base. The average fruit weight is 9.0 g and fruit size of 3.56cm x 2.10cm. It has pulp thickness of 0.64cm having 31.0 per cent TSS at *doka*. The fruits are sweet at full *doka*. The Halawy dates may be eaten raw. The fruits have a tendency to shrivel during the ripening stage particularly in case of insufficient irrigations. It is early in ripening as the fruit of this variety reach at full *doka* stage at Bikaner by second fortnight of July. Halawy has average yield potential of 100 kg *doka* fruits per plant per year.

Plate 5.2. Early fruiting in Date palm cv. Halawy

Shamran

The trees of this variety are semi-vigorous. The fruits are yellow in colour with slightly pink shaded base. The fruits are medium in size and oblong in shape with obtuse apex. The average fruit weight is 6.5 g and fruit size of 3.50 cm x 2.18 cm. It has pulp thickness of 0.56 cm having 41.0 per cent TSS at full *doka* stage. The fruits are astringent at *doka* stage. It is suitable for dry dates and soft dates. Its soft dates have typical aroma. It is medium in ripening as fruits reach full *doka* by the end of July at Bikaner. It is a prolific bearer and has average yield potential of 100 kg *doka* fruits per year (Plate 5.3).

Plate 5.3. Mature bunch of Date palm cv. Samran

Barhee

The trees of this variety are semi-vigorous and medium in height, having a good canopy size. The fruits are golden yellow in colour at *doka* stage and oval in shape. The average fruit weight is 7.5 g and fruit size of 2.90cm x 2.30cm. It has pulp thickness of 0.73cm having TSS of 32.0 per cent at *doka* stage. The fruits are sweet and non-astringent having very pleasant taste at *doka* stage. Fruits of this variety are non-astringent at *gandora* stage also. This variety is the most suitable for raw eating dates; however, it is not good for soft dates. It is late in ripening as fruits of this variety reach at full *doka* stage by mid August at Bikaner and hence it is moderately damaged by rains and high humidity. Average yield potential is 100 kg *doka* fruits per year.

Khalas

It has shown great promise for cultivation in India. The trees are semi-vigorous in growth. Its fruit colour is yellow at *doka* and golden brown at *pind* stage. The fruits are oblong in shape. The average fruit weight is 11.0 g and size of 3.2cm x 2.4cm. It has pulp thickness of 0.83cm having 28.0 per cent TSS. The fruits are sweet at full *doka* stage and suitable for raw eating as well as for soft dates. It is early in ripening as the fruits of this variety reach full *doka* stage in the second fortnight of July at Bikaner. Average yield potential is 75 kg *doka* fruits per tree per year.

Medjool

It is a popular variety and has potential of commercial cultivation in the country. The trees are vigorous and tall with erect growth habit. The fruits are large but variable in size having attractive golden yellow colour and more firm than Barhee and Khadrawy. The irregularities in shape are common and are associated with ridges on the seed. The average fruit weight is 15.0 g and fruit size of 3.90cm x 2.80cm. It has pulp thickness of 0.88 cm having 29.0per cent TSS at *doka* stage. The fruits at *doka* stage are highly astringent. It is suitable for dry dates (*chhuhara*) because of its large sized fruits and good pulp thickness. It is tolerant to rainfall and high humidity conditions. It is late in ripening as fruits of this variety reach full *doka* stage in mid August at Bikaner. Average yield potential is 75 kg *doka* fruits per year.

Sewi

It is vigorous in growth. The trees are tall and erect with moderate canopy. The fruits are yellowish green in colour at *doka* stage and oblong in shape. The average fruit weight is 6.1 g and fruit size of 2.9 cm x 1.90 cm. It has pulp thickness of 0.6cm having 40.0 per cent TSS at full *doka* stage. The fruits are sweet at full *doka* stage and therefore suitable for both raw eating as well as soft dates. It is late in ripening as fruits of this variety reach full *doka* stage by the first fortnight of August at Bikaner. The fully grown tree of Sewi has an average yield potential of 50 kg *doka* fruits per year.

Khuneizi

This is emerging as a potential date variety because of quality fruits. The trees are semi-vigorous having spreading growth habit. The fruit is dark red in colour at *doka* stage (Plate 5.4) with crispy pulp and dark brown at *pind* stage. The fruit is oval shaped with pointed apex. The average fruit weight is 10.2 g and fruit size of 3.50cm x 2.20cm. It has pulp thickness of 0.77cm having 43.0 per cent TSS at full *doka* stage. The fruits are sweet at full *doka* and thus suitable for raw eating. However, these are not good for soft dates. This variety is also early as fruits reach full *doka* stage by the end of first fortnight of July at Bikaner. Average yield potential is 40 kg *doka* fruits per tree per year.

Zagloul

The trees of this variety are vigorous in growth having erect growth habit. The fruits are red in colour and large in size. These are oblong,

Plate 5.4. Bumper harvest of Khunezi date

oblique and asymmetrical in shape. The calyx is small in size and slightly depressed at the base. The fruit bunches are bigger in size. The average fruit weight is 9.5g and fruit size is 3.90 cm x 2.20cm. It has pulp thickness of 0.64 cm having 28.0 per cent of TSS at *doka* stage. The fruits are astringent at *doka*. The *dang* and *pind* fruits are relatively less sweet but have pleasant taste. It is suitable for soft dates and for preparation of fruit products such as date juice and preserved products. It is medium in ripening as fruits of this variety reach full *doka* stage by the beginning of second fortnight of July at Bikaner. It is prolific bearer (Plate 5.5) and average yield potential is 150 kg *doka* fruits per year.

Plate 5.5. Fruiting palm of cv. Zagloul

Zahidi

The trees of this variety are semi-vigorous. The fruits are yellow in colour at *doka* stage (Plate 5.6) with smooth and hard surface, hence least affected. The fruits are ovate in shape with the broader end towards the apex. The average fruit weight is 10.0 g and fruit size of 3.00 cm x 2.20cm. It has pulp thickness of 0.68 cm having 30.0 per cent TSS at *doka* stage. The fruits are astringent at *doka* stage. The partial *dang* fruits are eaten as raw dates. It is a suitable variety for preparation of soft dates. It is late in ripening as fruits of this variety reach full *doka* stage in mid August at Bikaner. It is a high yielding variety and has average yield potential of 125 kg *doka* fruits per year.

Plate 5.6. Bunch of cv. Zahadi

Khadrawy

The trees of this variety are dwarf in growth. The fruits at *doka* stage are greenish yellow in colour and oblong in shape (Plate 5.7) with broad apex and a bit slanting base. The average fruit weight is 7.7 g and fruit size of 3.10 cm x 2.10 cm. It has pulp thickness of 0.66 cm having 30.0 per cent TSS at *doka* stage. The fruits are highly astringent at full *doka* stage and are badly spoiled due to rains at the time of ripening. This variety is suitable for preparation of both soft dates as well as dry dates. It is medium in ripening as the fruits of this variety reach full *doka* stage by end of July at Bikaner. Average yield potential is 40 kg *doka* fruits per year.

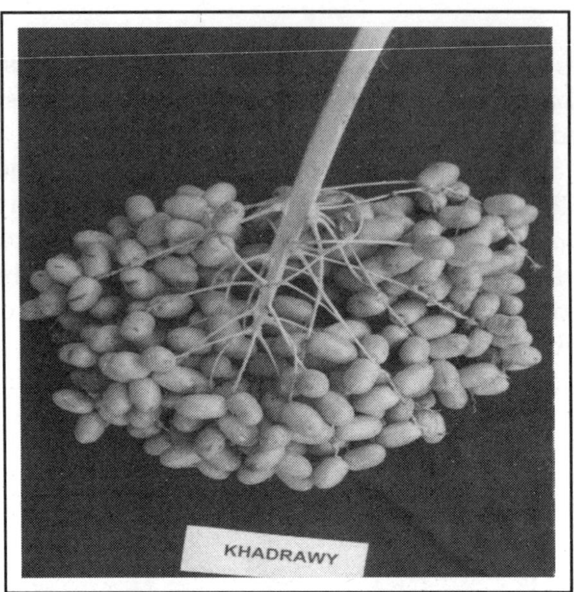

Plate 5.7. Fruits of cv. Khadrawy

Plant Propagation and Nursery Management

Date palm can be propagated through seeds as well as through naturally separating off shoots. It can also be propagated through tissue culture with caution. Owing to heterozygosity and dioecious nature of the palm, offshoots are always used for its propagation for commercial plantation. But 8-20 off-shoots can only be obtained between its fourth and tenth year of life and none thereafter. This is obviously a slow rate of multiplication. Absence of fast multiplication technique for date palm is thus a prominent bottleneck in the extension of the area under its plantation. Vegetative propagation in date palm is mainly done through off shoots, which after 3-5 years of attachment with the parent palm, produce roots and by then start producing second generation of off shoot. Aerial offshoots are not used for planting as these are either devoid of roots or have very poor root development. Ground off shoots of 8-10 kg average weight and well-developed root systems are selected for commercial plantation. Slow multiplication of date palm through off shoot is main cause of slow rate of its expansion in terms of area. Vishal Nath and Singh, (2003) reported that a healthy mother palm produce only 15-20 off shoots in its life time. After about 20 years mother palm do not produce any viable off shoot which can be used for planting. Therefore the necessity felt for rapid multiplication of this important species by tissue culture. In micro propagated plants, existance of soma-clonal variation particularly towards the production of male palms is a matter of concern and therefore efforts are on to produce genuine female plants through this technique. Besides private sector companies, some efforts have been made by Rajasthan Agricultural University, Bikaner to produce planting materials through tissue culture but the plants need field evaluation for their sex expression and fruiting behaviour. Looking to the present requirement of the plant materials, raising of seedlings in polyethylene bag has been tried by the Date Palm Research Centre, Bikaner. Seeds may be shown in polyethylene bags either in August or in March. The germination is more than 85 per cent and plants become ready for planting within a year but the genetic make up of such plant can not be predicted.

Planting

The suitable period for planting date palm offshoots is rainy season (July-September) or during spring (February-March). During this period, relative humidity is high and temperature is congenial for establishment of plants. The date palm offshoots are planted in the field in square system at a planting distance of 8m (Plate 5.8). Optimum distance between rows and within rows is very essential as date palm

Plate 5.8. View of the new plantations of Date palm

remain in bearing for at least 50-60 years and close distance may create problem in later years of bearing due to over crowding and shading. While planting precautions should be taken that the crown of the off shoot should not come in contact of water to be applied in the basins (Vishal Nath and Singh, 2003). Seedlings raised in polythene bags are transplanted at the distance of 8 x 2 m for maintaining the proper sex ratio in the field.

Cent per cent survivability of offshoots treated *in situ* with Bavistin (0.1%), Chloropyriphos (0.1%) and 1000 ppm Indole Butyric Acid (IBA) has been observed at Bikaner. The treatment of Bavistin (0.2%) and Chloropyriphos (0.1%) should be repeated after a month with alternate days irrigation of offshoots till three month period. To protect the crown portion from the damage caused by irrigation water, it should be at least 10-15cm above the ground level. The windbreak should be planted in north and west directions of the orchard, to protect date palm trees against hot and cold winds (Manohar and Chandra, 1995).

Training and Pruning

Being monocotyledonous, date palm trees have single stem and generally do not branch. Therefore, diseased, dried and damaged leaves must be removed during winter. Very old leaves must be removed, as these are less efficient in photosynthesis. Besides removal of leaves, dethroning is also essential close to fruit bunches as it help in pollination, thinning of bunches, spraying and harvesting of fruits. After fruit set, twisting of bunch is done during April and May, which helps to withstand bunch breakage due to high wind velocity. Removal of one third of their central strands after fruit set has shown better development of fruits and hastened ripening and fruit quality. It is also found that pruning date leaves leaving 6 leaves per bunch and also removal of one-third strands from centre have given good quality date production (Anon., 2002a). Therefore, it is suggested that nearly 70-100 leaves per palm are required for optimum productivity.

Orchard Management

Orchard management in commercial date palm plantations should be done very precisely. For production of quality fruits and having competition with other date producing countries in the export market recommended management practices should be followed.

Nutrient Management

Date palm requires proper nutrition particularly in sandy and gravely soils to achieve good growth and economical production. The amount of fertilizer needed by date palm tree depends on soil type; kind of inter crops grown under, as well as variety and age of the tree. In sandy soils of western Rajasthan, well rotten farm yard manure must be given every year at the rate of 20 - 40 kg per tree during September to December, to the bearing date trees. Besides this, each bearing tree should be given 0.5 - 1.0 kg nitrogen, 0.5 - 1.0 kg phosphorus and 0.5 kg potash per year (Vishal Nath and Singh, 2003). The entire quantity of phosphorus and potash and sixty per cent of nitrogen should be given 3 weeks before flowering and remaining quantity of nitrogen during the month of March and April after fruit set. Studies conducted at Bikaner centre revealed that the addition of 1500 g nitrogen per palm per year into two split doses viz. sixty per cent before flowering and remaining after fruit set in combination with 500 g each of phosphorus and potash as basal dose provided better vegetative growth, yield and quality fruits in cultivar Halawy (Anon., 2002c). At Abohar, application of 400g nitrogen per palm per year increased plant height, spread, leaf length and number of bunches per palm in cv. Khadrawy (Anon., 2002b).

Water Management

The date palm tree is thrust tolerant and drought resistance to a great extent. In spite of that, date palm trees are salinity resistance compared to other fruit trees. Irrigation is absolutely necessary in date palm, because it is grown in hot and dry, low rainfall areas. The water requirement of date palm is quite high, therefore under such circumstances date palm cultivation could not be extended in such areas where there is lack of irrigation facilities as in most of arid zone areas. Frequency of irrigation depends on soil texture and weather conditions. In light soils, date palm orchards are usually irrigated every 7 to 10 days intervals in mid summer and every 15 to 30 days during winter. Vishal Nath and Singh (2003) mentioned that one irrigation prior to spathe initiation is critical for normal fruiting. Irrigation after fruit set should be given regularly. Scarcity of irrigation during fruiting may result in premature fruit drop, shriveling of fruits and reduction in fruit size, weight and quality. Studies conducted at Bikaner on conservation of soil moisture by use of mulches revealed that mulching palm basins with black polythene sheet or 10cm thick layer of locally available dry weed, bui (*Aerva persica*) helps in better soil moisture retention, supresses weed growth (Plate 5.9) and improves fruit quality (Anon., 1998a). The drip irrigation method in date palm orchard is useful as it provides efficient utilization of irrigation water. If drip irrigation is used from an early age, the root distribution of date palm will be within the area of moist soil around the tree. The emitters should be moved out away from the tree as it grows so that a normal spreading root system will develop. It is also important to let the soil aerated between watering.

Plate 5.9. Moisture conservation in the palm basin through mulching (i) mulching with local weeds (ii) mulching with black polythene.

Orchard Floor Management

Time to time removal of weeds is very necessary in date orchard. Hoeing of newly planted date orchard should be done in the basin regularly. It not only helps in destruction of weeds but also provides better aeration to roots and moisture retention in the soil. In the well-established orchards also, removal of weeds is very essential to avoid competition of weeds for nutrients and moisture with date palm trees. Weeding in date palm is done mechanically however, synthetic weedicides have also been found effective. Vishal Nath and Singh (2003) reported that monocot and dicot weeds in date orchard can be managed effectively by post emergence application diuron (5 kg/ha) followed by glyphosate (3.75 l/ha). Intercropping in date palm orchard in feasible for quite long period. Leguminous crops, vegetables and small stature fruit crops can be grown in date palm orchard. Plants having strong competition should not be taken as intercrop.

Diseases Management

Graphiola leaf spot, *Alternaria* leaf spot, rotting of main shoot of offshoots and fruit rotting are major diseases of date palm in western Rajasthan. Amongst them, *Graphiola* leaf spot is the only disease, which cause economic losses by devitalizing the leaves to reduce their number on the palm affecting production. It has been observed to be severely affected by false smut of dates, *Graphiola phoenicis* (Moung). It is the most wide spread disease of date palm and probably occurs whenever the palm is cultivated under humid conditions, which prevail in many subtropical to marginal date growing areas. Leaves of date palm are spotted with yellowish dots with small black scales on worts having a gray brown to dark brown horny outer surface and long sterile flexous hyphae protruding from an inner membrane containing powder yellow or light brown spots. In case of heavy infestation, leaves dry and consequently die. Thus, it causes great damage to the plant growth and reduces the fruit yield. Studies on epidemiology, varietal screening and disease management were carried out at Bikaner centre. *Graphiola* leaf spot has been observed to be wide occurrence in the hyper-arid partially irrigated western plains of Rajasthan at Bikaner, Jaisalmer, Barmer and Jodhpur, which have hot arid climate. The disease progressively developed on older leaves as compared to new one. Highest infection was recorded in the month of December due to low temperature and higher humidity in the atmosphere. Incidence of the disease has been recorded to be varying from 2.50 to 76.25 per cent in the field. Thirty-four varieties were screened under natural as well as artificial conditions, out of which the varieties Sayar, Halawy and Khalas were found to be highly susceptible and variety Hatemi was found to be resistant. Two alternate spray of Bliotx-50 (0.4%) have been found followed by Bavistin (0.1%) effective to minimize the *Graphiola* leaf spot.

Four date varieties namely; Sayar, Khalas, Halawy and Hatemi were used to investigate the changes in green pigments (chlorophyll and carotenoids) at different levels of severity of the disease. The per cent reduction in chlorophyll and carotenoides in diseased plant was obviously much more in highly susceptible varieties in comparison to resistant ones. If the severity of the disease increased, the chlorophyll and carotenoides pigment decreased proportionately in all the varieties of date palm.

Pre-harvest fruit rot is another problem, since their incidence is governed by the occurrence of rain and high humidity during *Khalal* and later stages of ripening. Considerable loss observed by fruit rots caused by *Aspergillus niger, A. flavus, Rhizopus, Penicillium* sp. and *Botryodiplodia* sp., when humid and rainy weather occurs during the ripening season. Out of 31 date cultivars screened against fruit rot before and after rains, none of the cultivars was found immune, while cultivars Khuniezi, Halawy, Medjool, Kahals and Barhee were susceptible to the disease. Only one cultivar i.e. Zagloul was found resistant i.e. below 5 per cent disease incidence.Two spray of chemical Carbendazim (0.1%) have been found effective in controlling fruit rotting (64.98%) followed by copper oxychloride (0.4%).

Pest Management

Among the insects, *Parlatoria blanchardii* commonly known as scale insect seems to be one of the most injurious pests in the Thar desert of Rajasthan. The nymph and adult of scale insect causes damage to the trees by feeding on white succulent tissues at the base of the leaf stalk. Its infestation affects normal growth of the tree and productivity goes down. Spraying of Monocrotophos or Dimethoate (0.05%) has been found beneficial for the control of scale insect. The heavily infested leaves with scale insects should be removed and destroyed. Studies conducted at Date Palm Research Centre, Bikaner revealed that scale insect population might be kept below Economic threshold level with the introduction of predatory beetle *Chilocorus nigritus.*

The termites *(Odontotermes obesus)* workers occur on all the tree parts all the year round and considerably affect the palm. Termite's workers attack on root zone, feed on the roots and mine into the main trunk. The young establishing trees under its severe attack wither and die. The root zone of trees should be treated with chlorpyriphos or endosulfan periodically. The galleries of termites on the tree trunk should be scrapped and treated with above-mentioned insecticides.

A new insect pest identified as lesser date moth; *Batrachedra* sp. was also observed for the first time in Bikaner (Anon. 2002 a,c). Larvae of this moth attack the flowers and fruits. Infestation occurs during reproductive period i.e. between March to August. To control this pest, Malathion or Diazinon @ 0.05 per cent should be applied at about one week after fruit set and repeat again after 15 days of the first application, if required.

Bird Management

The period of fruit growth and development from May to July in date palm coincides with the period when there is almost no field crop available for birds for feeding in the areas where date palm can be grown in drier parts of Rajasthan. Thus, birds cause severe damage to date palm fruits. The date bunches at the time of ripening are very much damaged by myna, crow, parrot, sparrow, purple sunbird etc. The maximum damage to the fruit is caused at full *doka* and *dang* stages. Therefore, protection from bird attack is very much needed during this period. The protection from bird damage can also be done by covering fruit bunches with wire gauge (3x3 mm mesh) or nets (Plate 5.10) at the start of *doka* stage till harvesting.

Plate 5.10. Protection of date bunches from bird (i) Covering with wire mess (ii) bagging with net

Flowering, Fruiting and Yield

Flowering in date palm is started by spathe emergence. Spathe emergence takes place from late winter to spring, which is influenced by various factors such as growing site, climatic conditions mainly temperature fluctuations, cultivars and sex of the plant, etc. At Bikaner, it has been observed that generally, spathe emergence takes place much earlier in male palm as compared to female trees (Plate 5.11). The spathe initiation commenced early in cultivar Zahidi, Nagal, Khadrawy and Khuneizi (first week of December) while it was delayed in Umshok and Sakloti (first week of March) at Bikaner (Pundir and Porwal, 1999).

Plate 5.11. Female (L) Male (R) and inflorescence of date palm

Date palm being dioecious in nature, bearing male and female flowers on separate trees requires artificial pollination for good fruit setting, as pollination through natural means such as wind and insect is negligible. Hand pollination is commonly practiced in Western Rajasthan (Plate 5.12). It is done by dusting pollen through cotton balls on freshly opened female spathes in the early morning hours at

Plate 5.12. Manual pollination for better sets in date palm

least for 2-3 days followed by placing pollen impregnated cotton balls in the female flowers or by hanging strands of male inflorescence in an inverted condition on female spathes. For collection of pollen from male spathe, the mature male spathes are swept on paper. These collected pollens (6 hrs in sunlight followed by 18 hrs in shade) can be stored in airtight glass vials in cool place at normal temperature for 8 weeks or in a refrigerator at 9° C for a year.

In date palm fruit set takes place on the loss of two unfertilized carpels (*hababouk* stage) after completion of fertilization process. There after, fruit growth is commenced. The fruit development stages in date palm can be divided into four stages viz., *Gandora* or *Chimri* (fruit attain size and remain hard green); *doka* or *khalal* (fruit attain full growth, remain hard and their colour changes into yellow or red); *dang* or *rutab* (softening of fruits start from tip and finally the whole fruit is softened) and *pind* or *tamar* (fruit is fully ripened, fruit weight decreases as a result of dehydration of fruits). The fruits become edible from *doka* stage except of those cultivars, which are highly astringent. At Bikaner, pre-harvest application of ethrel (2 - Chloroethyl phosphoric acid) @ 1000 ppm on fruit bunches at colour break stage (when fruit starts changing there colour from green to yellow or red) hastens ripening of fruits and also increases size and weight of fruit in cultivar Medjool (Anon., 1991). Trial conducted at Abohar (Punjab) on Khadrawy date palm showed that spraying of 100-150 ppm GA at colour break stage enhanced ripening (Anon., 1998b).

Harvesting and Post Harvest Management

Date palm trees come into bearing after six years of planting. Initially, yield is less and it goes on increasing with the age of the trees. Fruit yield differs from variety to variety and also depends upon the age of the tree & orchard management. On an average, 50 kg of *doka* fruits per year can be harvested up to the age of 15 years and thereafter 75-100 kg per tree per year by using recommended package of practices. The dates are eaten at different stages of maturity depending upon the varieties and according to local demand, customs and climate. The bunches are harvested at full *doka* stage for fresh dates in those cultivars which are edible at *doka* stage. The fruits are separated from strands and graded. Shriveled, wrinkled, diseased and undersized fruits are discarded to make produce more attractive for the market. For the preparation of dry dates (*chhuhara*), the fruits are harvested at full *doka* and for soft dates (*Pind khajoor*), partial to full *dang* stage.

The post harvest losses in date palm fruits in India are as high as 32- 40 per cent due to heavy rains during its fruit maturity. The post harvest losses in date fruit can be minimized to a great extent by harvesting of date palm fruits at right stage of maturity. Determining the maturity standards of different varieties of dates is therefore, important for proper management, handling, harvesting, drying, packaging and storage (Plate 5.13).

Plate 5.13. Post arvest handling of date palm fruits at *Doka* stage (i) packing of separated berries in paper cortons (ii) sale of produce in the local marketing yard.

Harvesting stage influences the fruit weight, acidity, TSS, organoleptic rating and spoilage percentage of fruits. The weight of fruits in all the cultivars increased up to *doka* stage and then slightly decreased at *dang* stage. The total soluble solids in all the cultivars increased from *gandora* to *dang* stage, whereas acidity decreased. This study also revealed that for raw consumption of dates as well as for its better keeping quality, fruit should be harvested at *doka* stage. A positive correlation was observed between TSS and organoleptic rating, hence, it is suggested that TSS may be taken as an index of maturity of dates.

Storage studies of fresh *doka* fruits at Bikaner have revealed that at room temperature, fruits could not be stored for more than 4 days; whereas these could be stored up to 30 days under refrigeration and up to 50 days at freezing temperature. All fruits of Halawy turned to *dang* stage at freezing, whereas only 75 per cent *doka* fruits turned *dang* in refrigerator.

Processing and Value Addition

Date palm berries can be processed into dry dates and soft dates as commercial product. The berries can also be utilized for preparation of various value added products such as juice, pickle, date leathers etc.

For the preparation of dry dates (*chhuhara*), the full *doka* fruits are washed and dipped in boiling water for 5 to 10 minutes and dried either in air circulating oven at 48° to 52°C for 70 to 95 hours or through sun drying for 80-120 hours, if weather is dry. Variety Medjool has been found the best for preparation of dry dates. Soft dates (*Pind khajoor*) are prepared by dipping *dang* fruits in boiling water for 20 to 30 seconds and dehydrating then in the air-circulating oven, at 38-40 °C. Fully *dang* fruits can be converted into soft dates simply by drying them in air circulating oven. At Mundra (Gujarat), 60 seconds curing (dipping in boiling water) treatment was found better for making *pind khajoor* (Anon., 2000a). Beverage from local date fruits was prepared with ginger powder at Mundra and quality of juice was found good (Anon., 2000b). Halawy variety was found suitable for making dry dates (Anon., 2002d).

The juice and sugars can successfully be utilized as sweetening and flavouring agent in ice cream in bakery products. A satisfactory pickle can also be obtained after six weeks of pickling from green date fruits (*Chimri*) treated with 50 per cent sodium chloride and 2 per cent acetic acid and for use in broiler rations. There is possibility of utilizing date fruits harvested at *doka* or *khalal* stage for preparation of alcohol, vinegar and concentrated juice.

Future Research Needs

Western districts of Rajasthan such as Jaisalmer, Barmer, Bikaner and Jodhpur besides Nagaur, Churu and Sriganganagar being extremely dry, are considered to have the highest potential of date palm cultivation. Vast areas of above-mentioned districts can be effectively utilized for date palm cultivation including areas, where irrigation water is also saline. The plantation of date palm in western Rajasthan can also help in checking desertification and economic upliftment of rural masses by generation of employment besides giving net monitory income to the farmers. By adopting improved package and practices for its cultivation, net annual profit of approximately Rs. 75000 per hectare can be had from full bearing of date palm orchard.

Dates are high in quality due to their high sugar to acid ratio, self preserved fruit, good energy and mineral supplement, low moisture content, excellent skin integrity, natural dessert with minimum acidity and pleasant taste. Its paste is a natural extractable product and can easily be incorporated as fillers in bakery products. It can be considered as small and marginal cottage industry to prepare bakery, juice, pickles, and chutney to get some economic earning by recommended post harvest technology. Date growing has great prospect in India. The factors which decide the commercial possibilities of date palm growing are,

(i) Long dry summer and sufficient accumulation of heat unit for development and ripening of fruits.

(ii) Sufficient water resource for irrigation

(iii) Site-specific production technology suitable for different agro-ecological situations.

The heat units available at some locations in arid zone counted from the time of flowering (end of February) indicate that most of these regions of arid zone are suitable for date palm cultivation. For successful date maturation nearly 3000 heat units are required most of which accumulate by July in the arid zone. In the date growing countries with completely rainless period from February-September, date maturation up to *pind* stage takes about 180 days. The number of days available for fruit development, however is cut short to 125-150 days. In Western Rajasthan, dates growing for *doka* stage in most of the cultivars have the potentiality. Possibility of near complete fruit maturity on the tree has been indicated in best plantation in Jaisalmer district but commercial feasibility is to be exploited. Thus, the extremely dry areas comprising Jaisalmer, Barmer and western parts of Bikaner and Jodhpur districts seem to be best potential region for date growing in Rajasthan. In eastern part of Jodhpur, Bikaner, Barmer and western part of Nagour, Churu and Ganganagar districts, date fruits will have to be harvested at *dang* stage. Thus, a well thoughtout policy for date palm is essential on following aspects.

(i) The suitable plant material of varieties having the characteristics of early ripening, high yielding and high sugar content and resistant to rain and *Graphiola* should be imported, acclimatized, evaluated and promising types should be multiplied for commercial cultivation.

(ii) The plant material raised by tissue culture by even the private companies should be purchased and provided to the farmers for wider plantations.

(iii) The irrigation facilities under arid zone conditions must be developed on priority basis.

(iv) Date palm being tolerant to high soil salinity, its plantation can be promoted under IGNP area, where water table and salinity levels are high.

(v) Under waste land conditions of IGNP area, date palm can be grown through seedling. Though the quality of fruit will be poor to its parent and sex ratio will be the another problem but the plantation done under such conditions will reduce the degradation of soil, improve environmental conditions and it will also help in the development of cottage industries which ultimately will improve the economic condition of the rural mass.

Despite vast potential, the progress of research as well as area expansion in date palm has been very meagre. The following are the priority areas of work to support large scale cultivation of date palm in suitable region of the country.

(i) *Propagation technology:* For getting superior quality plant material, date palm may be propagated by offshoots or by micro propagation. The multiplication through offshoots is a natural phenomenon of the plant but the availability through natural development of shoots is very limited. Therefore, the offshoots may be initiated by using growth regulators, supplementary nutrition and providing sufficient irrigation, which require research support. For Micro propagation however, standardization of protocols and proper hardening of tissue culture plants before planting needs standardization.

(ii) *Improvement:* Selection of indigenous elite clones suitable for the regions and their multiplication and import of the elite varieties and their performance evaluation, acclimatization and selection should be taken on priority. Studies on the physiology of flowering and fruit set, breeding and physiological manipulation for delayed flowering for different uses including soft date preparation needs to be conducted.

(iii) The control of *Graphiloa* leaf spot through integrated management including manipulation of cultural practices, sanitation, host resistance and use of fungicide is required.

(iv) *Insect pest management:* Lesser date moth has been the newly recorded pest of the date palm, which damage the flowers and fruits of the date palm, therefore, the studies on behaviour of this insect and its control measures are also of prime importance.

(v) *Cropping system:* Date palm is a tall growing trees which has sufficient space between the rows, which can very well utilized for intercropping with other fruit crops, legumes, fodder and vegetables. If this system found suitable, it will increase the economic return to the grower from the same piece of land.

(vi) *Cultivation under marshy waste land:* There are sizable area with high water table and high salinity level in the country. Date palm being the salt tolerant and requiring high amount of water needs to be tried in such areas.

REFERENCES

Anon. (1991) Annual Report of AICRP on Arid Zone Fruits of Date palm Research Centre, Rajasthan Agricultural University, Bikaner.

Anon.(1998 a, b) Biennial Report. IX Group Workers Meet. AICRP on AZF, CCS, H.A.U., Hisar (Haryana).

Anon.(2000 a,b) Biennial Report. X Group Workers Meet. AICRP on AZF, Acharya N.G. Ranga Agril. Univ., Hydrabad (A.P.).

Anon.(2002 a, b, c, d) Biennial Report. XI Group Workers Meet. AICRP on AZF, Rajasthan Agricultural Univ., Bikaner (Rajasthan)

Chandra, Atul; Chandra, Anju and Gupta, I.C. (1992) Date Palm research in Thar Desert., Scientific Publishers, Jodhpur (Rajasthan).

Chandra, Atul; Swaminathan, R.; Chaudhary, N.L.; Manohar, M.S. and Pareek, O.P. (1990) Performance of dateplam cultivars in Thar desert. *Indian Journal of Horticulture,* 47 : 28-33

Chundawat, B.S. (1990) Arid Fruit Culture, Oxford and IBH Publishing Co., New Delhi.

Manohar, M.S. and Chandra, A. (1995) Date Palm culture in Rajasthan. Directorate of Research Rajasthan Agril. Univ., Bikaner.

Pareek, O.P. (1984) Date Palm growing potential of Indian arid zone. *Indian Horticulture.* 29(2-5): 8.

Pareek, O. P. and Vishal Nath (1996) Date Palm. In: Coordinated fruit research in Indian arid zone- Two decades profile NRCAH, Bikaner: 1-101.

Pundir, J.P.S. and Porwal, R. (1999) Performance of different date palm cultivars under hyper partially irrigated Western parts of Rajasthan (India). Proc. Int. Conf. Date palm held at U.A.E. University, Al-Ain (U.A.E.) on 8-10 March, 329-336.

Randhawa, G.S. (1980) A History of Agriculture in India. Vol. 1-4, ICAR, New Delhi.

Vishal Nath and R. S. Singh (2003) Boost up Date production in arid region. *Intensive Agriculture*. 41 (7-8): 3-6.

—x—

6

BAEL

Bael (*Aegle marmelos* Corr.) is an ancient native fruit of North India. The fruit has been known in India since the prehistoric era. It is a sacred tree dedicated to Lord Shiva and is mentioned in Vedic literature and Ramayana. Its tree is grown in almost every temple premises of Lord Shiva. It is very popular due to its medicinal and nutritional properties and it is regarded very beneficial in cure of diarrhoea and dysentery. All parts of this tree i.e., root, bark, leaves, flowers, fruits and seed oil are used in various *Ayurvedic* medicines. *Bael* fruit is one of the richest sources of Riboflavin. It contains 31.8 g carbohydrates, 1.8 g protein, 2.9 g fiber, 1.7 g mineral (mainly calcium and iron), 8-18 mg vitamin C; 0.03 mg riboflavin per 100g pulp. The pulp also contains mucilage which contains high level of Phenol. Its fruit and leaves are used to cure stomach disease and to improve digestive system. Fruits can be processed into various beverages and preserves. Wood is used to prepare various agricultural implements and as fuel. *Bael* plant has religious importance too.

Origin and Taxonomy

Bael is indigenous to India and known since ancient time. The fruit is mentioned in ancient literature viz., Vedas, Ramayana, Upavan Vinod, Brihat Samhita (Singh and Roy, 1984). Methods of ripening of the bael fruit are also given in Buddhist and Jain literature 800 BC - 325 BC. It has been mentioned that bael trees were found in the Chitrakut hills and Panchvati area during the Ramayana period.

Aegle marmelos Corr. belongs to the family Rutaceae. Other members of Rutaceae are *Citrus, Casimiroa, Clausena, Eremocitrus, Limonia, Feroniella, Fortunella, Poncirus, Triphasia,* etc. The generic name

Aegle is of Greek origin and the species *marmelos* is of Portuguese origin. The chromosome number is x=9 and 2n=36. The tree is a medium to tall, deciduous, slow growing and 5-10 m in height. Its leaves are green aromatic, trifoliate; branches have spines, and trunk is strong and stout. Mishra *et. al.*, (1999) have reported great variations in leaf characters, development pattern and shoot growth of 8 selected genotypes of bael under Tarai conditions of Uttaranchal. The bisexual flowers are nearly 2 cm wide and borne in clusters and greenish white in colour. The calyx is shallow with 5 short, broad teeth, pubescent outside. There are 5 petals, which are oblong oval, blunt, thick, pale greenish white, dotted with glands. Stamens are numerous, sometimes coherent in bundle. The ovary is oblong ovoid, slightly tapering, the axis being wide; cells are numerous 8-20, small and arranged in a circle, with numerous ovules in each cell. Botanically, the fruit is 'Amphisarca' which is a hard-shelled berry (Plate 6.1). Fruit shape of bael varies from round with flat ends to globosely oblong (Vishal Nath *et. al.*, 2003). Pericarp is smooth and hard with grayish yellow in colour. Seeds are numerous, arranged in seed cavity/sacs. Mucilage is present with seed in sacs/locules. The testa is white with wooly hairs. The embryo has large cotyledons (Reuther *et al.*, 1967 and Roy and Singh, 1979).

Plate 6.1. Mature fruits of Bael (Amphisarca) showing cross sectional view

Area and Distribution

There is no organized orcharding of bael in our country, hence exact data on acreage and production is not available. However, in recent years, some organizations have started the survey, collection, evaluation and establishment of germplasm blocks at their experimental stations. Some progressive farmers have also started planting bael varieties either in the form of small orchards or as boundary plantation of other fruits. As an estimate, about 27 ha area is under bael plantation with about 150 tones of production of improved varieties. Moreover, the fruits available mostly in the markets are harvested from the trees either naturally growing in the forest areas or from seedling plantations. Thus, the total production per year is estimated to be about one thousand tons. Bael is found in neighboring countries like Nepal, Sri Lanka, Pakistan, Bangladesh, Myanmar, Thailand and most of the Southeast Asian countries. In India, it is distributed throughout the country but concentrated area under bael is in Uttar Pradesh, Bihar, Madhya Pradesh, West Bengal, Orissa and Rajasthan. Apart from systematic orchard, bael trees are also planted in nutritional gardens, parks, temple gardens, roadsides for various purposes.

Soil and Climate

The bael tree is very hardy, it can thrive well even in saline, alkaline, and rocky soils having pH range from 5 to 10, where many other fruit trees fail to establish. It has wide range of adaptability to adverse soil and climate. The extent of hardiness of bael plants under Thar desert have also been observed that the plant even after being buried under sand for 2-3 month are capable of rejuvenating itself. It can tolerate salinity level up to 9 dsm^{-1} (Anon., 2001). However, well-drained sandy loam soil is ideally suitable for *bae*l orchards. It can grow easily in wasteland and in sandy soils of arid ecosystem, having low fertility status and poor moisture holding capacity. It can also be successfully grown in saline, sodic and sandy wastelands. In highly saline and sodic lands, plantations can be done successfully by treating the pits with gypsum and pyrite. *Bael* can tolerate the extremes of high temperatures by dropping its leaves in natural forests. However, young plants need to be protected from low temperature (7° C) and descicating summer winds. Marked reductions in the contents of leaf NPK and Ca were observed in response to increase in salinity and sodicity levels in the soil in which plants were grown. Salinity caused significant increase in leaf Mg, while sodicity decreased it. Leaf Na was at toxic levels in both saline and sodic soils.

Bael is a plant of dry tropical and sub-tropical region. However, the bael tree grows successfully and produces comparatively higher yield in sub-tropical climate where summer is hot and winter is mild. Further, the seedling *bael* tree is reported to be cold hardy is found to be grown up to an elevation of 1200 msl and is not affected by low temperature, but under arid conditions of Bikaner (Western part of Rajasthan), the leaves and tender twigs of improved budded *bael* cultivars were affected at a temperature <5 ^{0}C. However, plants tolerate high temperature and drought by shedding leaves during summer.

Plant Genetic Resource Management and Varieties

Bael plant is found naturally in Uttar Pradesh, Bihar, Jharkhand, Madhya Pradesh, Orissa, West Bengal and Chattisgarh with wide genetic variability. In Uttar, Pradesh, Deoria, Basti, Gonda, Faizabad, Sultanpur, Jaunpur, Pratapgarh, Mirzapur, Allahabad, Lucknow, Etawah, Agra are the districts where large number of promising genotypes of bael are available in natural condition or planted near the houses. Rai and Dwivedi (1992) and Vishal Nath *et. al.,* (2003) reported wide range of bael genetic diversity (Plate 6.2) in Northern and East Central India. Apart from the tree morphological characters,

Plate 6.2. Variability in Bael genotypes

wide variability exists in fruit size and shape, bearing habit, flesh colour, texture, fibre content, sugar content, mucilage content, etc. Some types have more number of seeds, gum locules and thick pericarp (Mishra *et al.*, 2000). However, some selections have been made at NDUAT, Faizabad and GBPUAT, Pantnagar, which are gaining popularity for commercial cultivation. At Central Institute for Arid Horticulture, Bikaner also collection of bael germplasm have been done, which are under evaluation. Besides this, some germplasm were also collected and maintained in the field repository at CAZRI, Jodhpur; NDUAT, Faizabad, GBPUAT, Pantnagar, HARP, Ranchi and TNAU, Arrupokottai. Recently, Lal (2002) evaluated 12 genotypes collected from Jaipur (Rajasthan) and found that 8 genotypes produce fruits of excellent quality. Table 6.1 explains the recently released/identified varieties of bael in India.

Table 6.1 : Bael varieties developed in India

Varieties/Cultivars	Organisations from where developed
Narendra Bael-5, Narendra Bael-7, Narendra Bael-9, Narendra Bael-16, Narendra Bael-17,	Narendra Deva University of Agriculture and Technology , Kumarganj, Faizabad. (UP)
Pant Aparna, Pant Urvashi, Pant Sivani, Pant Sujata	G.B.Pant University of Agriculture and Technology, Pant Nagar, Nainital (Uttaranchal)
Dhara Road	Central Arid Zone Research Institute, Jodhpur (Rajsthan)
CISH Bael-1, CISH B-2,	Central Institute of Sub-tropical Horticulture, Lucknow (UP)
Seedling selections with desired horticultural traits Thin skull, less seed, fiber and muscilage, high TSS, etc.	HARP, Ranchi, FRS, Basti, BAC, Sabour, MPUA T, Udaipur, CSAUAT, Kanpur and OUAT Bhubaneswar.

Out of the existing extensive biodiversity of bael in India, two distinct types *viz.* the small fruiting type having acrid pulp and high seed and mucilase and the large fruiting type with thin skull, less seeds and mucilage and more sweet pulp have been broadly identified. The former is used in medicinal preparations because of high marmelocin content while the later is used for preparation of *sharbat*, processed products and as a desert fruit (Singh and Vishal Nath, 1999). From the existing gene pool, some popular ecotypes have been selected in different variability regions. Notable among are Siwan in Bihar and Kagzi, Deoria Bada, Etawah, Gonda, Chakiya, Ayodhaya and Mirzapuri in Uttar Pradesh. Some recent selections are NB 5, NB 9, NB 4 and NB 7 at Faizabad, Pant Sujata, Pant Shiwani, Pant Aparna and Pant Suvarna at Pantnager and Dhara Road at Jodhpur. Vishal Nath *et al.,* (2003) and Pandey *et al.,* (2005) mentioned five improved seedling type bael germplasms (VN-DP 5/9 (IC 436483), VN-DP 5/ 12 (IC 436486), VN-DP 5/14(IC 436488), VN-DP 5/27 (IC 436501), and VN-DP 5/33 (IC 436507)) from Jharkhand and Bihar. Salient feature of some of these cultivars are given below:

Narendra *Bael* 5

This is an early maturing cultivar with average plant height of 3-5 m and canopy spread of 2.5-5.0 m^2 (Plate 6.3). Its medium sized (21 x 25 cm), round fruits with flat ends are sweet (25-27^0 Brix) and less seeded (40-50 per fruit). The average fruit weight range from 900-1000g (Plate 6.4).

Plate 6.3. Fruiting in Bael cv. NB-5

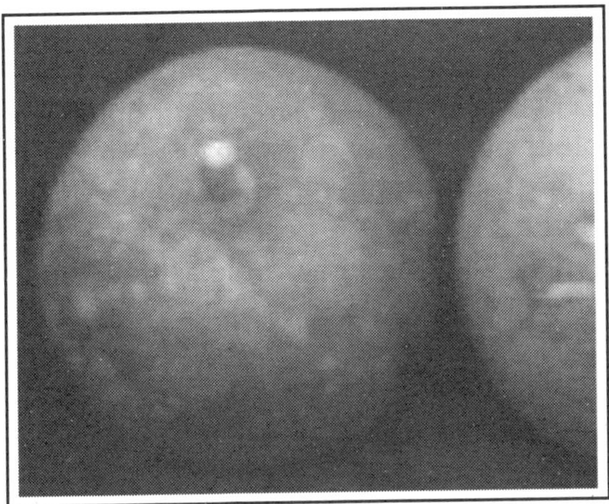

Plate 6.4. View of individual fruit of Bael cv. NB-5

Narendra Bael 7

This is a tall (5-7 m height) cultivar with medium (3-5 m^2) canopy spread. Fruit is round and sunken at both the ends and weighs more than 1500g. Its tree has been observed to have more vegetative growth in southern India.

Narendra Bael 9

Its trees are of medium height (4-6) and spread and produce large fruits (26x33 cm size) containing more pulp and less mucilage (Plate 6.5). TSS of fully ripe fruit range from 35 to 40^0 Brix and average fruit weight from 1000 to 1500 g.

Plate 6.5. Fruiting twig of Bael cv. NB-9

Pant Shivani

It is a mid season cultivar. Trees are tall, vigorous, dense, upright-growing, precocious and heavy bearer, fruit oblong in shape, weight 2 kg, lemon colour at maturity, medium fiber and mucilage content, rind medium thick, pulp light yellow in colour with very good taste and pleasant flavour. It has 69.% flesh, 34% TSS, total titortable 1.31% acidity and 15 mg / 100 g flesh ascorbic acid.

Pant Urvashi

It is a mid season variety. Trees are tall, vigorous, dense, upright growing, precocious and heavy bearer. Fruit is ovoid - oblong with average size 49.7 x 47.7 cm and weight 1.6 kg. Fruit is leman yellow colour, rind is medium thin and pulp is yellow. Fruit has 64% flesh with pleasant flavour. Seed and mucilage are medium, fibre content low, TSS 33%, acidity 1.15% and ascorbic and 12.6 mg / 100 g flesh.

Pant Aparna

Trees are dwarf with drooping sparse foliage, almost thornless, precocious and heavy bearer, leaves are large, dark green and pear shaped. Fruit has globose shape with average size and weight 0.6 - 0.8 kg. Fruit contains 34.1% TSS, 0.8%, total titratable acidity and 15 mg/ 100 g of ascorbic acid. Mucilage, seed and fibre contents are low. Flavour and taste of fruit is very good.

Pant Sujata

It is an early variety but has problem of fruit splitting. Trees are medium-dwarf with drooping and spreading foliage, dense, precocious and heavy bearer. Thorn are stout and bigger. Fruit is globose shaped depressed at both ends with average size 48.0 cm x 50.4 cm and weight 1.14 kg. Fruit and pulp are light yellow. Storage quality is better, rind is thin and seeds, mucilage and fibres are low. Its flavour is pleasant and taste is very good. Flesh content is 72%, TSS 30%, Acidity - 0.75% and assorbic acid 18 mg / 100 g flesh.

Plant Propagation and Nursery Management

Traditionally, bael is propagated by seeds, however on account of inherent limitations associated with the seedling progenies, seed propagation is limited only for the raising of rootstocks. For seedlings rootstocks, the seeds extracted from mature fruits preferably of seedling origin are sown in nursery bed. The seeds have no dormancy and therefore they germinate within 10-15 days after sowing. The young seedlings should be protected from intense radiation and frost. Generally, the freshly extracted seeds are used for sowing, however if required thse can be stored upto 132 days with proper treatment. For storage, three days after extraction, the seeds should be treated with fungicide such as capton, KH_2PO_4 and then dried back to their original weight and stored in alkathene bags at room temperature. Seeds can also be sown 12 days after treatment which gives more than 88 per cent germination. Stored seed also gives more than 90 % germination if treated with 1per cent KH_2PO_4 for 24 hour coated with Paste-1 (Agromin at 0.75 g 100 g^{-1} seeds + 100 ppm GA + 0.75 g KH_2PO_4 100 g^{-1} + 0.75 g KNO3 100 g^{-1}) (Hore and Sen, 1985).

No work on standardization of bael rootstocks has been done so far, which needs attention to assess vegetative compatibility and vigour, fruiting, fruit quality, pest and disease reactions, usefulness under degraded sites etc. of component plant. On these rootstocks bael can be propagated by budding and grafting. Generally, one year old seedlings are used for budding and grafting. To prepare budded plants, rootstocks are raised from seeds in polythene tubes or in nursery beds. 300 guage polythene tubes of 30x15 cm size, filled with FYM+clay+sand mixture (1:1:1 ratio), are placed in trenches of 6x1x0.3 m to raise the rootstocks. Raised nursery beds can also be used for raising rootstocks by sowing seeds at 20-30 cm apart. Sowing of seeds is done during June-July so that the rootstock attain the desired thickness for budding by next year. The most suitable period for budding is May-August under North Indian condtion (Kumar *et al.*, 1994).

The effect of sodicity on seed germination and seedling growth was studied by Pandey *et al.*, (1988). A pot experiment was conducted to study the effect of artificially maintained sodicity levels (18.5, 29.0, 38.0 and 49.5 ESP (Exchangeable Sodium Percentage)) on germination of seeds and growth parameters of bael seedlings (Shukla and Singh, 1996 a). Delayed and poor seed germination and reduced plant growth were observed in response to increased sodicity. However, satisfactory performance of bael with respect to seed germination and plant growth was observed in sodic soils up to 29.0 ESP without application of any chemical amendments. The media requirement for seed germination and seedling establishment has also been suggested by Chattopadhyay and Mahanta (1989).

The foliar spray of plant biological regulators, i.e. gibberellic acid (GA_3) and IBA (both at 250, 500, 750 and 1000 ppm), and potassium nitrate (250, 500, 750 and 1000 ppm) improves seedling vigour by means of improved growth of stem and top roots. Root and shoot growth studies indicates that shoot growth was maximum with foliar sprays of GA_3 at 1000 ppm, while the number of fibrous roots per seedling, length of tap root and dry weight of roots were found maximum with foliar sprays of IBA (1000 ppm). The shoot root ratio was found maximum with GA (1000 ppm) and minimum with IBA (1000 ppm). The correlation among most of the morphological traits was found positive and significant, but the correlation of the number of branches per seedling with dry weight of leaves; fresh weight of top with dry weight of leaves and stem; and dry weight of top with fresh weight of top, dry weight of leaves and stem, were more significant (Mishra and Jaisawal, 2001).

The orchard raised by seedlings is not true to type and exhibits variability. Therefore vegetative propagation techniques are recommended for commercial orcharding of bael. Among the vegetative techniques budding, grafting and suckers techniques are most successful.

Patch budding in the month of June or July proved to be the best. Singh (1954) reported that budding in *Aegle marmelos* in June with bud wood from one month old scion gave 80 per cent success. Moti *et al.,* (1976) also suggested patch budding in the month of June for propagation of bael. The effect of four scion genotypes (Pant Aparna, Pant Shivani, Pant Sujata, and Pant Urvashi) on patch budding of seedling rootstocks of bael was studied at G.B. Pant University of Agriculture and Technology, Pantnagar. Pant Aparna gave higher bud sprouting, lower unsprouted green and unsurvived buds, higher final survival, sprout length, number of leaves per budding and maximum sprout diameter while Pant Urvashi performed poorly in most of these characters. Pant Shivani had minimum sprout diameter while Pant Sujata had higher percentage of unsprouted green buds (Mishra and Jaiswal, 2001). Singh *et al.,* (1976) propagated bael by patch budding, T-budding at monthly intervals from July-October and from March-June. Bud take was 100 per cent with patch budding in June or July. Kumar *et al.,* (1994) reported that patch and modified ring budding showed more than 80 % success from mid of May to Mid of July under north Indian condition. Further the effect of genotype, method and time of budding was studied in *bael*. The results showed that more sprout length was found in Pant Aprana while leaf area was greater in Pant Shivani. Per cent survival and number of leaves per sprout did not differ significantly in genotypes. In all characteristics, patch budding excelled over 'T' and inverted 'T' budding. Among interaction, higher percentage of survival was recorded with patch budding performed on 30th June in Pant Aparna and patch budding on 8th July in Pant shivani (Rai *et al.*, 1999). Patch budding is ideal method with 90 per cent success rate and June-July is the best time for it (Chadha, 2001). Under arid conditions of Bikaner (Rajasthan), more than 90 per cent success was obtained through patch budding in July in poly bag raised rootstocks.

In arid regions, *in situ* budding is the most successful method for establishing bael orchard (Plate 6.6). This is done by sowing 2-3 seeds directly in the field or by planting seedlings. After one year, budding is done in the field. This method not only saves time but also ensures higher success. Generally 2-3 rootstocks sprouts developing from ground level are budded and ultimately only one of them is allowed to grow. Wild sprouts emerging from the rootstock portion should be removed from time to time. Top working of old and uneconomic trees can also be done by heading them back during March and budding on the new shoots during June-August.

Plate 6.6. Seedling plant of Bael ready for *in situ* budding

Maiti *et al.*, (1999) reported that bael can be propagated successfully by grafting and among different grafting methods tried; whip grafting had given good response (70% success). Recently softwood grafting has also been found successful with more than 80 percent success in bael under North Indian condition. It can also be grafted on the *Aegle fraeglegabonensis, Aeglopsis chevalier* and *A. paniculata.* However, it is not commonly used for raising of *bael* plants.

Root cutting and layers are also successful in bael. Separation and planting of root suckers can be done during monsoon. To ensure establishment, suckers are planted in nursery beds for about two years after uprooting and are then shifted to the main field. Some root suckers which, arise from the roots of the bael trees with profuse roots, can be separated during monsoon and directly transplanted into field. But, the success percentage is very low. It is not recommended for arid conditions.

Bael can also be propagated through cuttings and air layering on its own roots. Ray and Chaterjee (1996) reported that growth regulator and etiolation treatments were significantly effective in inducing roots in ringed stem cuttings of *A. marmelos*. An invigoration treatment (the production of water shoots following removal of large branches) accompanied by growth regulator and etiolation treatments significantly increased root quality of cuttings. The highest rooting rates of 75-80% were achieved by using 5000 ppm IBA, etiolation and invigoration. Air layering is very successful in bael (90% rooting and 77% survival of rooted layers) provided that mother trees are given invigoration treatment by heading back to few of the thick branches during April. Air layers are prepared in the second week of August by bark ringing and application of IBA at 10,000 pm in lanolin paste.

Recently, Micro-propagation techniques has also been found successful in bael. Arumugam and Rao (1996) reported that cotyledonary node explants, excised from 15-day-old seedlings of *A. marmelos* were placed MS medium supplemented with BAP [benzyladenine], IBA, IAA or NAA. BAP induced the best production of multiple shoots and subsequent plant regeneration. The highest number of shoots (75.2/explant) was observed on MS medium supplemented with BAP at 3 mg litre^{-1}. The number of shoots was further enhanced by (i) using nodal explants of in vitro regenerated shoots as microcuttings, and (ii) repeated subculture of the original explants on the same medium after excising the shoots. Several shoots were obtained from a single explant within 5 months. Regenerated shoots produced roots in 30% of cultures when transferred to a medium containing IBA at 4 mg litre^{-1}. Plantlets were transferred to soil, acclimatized and transplanted to the field. Rapid clonal multiplication of *Aegle marmelos* was achieved by enhanced auxiliary bud proliferation in young single-node segments of a 25-year-old tree cultured in MS nutrient medium. Bud break was dependent on cytokinin supply, but the synergistic combination of 2.5 mg 6-benzylaminopurine (BAP) [benzyladenine] and 1.0 mg IAA litre^{-1} induced the formation of 12.1 shoots of up to 5.2 cm in length in 48 per cent of the explants after 7 weeks of culture. Explants of in vitro-grown shoots (node, whole leaf, shoot tip and internodes) were subcultured in the presence of 0.05-2.5 mg BAP litre^{-1} to produce 11.3, 18.4, 5.3 and 3.2 shoots and shoot buds at a rate of 100, 70, 95 and 40 per cent, respectively, in 7 weeks. Different shoot nodes and leaves were equally regenerative and adventitious organogenesis in the latter was confined to cut petiole ends. Nodal explants responded most favorably at low BAP (0.05-0.1 mg/litre) and produced uniform (3.8-5.3 cm) shoots facilitating their simultaneous harvest for rooting. Repeated subculturing through five cycles of nodes and leaves of shoot cultures enabled continuous production of healthy callus-free shoots without any sign of decline. Shoot cuttings (3.0-5.2 cm) were best rooted in half-strength MS medium with 0.5 mg IAA/litre (70%) or 10.0 mg IBA/litre (90%). Some 85 per cent of the rooted plants were established in polybags after hardening (Ajithkumar and Seeni, 1998). Callus of *A. marmelos* was initiated from stem explants on MS medium supplemented with different concentrations of kinetin, 2,4-D and NAA. Meristemoids developed in the callus when subcultured on medium supplemented with 1 mg kinetin + 5 mg NAA litre^{-1}. In the presence of BA, alone or in combination with NAA, these callus showed shoot development. Multiple shoot induction from nodal explants was achieved on MS medium augmented with different concentrations of BA, kinetin and NAA. The shoot buds that developed from nodal explants were most numerous in medium supplemented with kinetin and NAA. Rhizogenesis of shoots was achieved in the presence of IAA (Varghese *et al.*, 1993).

A protocol for organogenesis from nucellar explants excised from fertilized ovules of immature fruits was developed. Adventitious buds were initiated on MS medium containing various combinations of benzyladenine (BA), NAA, IAA and gibberellic acid. Medium containing 4.4 μM BA and 2.7 μM NAA produced the greatest number of adventitious buds per explant. Shoots were elongated by transferring explants with shoot buds to medium with a low concentration of BA (0.44 μM). Rooting of in vitro-regenerated shoots was obtained in half-strength MS medium with 4.9 μM IBA. This is the first report of plant regeneration from nucellar explants of *A. marmelos* (Hossain *et al.*, 1993). Further, the protocol for organogenesis from leaf explants of in-vitro grown seedlings of *A. marmelos,* an adventitious buds were initiated on MS medium containing various concentrations and combinations of BA and IAA. The maximum frequency of shoot organogenesis (61.2%) and the highest number of shoots per explant (38.4) were obtained when 1.5 mg BA and 0.5 mg IAA/litre were used. Twenty-day-old seedlings proved to be the optimum source for explants. Shoots were elongated on the same basal medium without auxin and supplemented with 0.5 mg BA litre^{-1}. Elongated shoots were rooted on half strength MS medium supplemented with 0.1 mg IBA litre^{-1} (Islam *et al.*, 1993).

Planting

The planting of seedling bael is done at 10 x 10 m or 8 x 8 m depending upon the need and agro climatic conditions. Moreover, for vegetative propagated plants, 5 x 5 m spacing is sufficient. Though, planting space of 10 m has been recommended however, by development of dwarf rootstocks, spacing can be reduced accordingly (Singh and Roy, 1984). The pits of 2.5 x 2.5 x 2.5 ft. size and filled with top soil mixed with 10-15 kg well rotten manure and 50 g methyl parathion (2%) dust per pit is good for bael plantation. The ideal time of planting is July-August, however if irrigation facilities are available, it can be planted in February-March also. In arid region, *in-situ* planting is found suitable for orchard establishment.

No proper work has been done on planting geometry of bael orchard. In general, bael plantations are being done in square system. Seedling bael is also planted on the boundary of the orchard as wind break. While planting the *bael* in square system a spacing of 6x6 m and in rectangular system a spacing of 5x7 m has been recommended (Singh and Vishal Nath, 1999). Pits of 1x1x1 m size are dug during May-June which is kept open for 15 days. By the end of June, each pit is filled with a mixture containing top soil, 10-15 kg FYM, 100g diammonium phosphate (DAP) and 50g methyl parathion dust. In sodic soils, before filling, 5-10 kg gypsum should be added to each pit depending on the level of sodicity. The best time for planting is July. However, planting can also be done during February-March. After light irrigations must be done in the basin until the establishment of newly planted saplings. Bael can grow under rainfed condition if properly established with irrigations for 1-2 years. Timely irrigation, weeding, nutrient application, plant protection, etc. improve the establishment success. Shelter belts and windbreaks around the orchard protect the trees from hot and desiccating winds during summer and cool waves during the winter. For this 2-3 rows of fast growing, drought hardy tree species (*gonda, boradi, neem,* etc.) should be planted in staggered fashion. Newly planted saplings can be protected from hot and cool winds by close hedge/strip of fast growing plants like *dhaincha* and castor. Precautions must also be taken to protect from frost damage.

Training and Pruning

So far no work has been reported on training and canopy management of bael but training of young plant with clear trunk up to at least 1.0 m from ground level is essential in order to avoid difficulty in intercultural operations. The plants are trained to a straight central stem from which branches are not allowed upto 60-90 cm. Scaffold branches in all directions are permitted. Pruning for training in budded *bael* plants is usually not required because the branches of such plants are self oriented, however, in case of rosette growth, few branches should be clipped from its place of origin to have well spaced scaffolds. Bael plant is seldom pruned after start of fruiting. They do not require regular pruning also the fruiting takes place on older shoots. The trees are allowed to grow as such but dried, criss-cross branches, weak, diseased and broken twigs should be removed from time to time preferably after fruit harvest and before commencement of new flush when the plants are leafless during winter. Pruning of few growing branches becomes necessary, when plantation is raised under high-density planting for better management.

Orchard Management

Bael is a drought hardy fruit plant which requires least care after establishment. However, plants do responds to different orchard management operations during fruiting stage. Proper nutrient and water management and management of plant health and its normal physiology help improved fruit production from bael orchard.

Available information of various orchard management techniques have been reviewed and coleated to have a complete package of orchard management of bael plantation in the marginal and waste land.

Nutrient Management

No systematic work has been done on nutritional requirement of *bael* tree. However, growers apply some organic manure during rainy season. In wastelands, addition of organic matter will make cultivation successful. Application of manure is done during onset of rainy season according to the age of the plants. It is suggested to apply 10 kg FYM and 50:25:25g N, P, K /plant/year respectively. This should be increased every year in the same proportion up to the age of 10 years. Half dose of N and full dose of P and K should be applied in the beginning of flowering. However, remaining dose of N should be given in August-September. The fertilizers should be applied in and around active root zone areas. Singh and Mishra (2000) have reported seasonal variation in root distribution pattern of *bael*. Provision of green manuring has special significance for bael plantations established under degraded lands.

The effect of sodicity on mineral composition, chlorophyll content and injury symptoms in bael was studied by Pandey *et al.,* (1985). Shukla and Singh (1996 b) reported reductions in the contents of leaf N, P, K and Ca were observed in response to increase in salinity and sodicity levels in the soil in which plants were grown. Salinity caused significant increases in leaf Mg, while sodicity decreased it. Leaf Na was at toxic levels in both saline and sodic soils.

Doses of manures and fertilizers for *bael* depends on the fertility level of the soil. In sandy loam soils, besides 10-15 kg FYM, an annual dose of 100g nitrogen, 50 phosphorus and 75-100 g potash per plant should be given. The doses should be increased upto 7 years of plant age. After that fixed doses of manure and fertilizers are applied. July-August is the best time for the application of manures and fertilizers. If there is a drought spell, plants must be irrigated once after fertilizer application. Plant growth of *bael* is adversely affected by the deficiency of some micronutrients. Zinc efficiency is common in sandy soils of arid region. Three foliar sprays with 0.6% mixture containing zinc sulphate, borax and ferrus sulphate in equal proportion during July, October and December has been found beneficial.

Water Management

Since, the plant is very hardy in nature and once established can withstand even under intermittent drought conditions regular irrigation of plants are not required, however, young plants needs to be watered time to time until their establishment. For initial establishment 5-6 irrigations per year are considered enough. During summer period, when trees shed their leaves, requirement of water is considerably reduced because of dormant phase. In tropical and subtropical parts of the country, no seedling trees are irrigated. However, under arid conditions proper moisture and irrigation management are essentially required to make bael orchard productive and profitable. Maintenance of optimum soil moisture regime will also minimize the problem of fruit drop, cracking and damage of plants from frost. Use of mulches and inward sloping of tree basins also help increased soil moisture. Under semi-arid conditions of Agra (Uttar Pradesh), runoff collected during rainy season by developing micro-catchments had given good response in improving soil moisture regime for establishment of bael plants. Productivity and quality of fruits, however, are increased in arid region if irrigation is done during summer months. At least two irrigations in winter and three irrigations at fruit development stage are essentially required under sandy desertic conditions of hot arid region. Provision of microcatchment around the tree trunk for runoff concentration and mulching with the locally available organic waste or polythene materials conserved more moisture which leads better yield under rainfed condition.

Orchard Floor Management

During the early age of orchard establishment, inter space between tree rows can be utilized for raising suitable crops which not only enrich the soil but also generate additional income. In arid and semi-arid region, clusterbean and *moth* bean have been found beneficial and remunerative. Under irrigated conditions, however, high value crops like vegetables can be successfully grown.

Disease Management

Being a hardy plant, it does not require much protection but some diseases do infest the leaves causing leaf fall. **Bacterial blight** (canker) caused by (*Xanthomonas* sp.) produces brown spots on leaf surface resulting in leaf fall in young plantations. The disease can be effectively managed by 2-3 sprays of 500 ppm streptomycin at 15 days interval. Powdery mildew has also been noticed in the humid area. 2-3 sprays of sulphur 2 g/lit or karathane 2 ml /lit effectively controls the disease.

Pest Management

There is no serious pest in bael. However, **termite** attack in newly planted *bael* saplings is a serious problem. Application of chlorpyriphos @ 4-5 ml per plant in 10-15 liters water has been found effective to manage the termite. **Chafer beetle** or leaf eating caterpillars observed on *bael* plants can be controlled by 2-3 sprays of 0.01% monocrotophos at 15-20 day interval.

Physiological Disorders

Under hot arid conditions, the fruit cracking has been observed as a major physiological problem in bael cultivation. The cracking takes place twice in a year i.e i) during winter season (December-January) when developing fruits are immature and ii) during summer season (March-April) when fruits are mature and in ripening phase. The cracking at later phase (summer cracking) is more severe than former phase (winter cracking). The cracking can be minimized by maintaining optimum soil moisture regime and provision of windbreaks around the orchard. Fruit drop is another problem in bael orchards. Saini *et al.,* (2004) suggested that the fruit drop and cracking in bael can be minimized by either spraying of 0.1% borax or application of 100kg FYM/ tree as basal dose.

Problem of immature **fruit drop** (at pea stage) has also been observed which can be effectively minimised by adopting better orchard management practices and spraying of growth hormones like planofix (2 ml / 5 lit water) at pea stage. **Internal rotting** of mature fruits in the market is a serious problem which is mainly due to damage to the fruit during harvesting or due to the harvesting of unripe fruits. To avoid the problem, only fully mature fruits should be harvested.

Flowering, Fruiting and Yield

Vegetatively propagated *bael* tree starts bearing after 4-5 years of planting while the seedling plants take 7-10 years to commence fruiting. Commercial yield in budded plants comes after about 10 years of planting. In north India, the tree flowers during March-April and the fruits are harvested during February-March. Initially, fruit development is slow for about a month, which is followed by rapid development for next four month and thereafter the growth remains more or less static until the fruits are harvested. *Bael* is a climacteric fruit. Under north Indian conditions, fruit takes 8-10 months to mature and 10-12 month for ripening after fruit set. Under such condition maturity takes place in December-January and ripening in March-April, continuing up to June, because of low temperature prevailing during December-February. In south India, fruit starts ripening in February and the harvest extends up to April. Stray fruit can also be seen on the tree during rest of the period. When the fruit changes its colour from green to yellow during April-May, they should be harvested. A fully developed *bael* tree produces 200-800 fruits weighing between 2-4 quintal.

Harvesting and Post Harvest Management

Fully mature fruit slips from peduncle which indicates the correct stage of harvesting. Mature fruits along with about 5 cm stalk are harvested with the help of a fruit picker. Fruit skull becomes thick and stony on ripening. Ripe fruits are packed in gunny bags or wooden crates with dry grass as cushioning material to avoid cracking. Grading is done on the basis of size and shape of fruit. Fruits can be stored at 30 ± 2^0C for 2 weeks. However, this period can be increased up to 12 weeks if storage is done at 9 ± 1^0C.

Processing and Value Addition

Ripe fruit contains amber or honey coloured viscous sticky of glutinous, translucent sweet and fleshy aromatic pulp which is generally used for processing. Mature fruits can be processed into candy and preserve after slicing whereas various beverages (nectar, squash, RTS), cider, etc. are prepared from ripe fruit pulp (Plate 6.7) (Srivastava and Kumar, 2002) following the standarderd method as in case of jammun and phalsa. The bael cider is considered to be one of the important product having more herbal value. The quality bider from bael pulp can be prepared by the fermentation and staining process as suggested by Srivastava and Kumar (2002). Srivastava and Kumar (2002) described the recipies methods and procedures for developing various value added products from bael fruits. They described that the pulp should be extracted from ripe fruits involving the technique given in Fig. 6.1 for preparation of various beverages from bael fruit.

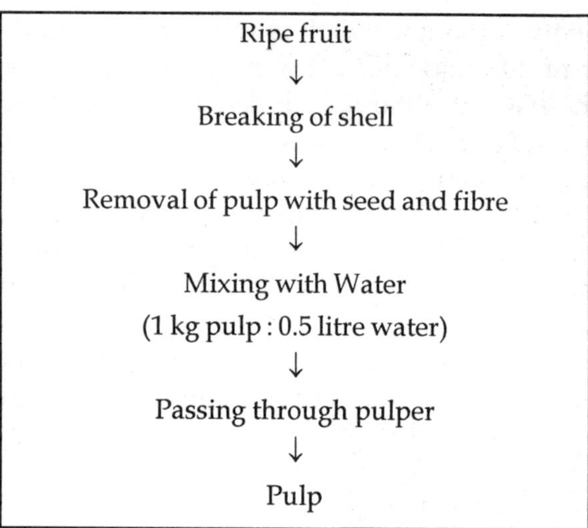

Ripe fruit
↓
Breaking of shell
↓
Removal of pulp with seed and fibre
↓
Mixing with Water
(1 kg pulp : 0.5 litre water)
↓
Passing through pulper
↓
Pulp

Fig. 6.1. Flowsheet for extraction of bael pulp

Plate 6.6. Value added products from ripe bael pulp. (i) RTS (ii) Squash

Future Research Needs

Bael is a fruit which is attaining importance due to its suitability to waste lands and high medicinal properties. Still there is research gap which comes in the way of its wide spread commercial cultivation. Bridging these research gaps will help increased bael production in the country. Some of the gaps has been identified.

1. A rich genetic diversity are available through out the country especially in the states of U.P., Bihar, Uttaranchal, Jharkhand, Chhatisgarh, M.P., etc. which needs to be exploited for selection of better genotypes.

2. Fruit drop and cracking of *bael* fruit are the major problems of cultivation. To overcome this problem, suitable genotypes should be identified with high yield potential and better quality fruits.
3. Development of low seeded variety is necessary to lure this fruit.
4. More emphasis on post harvest technology should be given for value added and export oriented processed products. Establishment of small scale processing units should be promoted.
5. Screening of genotypes/cultivars for problematic soils is also essential.
6. Development of complete production technology for *bael* fruit should be developed for various agro climatic zones.
7. Bael based farming systems should be developed for high income and proper land utilization.

REFERENCES

Ajith Kumar, D. and Seeni, S. (1998) Rapid clonal multiplication through in vitro axillary shoot proliferation of *Aegle marmelos* (L.) Corr., a medicinal tree, *Plant Cell Reports*, 17(5): 422-426.

Anon., (2001) Annual Report, AICRP on AZF, CIAH, Bikaner.

Arumugam, S. and Rao, M.V. (1996) *In-vitro* production of plantlets from cotyledonary node cultures of *Aegle marmelos* (L.) Corr. *Advances in Plant Sciences*, 9 (2): 181-186.

Banerjee, N. and Maiti, M. (1980) Some physico-chemical studies on a protein rich fruit seed (*Aegle marmelos* Correa). *Journal Indian Chemical Society*, 57(9): 945-46.

Bhadra, S. and Sen, S.K.(1998) Extension of storage life of bael (*Aegle marmelos* Correa.) var. Kalyani Selection-1. *Environment and Ecology*, 16 (3): 699-703.

Bhadra, S., Chakraborti, K. and Sen, S.K.(2000) Developmental physiology of bael (*Aegle marmelos* Correa) fruit var. Kalyani Selection-1, *Environment and Ecology*, 18(2): 470-471.

Bhargava, S.N., Shukla, D.N. and Singh, A.P. (1977) Stalk end rot of *Aegle marmelos* a new disease. *Indian Phytopathology*, 30(1):120-121.

Chadha, K. L. (2001) Bael. In: *A Handbook of Horticulture*, ICAR, New Delhi.

Chattopadhyay, R.K. and Mahanta, S.K. (1989) Media requirement for seed germination and seedling establishment. *Indian Horticulture*, 36(3): 27.

Chakrabarty, B.P., Sharama, B., Jain, K.L. and Prasad, C.K.P. (1984). A bacterial leaf spot of bael (*Aegle marmelos* Correa.) in Rajasthan and a revived name of the bacterium. *Current Science*, 53(9): 488-489.

Deepti Rrai., Misra, K.K. and Singh, V.P. (2002) Analysis of genetic divergence in bael (*Aegle marmelos* Correa) germplasm. *Progressive Horticulture*, 34(1): 35-38.

Garg, R.C., Joshi, B.P., Ram, H.B., Jain, S.P. and Tripathi, V.K. (1977) Effect of stage of maturity on the quality of bael fruit (*Aegle marmelos* Correa) and its preserve. *Progressive Horticulture*, 9(1):19-26.

Gopalan, C. B. M., Ramashastri, B. V., and Balasubramanium, S. C. (1985) Nutritive value of Indian foods, ICMR, Hyderabad.

Ghosh, D.K. and Gayen, P. (1990) Metroglyph grouping and association analysis on physical characters of bael fruits (*Aegle marmelos*), *Progressive Horticulture*, 22 (1-4): 6-11.

Haskar, C.N. and Kendurkar, S.G. (1961) Uses of bael gum. *Printindia*, 11(1): 135.

Hossain, M., Karim, M.R., Islam, R. and Joarder, O.I. (1993) Plant regeneration from nucellar tissues of *Aegle marmelos* through organogenesis. *Plant Cell, Tissue and Organ Culture*, 34: (2): 199-203.

Hore, J.K. and Sen, S.K (1985) Effect of seed treatments on longevity of bael (*Aegle marmelos* Correa.) seeds. *Haryana Journal Horticulture Sciences*, 14 (3-4): 204-210.

Islam, R., Hossain, M., Joarder, O.I.and Karim,M.R. (1993) Adventitious shoot formation on excised leaf explants of in-vitro grown seedlings of *Aegle marmelos* Corr., *Journal Horticulture Science*, 68(4): 495-498.

Jain, N.K. (1977). Antifungal activities of *Aegle marmelos* Correa, *Indian Journal Microbiology*, 17(1): 51.

Jaisawal, H.R. and Misra, K.K. (1996) Clonal variation in inflorescence and floral morphology of bael (*Aegle marmelos* Correa). *Progressive Horticulture*, 28(3-4): 119-123.

Jaisawal, H.R. and Misra, K.K. (1998). Studies on fruit set, retention and yield of bael (Aegle marmelos Correa) clones under Tarai conditions. *Progressive Horticulture*, 30(3-4): 164-167.

Kore, S.S. and Dhande, G.W. (1973) A new leaf spot of bael (*Aegle marmelos* Correa.) caused by *Fusarium roseum* Link from India, *Punjabrao Krishi Vidyapeeth Res. Journal*, 1(2): 223-224.

Kumar Dinesh, Pathak, R.K. and Ali, W. (1994).Studies on the effect of duration and methods of budding in bael.*Indian Journal Horticulture*,15(2):150-153.

Lal, G. (2002). Evaluation of bael (*Aegle marmelos* Correa) germplasm in semi-arid regions of Rajasthan. *Current Agriculture* Vol. 26 (1-2): 127-1129.

Madan, R.L. and Gupta, P.C. (1985) A leaf spot disease of *Aegle marmelos* L. caused by *Alternaria alternata* (Fr.) Keissler, *Indian Journal Plant Pathology*, 3 (2): 239.

Maiti, C.S., Nath, A. and Sen, S.K. (1999) Studies on the propagation of bael (*Aegle marmelos* Correa.) by different grafting methods in West Bengal, *Journal Applied Horticultural*, 1(2): 131-132

Misra, K.K., Sigh, R. and Jaisawal, H.R.(1999) Studies on leaf characters, development pattern and shoot growth in bael genotypes. *Progressive Horticulture*, 31 (3-4): 144-150.

Misra, K. K., Singh, R. and Jaiswal, H.R. (2000) Performance of *bael* (*Aegle marmelos*) genotypes under foothills regions of Uttar Pradesh. *Indian Journal Agriculture Science*, 70 (10): 682-683.

Misra, K.K and Jaisawal, H.R.(2001) Effect of scion genotypes on patch budding in bael (*Aegle marmelos* Correa). *Advances in Horticulture and Forestry*, 8: 61-66.

Misra, K.K and Jaisawal, H.R.(2001). Effect of plant bioregulators and potassium nitrate on seedling quality of bael (*Aegle marmelos* Correa). *Advances in Hort. & Forestry*, 8: 67-74.

Moti, Dhar, L. and Chaturvedi, O.P. (1976) Propagation of some tropical and subtropical fruits by budding. *Punjab Horticulture Journal*, 16 (1-2): 33-38.

Pal, P.K., Bhattacharya, D.K. and Ghosh, S. (1993) Chemical examination of seeds of *Aegle marmelos* (Bael fruits) and its medicinal importance. *J. Oil Technologists Association* India, 25(3): 55-59.

Pandey, D., Pathak, R. K. and Pramanik, P. K. (2001) *Shri Phal Bael. Udyan Rashmi* (In Hindi), 2(2):13-18.

Pandey, D.; Vishal Nath, S. K. Shukla and R. K. Pathak (2005) Promising bael genotypes. *Indian Horticulture*. 50 (1): 21.

Pandey, G. Pathak, R.K. and Om, H. (1985) Effect of sodicity on mineral composition, clorophyll content and injury symptoms in bael., *Progressive Horticulture*, 17(1): 5-8.

Pandey, G. Pathak, R.K. and Om, H. (1988) Effect of sodocity on seed germination and seedling growth in bael (Aegle marmelos), *Indian Journal Agrilculture Sciences*, 58(7) 577-578.

Prakash, A., Pasalu, I.C. and Mathur, K.C. (1983) Use of bael (*Aegle marmelos*) leaf powder as protectant against storage pests of Paddy. *Agrilculture Research Journal Kerala*, 21(2): 79-80.

Rai, M. and Dwivedi, R. (1992) Bael (*Aegle marmelos* Correa) diversity in India: *an Underutilized fruits of medicinal value, IBPGR Newsletter*, 10:13-14.

Rai, M., Dwivedi, R. and Gupta, P. N. (1991) Variability and potentials of identified germplasm of *bael* (*Aegle marmelos* correa.). *Indian Journal Plant Genetic Resources*, 4(2): 86-92.

Ram, D. and Singh, I.S. (2003) Physico-chemical studies on bael (*Aegle marmelos* Correa) fruits. *Progressive Horticulture*, 35(2): 199-201.

Ray, D.P. and Chaterjee, B.K. (1996) Effect of different concentrations of growth regulators, etiolation and vigoration treatments on the rooting of stem cutting of bael (*Aegle marmelos* Correa.) *Orissa Journal Horticulture*, 24(1-2): 36-41.

Reuther, W., Webber, H.J. and Bachlor, L.D. (1967) The *Citrus* industry, Vol. I, Pub. Division of Agriculture Sciences, University of California. Pp. 407-409.

Roy, S.K. and Singh, R.N. (1979) Preliminary studies on storage of *bael* fruits (*Aegle marmelos*). *Progressive Horticulture*, 11:3

Roy, S.K. and Singh, R.N. (1981) Studies on ripening of bael fruits (*Aegle marmelos*). *Punjab Horticulture Journal*, 21(122): 74-82.

Roy, S.K., Pandey, R.M. and Singh, R.N. (1972) Preliminary studies on the availability of *bael* fruit (*Aegle marmelos*) in off-season. *Indian Journal Horticulture*, 29 (3-4): 344-347.

Saini, R.S., Surender Singh and R.P.S. Deswal (2004) Effect of micronutrients, plant growth regulators and soil amendments on fruit drop, cracking, yield and quality of bael (*Aegle marmelos* Correa) under rainfed conditions. *Indian Journal Horticultural*, 61(2): 175-176.

Shukla, S.K. and Singh, G.N. (1996 a) Seed germination and growth of bael (*Aegle marmelos* Correa) seedlings as influenced by soil sodicity, *J.Tropical Forestry*, 12(4): 242-245.

Shukla, S.K. and Singh, G.N. (1996 b) Nutrient imbalances in bael (*Aegle marmelos* Correa) leaves as induced by salinity and sodicity, *Indian J. Plant Physiol.*, 1(4): 293-294.

Singh, K.V., Bhatt, S.K. and Sthapak, J.K. (1983) Antimicrobial and anthelmintic properties of the seeds of *Aegle marmelos. Fitoterapia*, 54(6): 261-264.

Singh, L. B. (1954) Propagating bael (*Aegle marmelos*) vegetatively. *Scince and Culture*, 19: 485-486.

Singh, A.K. and Misra, K.K. (2000) Seasonal variation in root distribution pattern of *bael* (*Aegle marmelos*). *Indian Journal Agriculture Science*, 70(10): 717-718.

Singh, R. N. and Roy, S. K. (1984) *The Bael*. I.C.A.R., New Delhi, p.25.

Singh, R. S.and Vishal Nath (1999) *Anupayogi Bhoomi mein – Upayogi Bael Vriksh Lagayein. Unnat Krishi*, 38 (3):3-8.

Singh, U. R., Pandey, I. C., Upadhaya, M. P. and Prasad, R. S. (1976) Propagation of bael (*Agele marmelos*) by budding. *Punjab Horticultural Journal*, 16:57-59.

Shrivastava, R.K.(1982) A note on phenological observations on *Aegle marmelos* Correa. *Indian Forester*, 108(80: 593-595.

Srivastava, K.K. and Singh, H.K.(2000) Floral biology of bael (*Aegle marmelos*) cultivars. *Indian Journal Agriculture Sciences*, 70(11): 797-798.

Srivastava, R. P. and Kumar, S. (2002) Fruits and vegetable preservation-Prinicipal and practices. 3rd Edition, IBDC, Lucknow.

Tokitoma, Y., Shimono, Y., Kobayashi, A. and Yamanishi, T. (1982) Aroma component of bael fruits (*Aegle marmelos* Correa.). *Agriculture Biological Chemistry*, 46(7): 1873-1877.

Varghese, S.K., Inamdar, J.A., Kiran Kalia, Subramanian, R.B., Nataraj, M. and Kalia, K. (1993) *Micropropagation of Aegle marmelos* (L.) Corr. *Phyto-morphology*, 43: 1-2, 87-92.

Verma, J.P. and Ahmed, S. (1958) Drying of bael (*Aegle marmelos* Correa) pulp. *Indian Food Packer*, 12(7): 7-9.

Vishal Nath, D. Pandey and B. Das (2003) Diversity of bael (*Aegle marmelos* Corr.) in East Central India. *Indian Journal of Plant Genetic Resources*. 16 (2): 85-89.

—x—

7

TAMARIND

Tamarind (*Tamarindus indica* L.) is found throughout the tropics, where it is grown unattended in backyards, roadsides or wastelands. However, these unimproved trees are commercially exploited in all producing countries. Tamarind is ideally suited for avenue plantings as a roadside tree, in and around villages as a multipurpose species for agro-forestry systems or as protective firebreaks for forest margins. It is valuable timber species widely used for making furniture, tool handles, charcoal, oil mills, rice ponders and fuel wood. The leaves are an important source of food and herbal medicine and the edible pulp of ripe fruits is used as a flavouring agent in cooking, soups, jams, chutneys, sauces, juices, etc. The fruit pulp of tamarind is the richest natural source of tartaric acid (8-18%) and is the chief acidulate used in the preparation of foods in India and many other Asian countries. The major industrial product of tamarind is the seed kernel powder (TKP) which is an important material used in sizing of textiles, paper and jute. Other industrial products are tamarind juice, tamarind juice concentrate, tamarid powder, pickles and paste (Shankaracharya, 1998). The seeds are gaining importance as a rich source of protein with favourable amino acid malnutrition widespread in many of the developed nations. It has also potential for substituting at least 30% of cereals in live stock rations.

Tamarind has an exceedingly wide range of domestic and industrial uses; yet this important tree remains under-exploited. In spite of its potential, which is well recognized, the unimproved wild trees are continuously exploited to meet the growing commercial demand. A tree that is easy to cultivate, free of any serious pests and diseases, with a long life span, deserves more focused research attention than it has so far received.

Tamarind is a nutritious fruit with a variety of uses. Tamarind has many valuable properties and virtually every part of the tree has been utilized by both rural and urban dwellers. Fruit pulp is the most valuable and commonly used part of the tamarind tree. The pulp constitutes 30-50% of the ripe fruit (Purseglove, 1987; Shankaracharya, 1998), the shell and fibre account for 11-30% and the seed about 25-40% (Chapman, 1984; Shakaracharya, 1998). Table-7.1 describes the average composition of tamarind fruit.

Table 7.1 : Average composition of tamarind fruit

Constituents	Amount (per 100g)
Water	17.8-35.8g
Protein	2-3g
Fat	0.6g
Carbohydrates	41.1-61.4g
Fibre	2.9
Ash	2.6-3.9g
Calcium	34-94mg
Phosphorous	34-78mg
Iron	0.2-0.9mg
Thiamine	0.33mg
Riboflavin	0.1mg
Niacin	1.0g
Vitamin C	44mg

Source : Coronel (1991); Feungchan *et al.*, (1996 a).

Fruit composition is variable, depending on locality (Table 7.2). Tamarind has low water content and a high level of protein, carbohydrates (60-72%) and minerals. The soluble solids content varies from 54-69.9^{0} Brix (Baragano de Mosqueda, 1980; Benero *et al.*, 1974). Ishola *et al.*, (1990) have reported that the pulp of Nigerian tamarind is poor in proteins (87.9 g/ka) and oil (25.3 g/ka). The oil which is greenish in colour and liquid at room temperature has high saponification and the low iodine value.

Table 7.2 : Proximate composition and food energy value of ripe tamarind

Constituents	Reports by various Workers/Organisation				
	A	B	C	D	E
Water	20.60	15.00-47.00	22.60	33.89	62.50-69.20
Protein	3.10	14.00-3.40	3.10	3.28	1.40-3.30
Lipids			0.40	0.50	0.71-0.81
Fat	0.40	0.90-1.00	0.40	0.50	
Carbohydrate (30-40% reducing sugars)	70.80	62.50	71.80	59.76	
Total sugars					21.40-30.85
Fibre	3.00	5.10	3.00	1.79	
Cellulose					1.80-3.20
Pentose					4.20-4.80
Total acidity					17.10-18.40
Free tartaric acid					8.40-12.40
Ash	2.10	1.50-4.20	2.10	2.57	1.16-1.72
Energy (k cal / 100g)			272.0	230.0	
Calcium		0.07			
Phosphorous		0.11			

Source : (a) Purseglove (1987); (b) FAO (1988); (c) Leung and Flores (1961); (d) Wenkam and Miller (1965); (e) Hasan and Ijaz (1972).

The most outstanding characteristic of tamarind is its acidity, which is due to tartaric acid (2,3-dihydroxybutanedioic acid, $C_4H_6O_6$, a dihydroxy-dicarboxylic acid), ranging from 12.2-23.8%, which is uncommon in other plant tissues (Ulrich, 1970). It is an unusual plant acid, which is formed from the primary carbohydrate products of photosynthesis and once formed, canned be further used in the plant due to the absence of the necessary enzymes. The tartaric acid is synthesized in tamarind leaves in the light and translocated to the flowers and fruits (Lewis *et al.*, 1961; Patnaik, 1974). It is high in young leaves and decreases with age and has been reported to show seasonal variations (Bueso, 1980). As reported by Lewis and Neelkantan (1964 b) the tartaric acid content in leaves decreased from 28% to 12%

suggesting that this is due to its transfer to the fruit during ripening. The content of tartaric acid, however, does not decrease during fruit ripening, indicating that it is not utilized in fruit development; but during this time, reducing sugar, increases up to 30-40% giving the sour fruit a sweeter taste. As the acidity does not disappear with ripening, but is more or less matched with increasing sugar levels, tamarind is known to be simultaneously the most acidic and sweetest fruit (Lewis and Neelakantan, 1964 a; Coronel, 1991).

In general, the dried tamarind pulp of commercial importance contains 8-18% tartaric acid (as potassium bitartarate) and 25-45% reducing sugars of which 70% is glucose and 30% fructose. Lewis and Neelakantan (1964 a) reported that one half of the tartaric acid was present as potassium bitartarate (cream of tartar) and to a lesser extent as calcium tartarate. The tender fruits contain most of the tartaric acid in free form (up to 16%), which can be easily extracted with hot water. The ascorbic acid content in tamarind is very small and varies 2-20 mg/100g (Ishola *et al.*, 1990; Lefevre, 1971). The other organic acids reported in tamarind fruit are oxalic acid, succinic acid, citric acid and quinic acid (Lewis and Neelakantan, 1962 a; Singh, 1973; Anon, 1976).

There is wide variations in the tartaric acid and sugar contents of tamarind pulp across the world. In Pakistan, Hasan and Ijaz (1972) found that in sour tamarind the tartaric acid content varied from 8.4 to 12.4% and the sugar from 21.4 to 30.9%. Sweet tamarind is found in Thailand where the tartaric acid content varied from 2.5 to 11.3% and the sugar from 5.0 to 40.0%. According to Feungchan *et al.* (1966 e), the tartaric acid content of sweet tamarind in Thailand was as low as 2.0 to 3.2% and the sugar was as high as 39.1 to 47.7%. The sweetness of the pulp is determined by the sugar content and the sugar content of the sweet tamarind types in Thailand varies between 39.06 to 47.71%, whereas the sugar content of sour tamarind is only 4.80 to 38.94%. About 55% of the total nitrogen in the tamarind pulp is non-protein N and 70% of this is contributed by free amino acids (Lakshminarayan Rao *et al.* 1954). The free amino acids like Proline, Serine, Beta-alanine, Phenylalanine, and Leucine have been identified in the tamarind pulp. Tamarind pulp is also rich in minerals; high in potassium (62-570 mg/100g); phosphorus (86-190 mg/100g); and calcium (81-466 mg/100g), and a fair source of iron (1.3-10.9 mg/100g). It also excels in riboflavin and is a good source of thiamin and niacin, but is poor in vitamin A and vitamin C (Leung and Flores, 1961).

The tamarind seed comprises 20-30% seed coat and 70-75% kernel (Coronel, 1991; Shankaracharya, 1998). It is the raw material used in the manufacturing of tamarind seed kernel powder (TKP), polysaccharide (jellose), adhesive and tannin. The seeds are also used for other purposes and are presently gaining importance as an alternative source of protein and some essential amino acids. Unlike the pulp the seed is a good source of protein and oil. Whole tamarind seed and kernels are rich in protein (13-20%) and the seed coat is rich in fibre (20%) and tannins (20%) (Table-7.3). Panigrahi *et al.*, (1989) reported that whole tamarind seed contains 131.3 g/kg crude protein, 67.1g/kg crude fibre, 48.2 g/kg crude fat, 56.2 g/kg tannins and trypsin inhibitor activity (TIA) of 10.8, with most of carbohydrate in the form of sugars. According to Ishola *et al.*, (1990), the seed also contains 47mg / 100g of phytic acid, which has minimal effect on its nutritive value. It also contains 14 to 18% albuminoid tannins located in the testa. According to Purseglove (1987), the seeds contain 63% starch and 4.5 to 6.5% of semi dried oil. Both pulp and the seeds are good sources of protein (269.3 g/kg), oil (109.1 g/kg) and calcium (Ishola *et al.*, 1990).

Table 7.3: Composition of tamarind seed, kernel and seed coat (%)

Constituent	Whole seed	Seed kernel	Seed coat
Moisture	9.4-11.3	11.4-22.7	11.0
Protein	13.3-26.9	15.0-20.9	-
Fat / Oil	4.5-16.2	3.9-16.2	-
Crude fibre	7.4-8.8	2.5-8.2	21.6
Carbohydrates	50.0-57.0	65.1-72.2	
Total Ash	1.60-4.2	2.4-4.2	7.4
Nitrogen-free extract	59.0	-	-
Yield of TKP	50.0-60.0	-	-
Calories / 100g	340.3	-	-
Total sugar	11.3-25.3	-	-
Reducing sugars	7.4	-	-
Starch	33.1	-	-

Source : Anon. (1976), Morad *et al.*, (1978); Ishola *et al.*, (1990); Bhattacharya *et al.*, (1994).

Tamarind is a versatile fruit, which can be used for many purposes. The unique sweet and sour taste of the pulp is popular in cooking and flavouring. Virtually every part of the tree has some value in commerce and particularly in the subsistence of rural people. Tamarind plants are used in agro forestry systems in many parts of the tropics due to its multiple uses. In India, many farmers integrate several species, including tamarind, with their agricultural crops and livestock. The increasing integration of tamarind with other trees and crops on farmlands offers a strategy to minimize the risk of crop failure. Since ancient times, in India and the Far East, members of flora and fauna were identified with particular personalities of the Hindu pantheon and worshipped. Tamarind is one of them which is highly sacred and worshipped tree. These beliefs and practices continue in certain sections of society, particularly among the rural folk and aboriginal tribes living in the forest. Nature worship and the faith of these peoples have helped to conserve many natural ecosystems in India. These people have preserved many forest trees such as tamarind and scared groves are conserved in their original form (Singha, 1995).

Tamarind is valued mostly for its fruit and pulp, which is used for a wide variety of domestic and industrial purposes (Kulkarni *et al.*, 1993). Tamarind is not a desert fruit, although the sweet tamarind is often eaten fresh directly from the pod. The pulp is usually removed from the pod and used to prepare juice, jam, syrup and candy. In Asia however, the immature green pods are often eaten by children and adults and dipped in salt as a snack. The acidic pulp is used as a favorite ingredient in culinary preparations such as curries, chutneys, sauces, ice cream and sherbet in countries where the tree grows naturally (Dalziel, 1937; Eggeling and Dale, 1951; Little and Wasworth, 1964). In some countries, the fruit pulp after removing the fibers is mixed with sugar, wrapped in paper and sold as toffees, which are a common sight in front of schools and on urban roadsides.

Tamarind seed is a by-product of the commercial utilization of the fruit; however, it has several uses. In the past, the seeds have been wasted whereas in 1942, two Inidan scientists, T. P. Ghose and S. Krishna, announced that the decorticated kernels contained 46-48% gel-forming substance which was patented in the name of jellose. (Savur, 1948). It has been recommended for use as a stabilizer in ice cream, mayonnaise and cheese and as an ingredient or agent in a number of pharmaceutical products (Morton, 1987). Presently, it is used in food and in industrial applications. Tamarind jellose has not been fully exploited, but due to its abundance and cheapness, seed jellose has great potential for replacing fruit pectin in many industries. In Gambia, seeds are pounded and eaten in times of famine. In India, seed kernels are used in times of food scarcity in the districts of Chennai, Andhra Pradesh and Madhya Pradesh and elsewhere, either alone or mixed with cereal flours (Shankarachrya, 1998). The seeds are also eaten after removal of the testa after roasting or boiling. In India, village women roast the seeds and chew them instead of betel nut (FAO, 1988). The seed is also used in the vegetable and food processing industries. Tamarind xyloglucan, commonly known as "tamarind gum", is the major component of TKP, forms a stiff gel and is used for thickening, stabilizing and gelling in food. It is commercially available as a food additive for improving the viscosity and texture of processed foods (Sone and Sato, 1994). Bhattacharyya *et al.*, (1994) reported that the incorporation of TKP decreased the specific volume and springiness of bread with increased TKP level, in contrast to increased hardness, while cohesiveness remained unaffected.

The leaves, flowers and immature pods of tamarind are also edible. The leaves and flowers are used to make curries, salads, sweets and soups in many countries, especially in times of scarcity (Benthall, 1933). In India, leaves are made into a dish called "Chindar". The seedlings are also eaten as a vegetable. Young leaves of tamarind are used as a seasoning vegetable in some Thai food recipes because of their sourness and specific aroma (Coronel, 1991). In Zimbabwe, the leaves are added to soup and the flowers used in salads. The children in Gambia mix the acid leaves with gum from fig trees to make a "chewing gum" (Sozolnoki, 1985). Flowers are an important nectar resource for honey bees in South India.

Tamarind wood is used for many purposes including making furniture, wheels, mallets, rice pounders, mortars, pestles, ploughs, well construction, tent pegs, canoes, side planks for boats, cart shifts and axles and naves of wheels, toys, hubs, oil pressers, sugar presses, tools and tool handles, turney, etc. (Coates-Palgrave, 1988; Troup, 1909). Tamarind heart wood is considered as a very durable timber and used is in furniture making as it takes on a good polish (Jayaweera, 1981). The wood also makes good fuel with a calorific value of 4850 k cal/kg, producing a great heat, which is required in brick making. The wood makes excellent charcoal, is valued for making gunpowder (NAS, 1979; Charturvedi, 1985; Purseglove, 1987).

The medicinal value of tamarind is mentioned in traditional Sanskrit literature. The laxative properties of the pulp and the diuretic properties of the leaf sap have been confirmed by modern medical science (Bueso, 1980). Tamarind fruits were well known in Europe for their medicinal properties, having been introduced by Arab traders from India (Rama Rao, 1975). Tamarind products, leaves, fruits and seeds have been extensively used in traditional Indian and African medicine (Jayaweera, 1981; Parrotta, 1990). Several medicinal properties are claimed for preparations containing tamarind pulp, leaves, flowers, bark and root (Bueso, 1980).

Tamarind pulp is used as a raw material for the manufacture of several industrial products, such as Tamarind Juice Concentrate (TJC), Tamarind Pulp Powder (TPP), tartaric acid, pectin, tartrates and alcohol (Anon., 1982 a; 1982 b). Pulp can also be processed in to various value added products such as Tamarind Kernel Powder, Pectins, Tamarind Pulp Powder, tamarind pickle, jam, syrup and candy.

Origin and Taxonomy

The origin of tamarind is not confirmed so far; however the species name "*indica*" gives the illusion of Indian origin (Morton, 1987). Parotta (1990) also suggested that the Arabs and Portuguese introduced tamarind to Africa from Asia. Von Maydell (1986) belived Madagascar to be the origin of tamarind, where today it dominates the reverine forests in the southwest and the dry forests further north. Evidences however also indicate that Central Africa particularly Abyssinia is the origin of tamarind because of high genetic diversity in the drier parts of Savannha from Sudan to Senegal (Irine, 1961, Torup, 1921). A high level of genetic diversity is also represented in the south and southeast Asian regions where tamarind is thought to have been introduced (Brenan, 1967) and where it was certainly naturalized a long time ago (Simmonds, 1984).

Tamarindus indica L. is a dicotyledonous plant belonging to family Fabaceae (sub-family Caesalpinioideae), which is the third largest family of flowering plants with 700 genera and approximately 17,000 species (Chant, 1993). Caesalpinioideae can be divided into five to nine tribes or groups of genera based on the morphological characters. According to Leonard (1957), tribe Amherstieae comprises 25 genera comprising of 21 from tropical Africa, 2 from tropical America and 2 from Asia. *Tamarindus* has zygomorphic and showy flowers and stamen filaments connate in a sheath, but have many differences in the floral structure, leaves, fruits and seeds. The genus *Tamarindus* is monotypic genus contains only one species. It is a diploid species with a chromosome number of x = 12 and 2n = 24 (Purseglove, 1987).

Distinction has made between the tamarinds of the West and East Indies. The tamarinds from the West Indies belonges to *Tamarindus occidentalis*, in which the pods are up to three times longer than their width and contain 1-4 seeds per pod. This is a tall tree, leaflets oblong, obtuse or rounded at the tip, deciduous and naturalized in all the West Indian islands (Grisebach, 1864). The East Indian tamarinds are recognized as *Tamarindus officinalis*, in which the pods are 6 times longer than the width and contain 6-12 seeds per pod (Coronel, 1991). However these differences have not been considered sufficient for separation into two species and the names *Tamarindus occidentalis* and *Tamarindus officinalis* presumed as synonyms of the Linnaean name *Tamarindus indica.* Tamarind is known by several vernacular names in India such as Indian tamarind, ambli, amli, imli (Hindi), amalika (Sanskrit), tintiri, tintul, tetul (Bengali) chinch, chitz, amli (Marwad), chis, hunchi (Hyrerabad), Karanji, Kamal, asam, hunse (Mysore), hunase, unsi, hulimara (Kannada), pulinje (Coorg), koya, tentuli (Oriya), chita, hitta, sitta (Gond), chinta (Telegu), puli, pulian (Tamil), tetili (Assam).

Tamarind is a long lived, magnificent, large, semi-evergreen tree, up to 30 m height. The trunk can be up to 8 m in circumference, with a spreading, rounded crown, up to 12 m in diameter. The trunk has light grey to brown, very rough and scaly outer bark; the inner bark is 1-2 cm thick and green-white in colour. The branches are often crooked, thick, wide-spreading and dropping at the ends to form a dense crown (Storrs, 1995). A blood-red-coloured gum exudes from the bole and branches when they are damaged. Plant produces a deep tap root and has an extensive lateral root system (Chaturvedi, 1985; Chaturvedi *et al.*, 1986). The main root is wiry and flexuose, while the lateral roots are numerous, moderately long and distributed down the main root. Leaves are alternate, even and peri-pinnately compound, 7-15 cm long with a pulvinus at the base and 6-20 pairs of opposite leaflets. The leaflets are 12-32 x 3-11 mm in size, narrowly oblong, unequal and entire, rounded or slightly emarginated at the apex and rounded the base. Leaves are sessile to sub-sessile glabrous, glaucous beneath and darker green above with reticulate venation. The petiole is up to 1.5 cm long with glabrous petiole and rachis. The apex is rounded to almost square, slightly notched and asymmetric with a tuft of yellow hairs.

Stipules are present, 0.5-1.0 cm, falcate, acuminate and pubescent, falling very early (Salim *et al.*, 1998).

The inflorescence racemes are small, 5-10 cm long, terminal and lateral drooping and often panicked (Plate 7.1). The flowers are bisexual and 2-2.5 cm in diameter. Pedicels are about 5 mm long, noosed

Plate 7.1. Flowering twig of tamarind

and joined at the apex. The bracts are caduceus, fragrant and zygomorphic, nearly as long as the flower bud. There are 2, boat-shaped bracteoles, 8 mm long and reddish in colour, four sepals, up to 1.5 cm long, unequal, ovate and pink, cream or pale yellow. Five petals in which the posterior and lateral ones are large and showy, slightly exceeding the clays, they are obovate to oblong in shape, white, cream, pale yellow, or pinkish in colour, streaked with red, the anterior ones are much reduced. There are 3 fertile stamens; the filaments are connate and alternate with bristle-like stamenoides to form a sheath, open above and inserted peri-gynously on the anterior part of the mouth of the calyx tube. Anthers are transverse, reddish brown and dehisce longitudinally. The ovary is superior uni-carpellary on a sheath coming from the posterior part of the calyx tube, it has 18 ovules, curving upwards and is green with a long, hooked style (Purseglove, 1987). The fruit is pendulous, pods are 5-10 cm x 2 cm, oblong or sausage-shaped, curved or straight with rounded ends. The shell is light grayish or brownish, indehiscent, often irregularly constricted between seeds, brittle and easily broken. It is filled with firm soft pulp surrounding the seed cavities, the outer surface of the pulp has 3 touches, woody, branching fibrous cords from the base to the apex and the pulp is thick and blackish brown in colour (Coronel, 1991; Purseglove, 1987). The indehiscent pods ripen about 10 months from flowering and many remain on the tree until the next flowering period (Chaturvedi, 1985). Pods contain 1-10 seeds, each 3-10 x 1.3 cm, irregularly shaped, flattened, rhomboid, with the centre of each flat side of the seed marked with a large central depression. The seeds are very hard, shiny, reddish or purplish brown. They are embedded in the pulp, lined with a tough parchment-like membrane and joined to each other with tough fibers (Purseglove, 1987). There are about 720 seeds per kilogram of fruit (Hong *et al.*, 1996) which may vary depending on the conditions under which the trees are grown.

Area and Distribution

At present tamarind is cultivated in about 52 countries of the world but the major areas of production are in the Asian and American continents. In most of the countries, it is a subsistence tree crop mostly meeting local demands, although some of them are also exporting it. The fruit in most of the countries are used for culinary purposes, making juices, chutneys, sauces, etc, while the seed is the main component used in industrial applications. India is the world's largest producer of tamarind although, the precise and accurate data on production and export of the tamarind are not available. The tree mostly grows wild, although it is cultivated to a limited extent. It is abundant in the Madhya Pradesh, Bihar, Andhra Pradesh, Karnataka, Tamil Nadu and West Bengal. In Bejapur, in the Deccan Plateau, the tree is famous for its fine varieties and is cultivated extensively (Plate 7.2). The

Plate 7.2. Tamarind plant in full bearing

production in India is mainly concentrated in the drier southern states. It is naturally regenerated on wastelands and forestlands. India has been exporting processed tamarind pulp to European and Arab countries from these scattered plantations. Recently, Thailand has become a major producer of tamarind, with sweet and sour varieties in production. Total area of tamarind in Thailand is about 105,785 ha. Mexico also produces tamarind commercially, with over 4400 hectares producing over 37,000 tons of pulp.

Soil and Climate

The tamarind tree can grow in a wide range of soils (Chaturvedi, 1985) however, thrives best in loamy, deep, well drained alluvial soil, which favors the development of a long tap root (Galang, 1955). With little cultivation it can flourish in poor soils and on rocky terrain. It tolerates sodic and saline soils where it grows in ravines and on degraded land (Anon, 1991). Dwivedi *et al.*, (1996) reported that tamarind could grow in soil containing up to 45% exchangeable sodium. The tree does not tolerate water-logging (Relwani, 1993; Vogt, 1995).

Tamarind is adapted to a wide range of climatic conditions due to wide geographical distribution in the sub and semi-arid tropics. It grows well with an evenly distributed mean annual rainfall of 500-1500 mm (Hocking, 1993). Dry weather is conductive for flower initiation and if heavy rains occur during flowering tamarind does not bear fruit. Dry weather is also very important for fruit development because regardless of total annual rainfall, tamarind produces more fruit when subjected to a fairly long annual dry period (Allen and Allen, 1981). Due to its deep and extensive root system tamarind can also be grown in dry areas where plant can withstand up to 6 months without rainfall (Coronel, 1991). Tamarind thrives under a maximum annual temperature ranging from 33-37^{0}C and a minimum of 9.5-20^{0}C. The plants are very sensitive to high temperature and frost particularly when young (Troup, 1921). Older trees are more resistant to extremes of temperature and can withstand as high as 47^{0}C and as low as -3^{0}C temperature without serious injury. It has strong and pliant branches and a deep and extensive root system, which anchors it to the ground (Coronel, 1986) and therefore can resist strong winds. Due to the diversity of habitats and soil types tamarind can grow in association with a number of different species under agro-forestry system.

Plant Genetic Resource Management and Varieties

The genetic variation of tamarind has been based on the phenotypic variation. This is primarily based on fruit characteristics such as length of pod, pod weight, seed number, pod colour and sweetness of pulp. The variation in sweetness and colour of the pulp in tamarind germplasm hold promise for genetic improvement. Variation also exists in fruit pulp characters where some are less acid or sweet fruits and others are very acidic. The variations have also been reported for tolerance to drought, wind, poor soils, water logging, high and low pH and grazing. Phenological diversity also exists in tamarind with respect to flowering and maturity of fruits (Mahadeven, 1991). Pathak *et al.* (1992) reported the considerable variation in fruit yield (200 to 800 kg/tree/year) of tamarind in different countries showing the potential for improvement in fruit production. Wide phenotypic variation in tamarind germplasm has been attributed to geographic location and gene mutation. The origin of sweet tamarind is one of the examples of geographic isolation and gene/point mutation (Feungchan *et al.*, 1996 a). It is also assumed that the sweetness in tamarind is a rare trait, which may be governed by recessive genes (Fuengchan *et al.*, 1996a). Occasionally, isolated branches on a tree may bear sweet fruits, while the other branches have the normal sour type which may be due to bud sport and camera. These morphological aberrations will provide ample opportunities for genetic improvement of tamarind.

A large variability in germplasm exists in terms of acidity of the pulp, pod bearing ability and pod size (Plate 7.3) world wide. Therefore, in order to incorporate such traits in the improvement programme,

Plate 7.3. Variability in Tamarind pods

germplasm collections and evaluation have to be made for each region. Although there is a wealth of tamarind germplasm across the regions where it is naturally grown, no systematic germplasm collection and evaluation has been attempted to date in spite of its future potential. Most countries have not considered it as a priority species for conservation. Plants with sweeter, juicier pulp and other desirable characteristics could be selected from such diverse germplasm. In recent years, less acidic, sweet tamarind types have been selected in the Philippines and Thailand which are becoming popular. It is important that germplasm is collected systematically, evaluated for important traits and conserved. This will enhance tree improvement and sustainable utilization of fruits and other products.

Whereas various sour and sweet type tamarind accessions are available in tamarind growing countries of world. Feungchan *et al.*, (1996 a) reported more than 50 sweet tamarind cultivars in Thailand. Knight (1980) developed a variety of sweet tamarind called 'Manila Sweet' in Florida out of the introduced material from Phillipines. Mentions of 46 accessions are from Philippines (Gunasena and Hughes, 2000) have been made. In India, however a wide range of variability available owing to natural regeneration through seeds, are being maintained by selecting the trees at different research Stations. Research evaluation and characterization of collected plus trees has enables to identify a good number of varieties with desirable horticultural traits such as pod colour, pulp yield per pod and fruit yield per tree. Some promising selections made on the basis of high yield of large and straight pods with extra white endocarp and high pulp content and TSS are PKM-1, PKM-2, PKM-3 and Urigam in Tamil Nadu; Pratisthan, No. 263, Yogeshwari, T-9, T-10, T-11, T-12, T-13, in Maharashtra and DTS-1 and DTS-2 in Karnataka. PKM-1 which also known as Periyakulam-1 is a sour type cultivar developed through clonal selection from a local variety by Collage of Horticultural and Research Station, Periyakulam. This starts early bearing with high pulp recovery (39%). A grafted plant of PKM-1 produces 200-260 kg fruit every year.

Although tamarind has commercial potential, little attempt has been made to improve it as a crop plant. Tree improvement programme in tamarind is time consuming. Indigenous farmers have selected cultivars from natural populations for desirable characterstics. Recently, genetic improvement using superior clones has been described by Kulkani *et al.*, (1993). Since, tamarind has a relatively long gestation period and heteroginity, conventional breeding approaches would require considerable time and money. Fortunately, there are trees with very high yielding potential exceeding 800 kg / tree / year in the South Asian region. In the short term, those trees having superior characteristics could be selected and vegetatively multiplied to produce fast growing trees for local (home gardens) and commercial (orchards) use. The selection of superior trees and their clonal development may be faster and may have greater impact than the conventional breeding.

Plus trees can be selected using the characteristics such as acidity of the pulp, content of tartaric acid and sugar, Real Value of Pulp, pod bearing ability (flowering and fruit maturing), pod size, pulp, fiber and seed weights and number of seeds. For improvement in tamarind, it is essential to identify ideotypes for different purposes, localities, environments and cultural practices. There is no universal ideotype suitable for all sites and end users. Chuntanaparb and Ranganathan (1990) stressed that ideotype identification should be focussed on specific purpose and products that could be marketed, either locally or internationally. The proposed characteristics for general tamarind ideotype has been in Table-7.4.

Table 7.4 : Proposed characteristics for tamarind ideotype

Tree Parts	**Characters for desirable Ideotypes**
Stem / Root	Strong, short, 3-5m, multiple branches for sweet types, Deep tap roots, normal tall trees with strong branches for sour types
Bearing habit	Early, 4th or 5th year, flowering period long, regular
Fruit	Heavy and fleshy, sweet / acidic pulp depending on market, high pulp recovery, high in nutrients
Yield	High, 600-800 kg/tree/year
Adaptability	Wide adaptability, resistance to pests and diseases

Thus, the improvement work in tamarind should be focused on

- Collection of germplasm from all naturalized areas and maintenance in the field gene banks.
- Complete characterization and evaluation of available germplasms based on International standards.
- Chemical characterization of fruits of promising cultivars for various components.

- Selection of high yielding cultivars based on phenological characters, including early, regular and year round production, high pod number, desirable fruit characters such as weight, pulp and their ability for vegetative propagation, quality, high nutritive value and resistance to pests and diseases.
- Identification of genetic variation particularly for the commercially important traits among cultivars and determination of molecular markets in selection of traits with low heritability.

Plant Propagation and Nursery Management

Tamarind is mainly propagated from seed, though vegetative and tissue culture propagation methods have been developed. The seed of tamarind is orthodox and fresh seeds retain viability for at least 6 months when kept at ambient temperature in dry conditions. Seed germination begins within a week after sowing and takes a month to complete. Tamarind has a hard seed coat, hence, delayed and poor germination is common therefore, seed scarification should be done for better germination. Generally, the germination capacity of fresh or well stored tamarind seed is about 65-75% (Coronel, 1991). Germination in tamarind is epigeal. The radicle emerges from one end of the seed and descends rapidly. The hypocotyl elongates arches slightly and raises the cotyledons enclosed in the testa. Soaking of seed in normal/cold water for 24 to 48 hours has been reported to hasten germination (Von Carlowitz, 1986). In Malawi, seeds of tamarind thoroughly cleaned and soaked in water overnight to encourage germination resulted in more than 80% germination (Prins and Megehembe, 1994). Seed treatment with carbendazim, Mancozeb, Thiram, Copper oxychloride or organo-mercury fungicide @ 3 g/kg of seed have been recommended for control of seed borne fungi (Purohit *et al.*, 1996). Seed should be collected from the mature ripe pods in March-April. After proper extraction seeds should be shown in beds/bags during June-July. The best medium for seed germination is sand or soil mixed with FYM. Chattopadhyah and Mohanta (1988) reported that seed germination could be encouraged by using cow dung or cow dung and sand in the propagation medium. However, a normal nursery potting mixture containing three parts of top soil, one part of sand and one part of compost, can be successfully used for germinating tamarind seed.

Seedlings grow rapidly in the early stages and produce a long tap root which may reach 30 cm or more within 2 months (Troup, 1921). Seedlings attaining a height of about 80 cm should be transplanted to the field at the beginning of the rainy season.

Vegetative propagation is preferred over seed propagation, as later does not produce true-to-type progenies. Vegetatively, tamarind can be propagated through stem cuttings and air layering on their own roots and by attach (inarching)/detach grafting (cleft or vineer)/ budding (shield or patch) techniques on the seedling rootstocks. Vegetative propagation of tamarind is also possible through root suckers at limited scale.

Though the success is less, stem cutting is the easiest and the cheapest method of propagating tamarind. Srivasuki *et al.,* (1990) reported that cuttings of shoots bearing new flushes of fully turgid leaves are collected in the morning. They are immediately dipped for 10 seconds in 1000 ppm of indole butyric acid (IBA) and in 50% isopropyl alcohol, before being plated in polypropylene tubes containing vermiculture/perlite (1:1) and placed in a mist propagator with 70-80% humidity. The use of IBA increased the rooting of cuttings to over 94% compared to the control, which recorded only 25%. Furthermore, the time taken to initiate rooting was only 10-15 days with IBA, while the control took 40-50 days. The terminal cuttings have an advantage over mid stem cuttings because there is only one cut end. This reduces the possibility of infection by disease causing organisms during or after the rooting phase. 15-20 cm long semi hard stem cuttings, taken from 1-2 year old branches can also be rooted (Swaminath *et al.,* 1990) in sand bed under mist with the treatment of 1000 ppm IBA. In this method, the cuttings are wrapped in moist cloth after removal from the trees to prevent moisture loss. They are then dipped in a rooting hormone, preferably IBA at 1000 ppm, and placed in a sand bed in a mist chamber. In these cuttings, bud initiation and root formation occurs after about 10 days and new leaves are formed after about 45 days. Hard wood cuttings do not root in tamarind.

Layering is a widely practiced propagation method for tamarind. In ground layering, a flexible branch is bent down and pegged to the ground and the point of contact covered with soil. A small cut is made in the lower side of the stem where it touches the ground to restrict sap circulation and encourage rooting. After 3-6 weeks, roots are formed at the point of contact with the soil and a new plant can be obtained by severing the branch above the place of rooting. For air layering a healthy branch is selected and the bark of 2.5 cm piece at 45-80 cm distance below the tip is removed. This area is covered with moist soil or moist *sphagamum* moss kept in position by wrapping with low gauge polyethylene film. When large quantities of roots are observed through the polyethylene film, the branch with its roots is severed from the parent tree and potted up for later planting. The appropriate time for air layering is at the highest physiological activity period of the parent plant. Navaneetha *et al.*, (1991) reported that May-June is the best time for air layering in tamarind. The application of IBA at the cut end increases the number and the length of roots, and reduces the time for rooting from about 10 weeks to 6 weeks in air-layers (Gowda, 1983). Navaneetha *et al.*, (1991) reported that IBA at 1000 ppm gave 75% rooting and resulted in the highest subsequent survival.

Shield and patch methods of budding are more successful in tamarind. For patch budding, seedling rootstocks should be grown in raised beds and budded when they are 6-9 months old. This is a suitable method for large-scale multiplication of tamarind. Pathak *et al.*, (1991) reported 96% and 94% success, respectively of patch budding and modified ring budding of 9-month-old seedling rootstocks. Shield can also be used for producing plants at large scale on 6-9 month old rootstock.

Approach grafting is a very reliable method for propagation of tamarind up to 95% success (Swaminath and Ravindran, 1989). For veneer grafting, 6 months old rootstocks having uniform size are selected. The scions should be defoliated before grafting (Plate 7.4).

Plate 7.4. Scion stick of tamarind for grafting

Veneer grafts are made at 8 cm high on the root stock and immediately after grafting, the stock is cut above the graft union. Where as other grafting methods gives about 50% success (Amin, 1978). Purushotham and Narasimharao (1990) mentioned that grafting methods are expensive and time consuming. For softwood grafting, the rootstock seedlings are defoliated and their tops cut off at 15 cm height immediately before grafting. A vertical downward cut of about 4 cm depth is made in the centre of the cut stem and the scion stocks are cut into wedge shapes, inserted into the stock and wrapped using 2 cm wide 200 gauge polyethylene strip. Soft wood grafting is considered to be the best grafting method in terms of successful unions and survival rates (Navaneetha *et al.*, 1990).

Tamarind can also be propagated by tissue culture techniques. Shoot tips, cotyledons and cotyledonal nodes have been used successfully as explants for tamarind tissue culture (Splittstoesser and Mohamed, 1991). The explants were cultured in a MS (Murashige and Skoog, 1962) medium supplemented with thidiazuron (TDZ) / IBA (indole butyric acid) or BA (butyric acid)/NAA and incubated at 28^0C for 5 weeks under an 18 hour photoperiod. The shoot tips of tamarind regenerates only one shoot while the cotyledonal nodes produces multiple shoots. Multiple shoots were however induced on shoot tips and nodal explants on MS basal medium supplemented with BAP Benzyl-amino-purine (5×10^{-6} M). All shoots were rooted in MS with IAA (indole acetic acid) (5×10^{-6}M) and the micro-propagated plants were successfully transferred to the soil (Jaiwal and Gulati, 1992). Shoot tips from *in vitro* grown seedlings have been regenerated into plantlets in MS medium containing IBA or IAA. Higher concentrations of growth hormones, (2 and 5 mg/l) induced better rooting, but a media supplemented only with coconut water or BAP supported growth of shoots do not produced that much of roots (Kopp and Nadaraja, 1990). Tissue culture raised plants have been reported to show better growth and height, branching habit, spread of branches, early flowering and fruiting cycle, with the start of flowering at an average height of 3.7 m whereas the seedlings raised plants flowered after attaining an average height of 8.0 m. Although the micro-propagation technique have shown promise, none of them has reached to the stage of commercialization. Tissue culture protocols have yet to be developed and made available for use as an effective tool in the large-scale multiplication and propagation of selected tamarind cultivars.

Planting

Tamarind is a multi purpose tree which can be grown in various ways such as an orchard crop in a pure stand, as an agro-forestry species in mixed cropping systems including in home gardens, or as hedgerow tree. In each of these systems, there are several methods available for planting the trees.

Tamarind can be directly established in the field by sowing seed at stake (Chaturvedi, 1985). Seeds may be sown directly to establish plantations, hedgerows, or home gardens. For the purpose, pits should be prepared in advance and filled with well-decomposed manure or compost. The seed should be planted 1.5 cm deep in these pits prepared at 5 m apart. For planting on the square, on the triangle, or in lines for orchard establishment, a few seeds are dibbled in 5 cm apart in each pit. This method is practiced in the state of Tamil Nadu and Andhra Pradesh. After germination, they should be thinned to have one-best seedling at each planting point. For seed planting, approximately 20 kg of seed is required per hectare.

Transplanting of nursery grown seedlings or vegetatively propagated plants in the field have advantages over planting seed at stake with respect to its establishment and assured quality of plants. The *in situ* plants on the other hand have deep, extensive and undisturbed root system but not true-to-type to the variety. Nursery raised plants can be transplanted in the field after about 12-14 months. Seedlings or grafted plants ranging from 40 cm to 2 m in height are commonly planted (Jambulingam and Fernandes, 1986) however, if growth is poor, the seedlings should be retained in the nursery for another one year but it will be difficult to handle these older seedlings as their tap roots which may have grown deep into the soil. These seedlings will require root pruning and after care. The pits of at least 30 to 50 cm^3 size should be prepared for transplanting.

The spacing between plants varies between 5 m and 13 m apart either way depending upon the soil type and climate. In fertile soils where the trees grow larger, spacing may be wider. When establishing a pure orchard, the spacing should be at least 10 x 10 m in square system. There can be an advantage in planting the saplings initially at higher density (5 x 5 m), followed by thinning alternate row either side, to obtain the appropriate final spacing. For Sweet tamarind orchards in Thailand, the recommended tree to tree spacing is 10-12 m (Yaacob and Subhadhrabhandu, 1995) whereas in India, tamarind is established at 8 m x 8 m, 8 m x 12 m or 12 m x 12 m (Jambulingam and Fernades, 1986). Tamarind is frequently planted along the land borders, canal banks, on field boundaries, in village groves or in home gardens as individual trees. In India, tamarind is used to make firebreaks, as the soil beneath is almost bare due to heavy shade. When creating firebreaks, the spacing adopted is 2.5 m x 2.5 m or 3 m x 3 m. If the spacing appears to be too close some plants can be removed later. In Mysore, where tamarind is used commonly in fire lines, seedlings are raised in tiled pots and planted in the field at a spacing of 3 m x 3 m in 1 m^3 pits. After care is minimal for these plantings, the only requirement being to hoe the soil around the plants once a year to remove weed growth (Troup, 1921).

The best time for planting is rainy season. This will reduce the need for frequent watering until the plants are firmly established in the soil. If irrigation is available, planting can be undertaken at any time of the year, even in the dry season. If due to unforeseen circumstances, planting in the rainy season is delayed, the seedlings should be cut back and the tap root should be trimmed.

Training and Pruning

Tamarind requires minimal care after plant establishment. It produces a compact tree with symmetrical branches therefore; the young trees should be trained properly to allow 3 to 5 well-spaced branches which develop as the main scaffold branches of the tree. Bearing trees require very little pruning other than maintenance pruning to remove dead, weak and diseased branches and water sprouts (Salim *et al.*, 1998). In closely planted orchards regular pruning is needed to rejuvenate fruiting wood and control the size of the trees.

Orchard Management

Tamarind is cultivated commercially in southern India. Systemetic orchards of improved varieties are now coming up very fast. Orchard management techniques for tamarind orchard in India and abroad has been reviewed to develope suitable management practice.

Nutrient Management

Tamarind is a leguminous tree species. The leguminous species are associated with rhizobium bacteria in root nodules which fix atmospheric nitrogen. Very little information is available on nitrogen fixating bacteria in tamarind compared to other cultivated legumes.

Tamarind was initially considered to be a non-nodulating species (Allen and Allen, 1936, 1981). However, some evidence suggests that it does a symbiotic association with rhizobium bacteria (Postgate, 1979; Athar and Mahmood, 1982; Quiniones, 1983) enabling the tree to fix atmospheric nitrogen under appropriate conditions (Quiniones, 1983; Ding *et al.*, 1986). Chiemsombat *et al.*, (1996) reported that the roots of tamarind do not develop nodule-like structures and have no bacteria or rhizobium association commonly observed in other leguminous plants. The acetylene reduction assay for nitrogen-fixing activity was unable to demonstrate any symbiotic nitrogen fixation in tamarind roots and nitrogen-fixing bacteria were absent in the roots. The most outstanding feature noted in the roots were finger-like projections called "microvilli," whose function is unknown, but it has been suggested that they assist in water absorption. However, tamarind seedlings inoculated with 13 vesicular-arbuscular mycorrhiza (VAM) fungi from various sources around the world have demonstrated positive results such as increased leaf number, plant height, stem girth, biomass, phosphate and zinc content. The number of VAM spores in the soil, percentage root colonization and external hyphae measured by soil aggregation were also higher. Tamarind responded best to inoculation with *Gigaspora margarita* and *Glomus faciculatus* (Reena and Bagyaraj, 1990). Maksoud *et al.*, (1994) also observed significant increases in growth parameters and leaf P and K contents when seedlings were planted in pots inoculated with VAM. Application of inoculum to the pits before transplanting the seedlings would therefore be an effective method of introduction of the VAM fungus.

Tamarind tree is seldom manured and therefore the information on fertilizer application is scarce and the nutrient requirements for tamarind have not been worked out. After establishment, trees are known to fruit well even without fertilizer application, due to their deep and extensive root system. But during early stage 25g urea and 25g single super phosphate per plant have been found optimum (Gupta and Mohan, 1990). At the time of pit filling, each pit should be filled with 5 kg FYM along with 1-1.5 kg neem/karanj cake thereafter 5 kg FYM and 0.5 kg of neem cake are applied per tree in the months of March and April. However, the commercial sweet tamarind growers use urea @ 100-200 g /tree to supports high yields. The amount of fertilizer is gradually increased as the trees grow. When the tree begin to bear fruits, about 500g of a complete fertilizer containing high amounts of nitrogen and potassium is applied per tree twice a year. A full bearing tree should receive 2-3 kg of NPK fertilizer each year.

Water Management

Tamarind is a hardy plant and grows well under rainfed conditions but responds to the application of irrigation water. Application of water during the initial establishment stage and during flowering and fruit development phase is beneficial. However, under rainfed conditions, soil and moisture conservation practices like mulching, water harvesting, cover cropping, etc., help in early establishment and development of vigorous plants. For water harvesting, 5 per cent slope to the land around the tree is provided. Mulching of the tree basins with locally available organic materials such as paddy straw, paddy husk, castor shells, sugar cane trash, coir pith or black polythene helps in conserving soil moisture.

Irrigation which is normally not practiced in tamarind cultivation, promotes better growth during establishment and the early stages of growth, especially during the dry seasons (Yaacob and Subhadrabhandu, 1995). If irrigation facilities are available, watering should be done as the need arises in the early stages of growth. In later years, as the deep tap root system develops, the need for watering becomes less. Flowering and fruiting is promoted by irrigation. In dry areas the use of water harvesting techniques during the rainy season should be considered as it encourages subsequent growth and fruiting (Chundawat, 1990). Mulching during the dry season also reduce water losses due to evaporation.

Orchard Floor Management

There is a wide choice of intercrops which can be grown depending on the soil, climate and the local market demand. Intercropping may be practiced in order to obtain some income during the early stages of tree growth. The intercropping period is limited to 3 or 4 years in vegetatively propagated trees and 10-12 year in seedling planting. It is advisable grow short duration crops preferably legumes during rainy season.

Disease Management

Several leaf spot diseases have been reported on tamarind in India. The major causal organisms of spot are *Bartalinia rabillardoides* Tassi., *Exosporium tamarindi* Syd., *Hendersonia tamarindi* Syd. *Pestalotia poonensis* Rao., *Phyllosticta tamarindicola* Rao., *P. tamarindina* Chandra and Tandon, *Prathigada tamarindi* Muthappa, *Xanthomonas tamarindi* Cook, *Sphaceloma* spp. and *Stigmina tamarindi* Munjal and Kulshreshta.

Powdery mildew (*Erysiphe polygoni* DC. and *Oidium* spp.), sooty mould (*Meliola tamarindi* Syd.), root and wood rot (*Ganoderma lucidum* (Leyss) Karst), stem rot (*Pholiota gollani* Henn.), trunk and root rot (*Stereum nitidulum* Berk.) and collar rot (*Phytophtora nicotianae* var. *nicotianae*); (Mukherji and Bhasin 1986; Siddaramaiah and Kulkani., 1982; Morton, 1987; Parrotta, 1990) are also causing loss to tamarind plants. Stony fruit disease caused by *Pestalotia macrotricha* Syd. (Lokesha and Shetty, 1991) makes the fruits hard and stony with fibrous structures in Karnataka State.

In the nursery, mildew caused by *Oidium* sp. is a common disease (Plants 7.5). The disease causes defoliation and retarded early growth. In Uttar Pradesh, tamarind has been attacked by *Cerospora* leaf disease resulting in severe defoliation (Khan *et al.*, 1988). For effective control 3-4 sprays of wetable sulphur at 15 day intervals is the most economical method (Siddaramaiah and Kulkarni, 1982). It can also be controlled by spraying Maneb 80 WP, Benzyl 50 WP or Captan 80 WP.

Plate 7.5. Incidence of powdery mildew disease in tamarind

Pest Management

Tamarind trees are attacked by a large number of insect pests. About, 40 insect species have been recorded on tamarind, which are causing severe economic losses (Joseph and Olmmen, 1960). The major pests which attack tamarind include shoot hole borers, toy beetles, leaf feeding caterpillars, bag worms, mealy bugs and scale insects (Coronel, 1991). Among the worst are the sap suckers (white flies, thrips, coccids and aphids) which affect the tender shoots and fresh foliage. The defoliators, which include caterpillars and the chafer beetles, also cause considerable damage. Some of these insects attack the flower buds and the developing fruits and seeds, while others damage the fruits during storage. The important pests of tamarind are:

Scale insect

The oriental yellow scale (*Aonidiella orientalis* Newst.) is a polyphagus pest which attacks tamarind. Some of the most serious pests in India are the hard scale insects species (*Aonidiella orientalis* Newst., *Aspidiotus destructor* Sign. and *Saisetia oleae* Ol.). Timyan, (1996) reported that female lays eggs under a protective covering such as cracks in the bark of the trees. After hatching, the nymphs crawl in search of suitable succulent parts of the plant. The scale insects can be controlled by removing the affected parts in the initial stages. When the infestation is serious, chemical spraying is required and pesticides such as Diazinon or Carbosulphan at 0.1% solution provide effective control (Bhutani, 1979).

Mealy bug

Mealy bugs (*Nipaecoccus viridis* Newst. and *Planococcus lilacinus* Ckll.) also attack tamarind trees. Nymphs and the females suck the sap on the ventral surface of the leaflets, the base of leaf petioles, tender shoots and even the mature shoots. *P. lilacinus* also attacks the roots. The leaflets become chlorotic and defoliation is common. Immature fruit fall is sometimes observed. Mealy bugs can be controlled by mechanical removal of infected shoots on a small scale. In serious infestations, chemicals such as Diazinon or Carbosulphan at 1% solution can be sprayed. *P. lilacinus* populations living in the soil can be displaced by raking or be killed using 5% aldrin at 500g per tree (Bhutani, 1978).

Thrips

Thrips (*Scirtothrips dorsalis* Hood) is a polyphagus pest and the adults live for 10 or 15 days and complete 25 overlapping generations per year (Raizada, 1965). Thrips can be controlled by spraying with dimethioate at 20-30 ml / 10 litres of water or fenthion at 10-15 ml / 10 litres of water.

Coccids

Several coccids have been reported as polyphagus pests of tamarind. These are *Aspidiotus destructor* Signoret, *A. ramarindi Green, Cardioccous castilleae* Green, *Hemiberlesia lataniae* Signoret, *Howasdia biclavis* Comstock, *Pinnaspis strachani* Cooley, P. *temporaria* Ferris. *Unaspis articolor* Green and *Saissetia oleae* Bernard. Spraying Carbosulphan at 0.1% solution or Diamethioate at 0.3% solution (20-40 ml in 10 litres of water) can control all of these coccid species.

Aphids

Aphid is serious pest where as whitefly is minor pest of tamarind. Aphid sucks the sap of tender shoots and leaflets causing them to become distorted and covered with moulds growing on the secreted honeydew. Aphids and white flies can be controlled by spraying Endosulphan at 1 ml in 10 ml of water or Dimethioate at 0.3% solution (20-40 ml in 10 litres of water), repeated every two weeks until the pests are no longer observed.

Storage pest

The ripe pods of tamarind are attacked by *Aphomia gularis* Zeller whereas stored fruits are infested with *Paralipsa gularis Zellar* and *Corcyra cephalonia* Stnt. *Assara albicostalis* Walk in India (Bhutani, 1978). The larvae are stout, grayish in colour with brownish spots which bore inside the fruit and eat the pulp and sometimes bore into the seed. Larvae of the tamarind weevil (*Caryedon serratus* Olivier) are also a serious pest which attacks the pods, seeds and fruits in the field and fruits and roasted seeds in storage in India (Bhutani, 1978). Mital and Khanna (1978) reported that Neem oil at the rate of 5 to 20 ml per kg of seed, has been used effectively to control bruchid beetles attacking tamarind pods and seeds. This pest can also be controlled by fumigation with phosphine from aluminium phosphide tablets weighing 3g each, applied at 0.5 tablet / ton. Ethylene bromide, carbon tetrachloride and acrylnitrile could also be used for fumigation of stored tamarind products.

Flowering, Fruiting and Yield

Seedling plants of tamarind take 10-14 years to start fruiting (Morton, 1987) whereas vegetatively propagated plants start bearing within 6 years and yield of the plant establishes after 10-15 consecutive bearing years. In north India, new flush in the plant appears in March-

April and flowering takes place during May-June and the fruits take 10-11 months to ripen. Generally, flowering and fruiting in tamarind takes place in dry season. Thimmaraju *et al.*, (1977) reported that flower bud development takes about 20 days from first visible initiation. Peak anthesis takes place at 6.00 a.m. and peak anther dehiscence at 10.30 a.m. Maximum stigma respectively is on the day of flower opening. Usha and Singh (1996) reported that cross-pollination results approximate 20% higher fruit set and retention in tamarind than open or self-pollinated. Bees are known to pollinate tamarind flowers and therefore bee farming could provide additional income to farmers. Feungchan *et al.*, (1996 d) attempted to increase fruit set of sweet tamarind by the application of growth regulators at the time of flowering. The growth regulators such as 4-CPA at 15 ppm, NAA at 100 ppm and GA 3 at 15 ppm have been found effective in increasing fruit set of sweet tamarind (Feungchan *et al.*, 1996 d). The growth regulator 4-CPA at 15 ppm was the most effective. Spraying growth regulators could be an important cultural practice for year round fruit production in sweet tamarind. Fruit development in tamarind takes place in three distinct stages of growth and development. Maturation and ripening takes about 245 days from first set to harvest (Gunasena and Hughles, 2000). The yield of tamarind varies considerably depending on genetic makeup and environmental factors. Feungchan *et al.*, (1996a) reported that the fruit yields are influenced by environmental and genetic factors. The size of the tree also affects the yield of the tree. Pod yield is sometimes cyclic, with bumper yields in every third year (Jambulingam and Fernandes, 1986) due to the high demand of flowers absorbing much of the available nutrients and thus weakening the physiological condition of the tree. Under such circumstances tree needs recovery in the next years before reaching peak performance again. A young tree yields 20-30 kg fruits per year, while full grown trees of 20 years age produces 150-200 kg/tree/year or 12 to 16 t/ha (Chapman, 1984). A full-grown tree is reported to yield about 180-225 kg of fruits in every season (FAO, 1988). In India, the average production of tamarind pods per tree is 175kg and processed pulp is 70kg/tree (Kulkani *et al.*, 1993). Rao (1995) reported 59% more yield (263 kg/tree) in cultivar Periyakulam-1, an improved cultivar developed in Tamil Nadu. Well managed tamarind tree can yield 200-300 kg fruits whereas 40-50 kg fruits can be obtained from 6-7 years old trees. A well managed tree continues to produce economical yield up to 60 years of age.

Harvesting and Post harvest Management

Ripe fruits are harvested during February-April by shaking the branches. Fruits are also harvested at half ripe stage. Sweet tamarind is harvested by clipping individual fruits to avoid damage (Coronel, 1991). The fresh fruits are dried using small-scale dehydrators; however in rural areas pods are dried in the sun. The shells, fibers and seeds are then removed and the pulp stored in plastic bags. The dry ripe pods can easily be cracked (Plate 7.6) and the pulp and fibers separated from the broken shells (Plate 7.7). The pulp is then processed by peeling

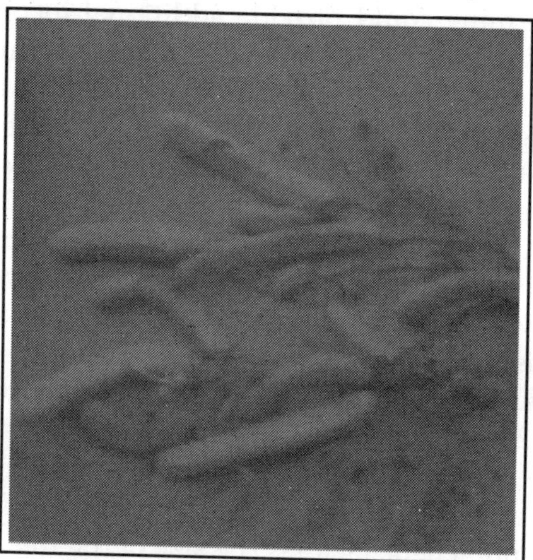

Plate 7.6. Mature pods of tamarind

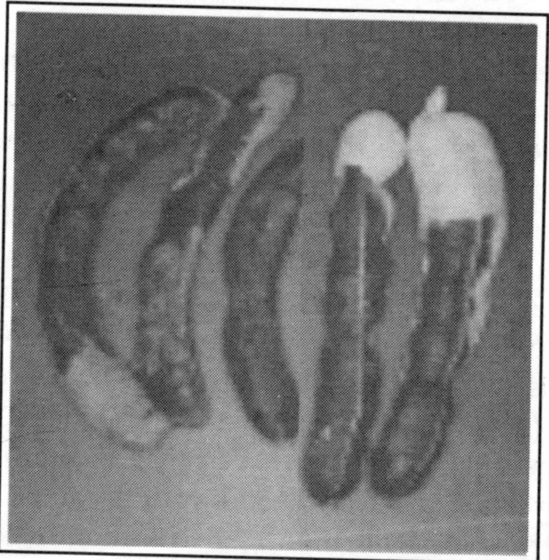

Plate 7.7. Ripe pods showing red and brown pulp

and removing the fiber stands from the pulp. After separating the pulp from the fibers, the seeds and shells, it is then compressed and packed in baskets, jute bags or plastic bags for storage and marketing (Plate 7.8). The pulp is pressed and preserved in large masses and sold in small shops and bazaars by weight. In East India, the pulp is covered with salt, rolled into balls, exposed to dew and stored in earthenware jars (Chapman, 1984; Morton, 1987). These methods are manual and crude and tend to be very labour intensive and inefficient. Mechanical methods of extracting pulp have been reported (Benero *et al.*, 1972) and a tamarind dehuller has also been designed and developed at UAS, Bangalore, India. The machine has a hulling capacity of 500 kg/h, with hulling efficiency of 80% for large fruits and 58% for small fruits (Ramkumar *et al.*, 1997). The quality and condition of the pulp and the selling price in the market are often related to the care taken during storage. The freshly prepared pulp is light brown in colour, but darkens in storage. Generally, under dry conditions, the pulp remains good for about one year, after which it becomes almost black. The pulp also becomes soft and sticky as pectolytic degradation takes place and moisture is absorbed, especially in humid weather (Lewis and Neelakantan, 1964a).

Plate 7.8. Matured pods of tamarind for marketing

Feungchan *et al.*, (1966 f) attempted several methods to prevent change of pulp colour including powdered salt, steam, sodium meta bisulphite, sun drying, hot air incubation and cold storage. Cold storage of pulp in transparent containers at 0°C and mixing powdered salt at 10% by weight in a transparent container is a satisfactory way to prevent the colour change in tamarind. The use of salt and storage in a transparent container is considered best due to its ease of handling and is more cost effective than cold storage. The pulp can be stored in a cool, dry place for three to six months without much deterioration. At some places in India, salted (10%) pulp is trodden into a mass and made into balls and exposed to the sun or steamed for a short time and then exposed to the sun and dew for about a week (Shankaracharya, 1998).

Processing and Value Addition

Tamarind can be processed in to different kinds of products. Some of the major products from tamarind have been enumerated.

Tamarind Kernel Powder

The major industrial use of the seeds is in the manufacture of Tamarind Kernel Powder (TKP). It is prepared by decorticating the seed and pulversing the creamy white kernels. The decorticated seed is ground to the required mesh size by machines to obtain a yield of 55-60%. The powder tends to deteriorate during storage under humid conditions; hence storage in a dry place in moisture proof containers is important. Mixing with 0.5% of sodium bisulphate before packing will prevent enzymatic deterioration. The TKP will become rancid and brown on storage and the storage ability and colour will be better if it is defatted (Sivarama Reddy *et al.*, 1979). The general characteristics of a good tamarind seed powder is that it should have flavouring characteristics when dissolved in water and be free of any burnt or other undesirable flavours, it should have good keeping quality and be free from any insect pests, fungal growth or extraneous materials.

Pectins

Polysaccharids obtained from tamarind seed kernels from mucilaginous dispersions with water and possess the characteristic property of forming gels with sugar concentrates, like fruit pectins. However, unlike fruit pectin, tamarind polysaccharide can form gells over a wide pH range, including neutral and basic conditions. Tamarind polysaccharides are also not affected by boiling in neutral aqueous solutions, even if boiled for long periods. Fruit pectins undergo degradation on boiling and fall to one-third of the original value after one hour of boiling. Therefore tamarind polysaccharide can be useful as a gel information agent and may be substituted for fruit pectins. Tamarind polysaccharide does not contain galacturonic acid and methyluronate and is therefore not regarded as true pectin; is termed as *"Jellose"* (Rao, 1948). Jellose is prepared on a large scale by adding TKP to 30-40 times its weight of boiling water, containing citric or tartaric acid at a concentration of 0.2%. It is then stirred vigorously and boiled for a further 30-40 minutes. The resultant solution is kept overnight for setting and the supernatant liquid is siphoned off and concentrated under vaccum, passed through a filter press and then dried in a drum drier. The resultant product is pulverized in a ball mill. A good sample of jellose should have a relative viscosity of 5.0% at 35°C in 0.5% solution, which is somewhat higher than corn starch. Jellose is much cheaper than corn starch and is required in smaller qualities (Gunasena and Hughes, 2000).

Tamarind Juice Concentrate

Tamarind juice concentrate (TJC) is a convenient product, as it is easy to dissolve and reconstitute in hot water (Anon., 1982 b). It can also be stored for long periods. TJC is prepared by extracting cleaned pulp with boiling water using the counter current principle, where dilute extracts are used for extracting fresh batches of the pulp. Using this process, an extract is obtained containing 20% soluble solids. The extract is separated from the pulp by sieving and is concentrated under vacuum in a forced circulation evaporator. When the soluble solids reach 68%, the pulp is placed in cans or bottles. It sets like jam on cooling. The yield of the concentrate is about 75% of the pulp used. The quality of the product is assured by using recently harvested fruits free from insect and rodent damage (Lewis *et al.*, 1964 a). An Indian standard is available for TJC (IS: 5955; 1933) developed to maintain the quality of the product.

Tamarind Pulp Powder

Tamarind pulp powder (TPP) is prepared by concentrating, drying and milling the pulp into a powder form. It is one of the convenience food products produced commercially by several manufactures in India. Depending on the manufacturing process, wide variations in the physico-chemical characteristics in TJC and TPP have been reported (Table 7.5). Manjunath *et al.*, (1991) reported that on an average the total soluble solids content of tamarind pulp powder varied from 18.6-25%, acidity from 8.7-11.1% with an average value of 9.9% (as tartaric acid). The moisture content ranges from 3.5-8.8%. Among the minerals, the calcium and potassium contents were high and ranged from 74-143mg and 23.8-27.7mg. Starch is the major ingredient in tamarind pulp powder (20-41%). The extraction and processing techniques of the pulp for the preparation of canned tamarind syrup, clarified tamarind juice and other soft drinks have been reported by Bueso (1980). The pulping operation consists of breaking the shells by hand and agitation in water to separate the pulp and the seeds. The industrial method used in India for the manufacture of tamarind juice concentrate is by extraction of all the solubles from the fruit pulp in boiling water (Nagaraja *et al.*, 1975). The recommended concentration of tamarind pulp in syrup is between 20-24%, so as to produce a beverage with distinctive flavour and acidity (Tarber, 1915; Bueso, 1980). Such syrup contains 56.7% total soluble solids, 43.8% reducing sugars and a total acidity of 1.11% as tartaric acid.

On a small scale, the fruit pulp is made into a refreshing drink after dissolving in water and squeezing by hand. After removing the extraneous matter, water is added to dilute the drink as required. Modern food technology has greatly advanced the preservation and utilization of tamarind fruit. Tamarind drinks in poly packs are commercially available in Thailand, Indonesia and in many countries of Africa. In Mexico and Guatemala, the drink is carbonated. It has a high content of vitamin B (Thiamine and Niacin) as well as small amounts vitamin C (ascorbic acid) and carotene (Sozolnoki, 1985). Tamarind fruit juice is most valued for its ascorbic acid content, which is about 5.1-6.1 mg/100ml.

Table 7.5 : Chemical composition (%) in tamarind juice concentrate and tamarind pulp power

Constituent	Tamarind juice concentrate (TJC)	Tamarind pulp powder (TPP)
Moisture	30.0	3.5-8.8
Tartaric acid (total)	13.0	8.7-11.1
Invert sugars	50.0	15.8-25.0
Proteins	2.0	1.7-2.4
Starch		20.0-41.3
Ash		2.1-3.2
Crude fibre	2.0	

Source : Nagaraja *et al.*, (1975); Manjunath *et al.*, (1991)

Tamarind Pickle

Pulp is used commercially to prepare tamarind pickle (Plate 7.9). The pickles are commonly used in Asia as an accompaniment to curries or other main meals. Tamarind pickles of hot, spicy and salty-sour taste can be preserved for several months. Preservation is due to the presence of salt, increased acidity and spices. The preparations of pickles are simple and can be done at a household level. For making pickles, fresh mature tamarind fruits are selected and soaked in clean water for 12 hours and pulp is separated. For each kilogram of tamarind, one kilogram of sugar is added and boiled, stirring the mixture continuously. Then mixtures of spices are added. Spice mixture for one kilogram of tamarind should include the following : 40g coriander, 50 g cumin seed, 30 g black cumin, 3-4 cloves, 3-4 pieces of cardamom, 3-4 pieces of cinnamon, 10-12 chillies, 30g salt, 250 ml mustard oil and 30 g black pepper. These are ground into a paste and added to the boiling tamarind and mixed thoroughly. The product is then packed into pre-sterilised glass jars or plastic bags and allowed to cool at room temperature. The pickle can be store in small polythene bags or in clean jars and sealed. It will keep well if stored in a cool place. Care should be taken to prevent air bubbles in the pickle and to ensure that a laver of oil covers the product.

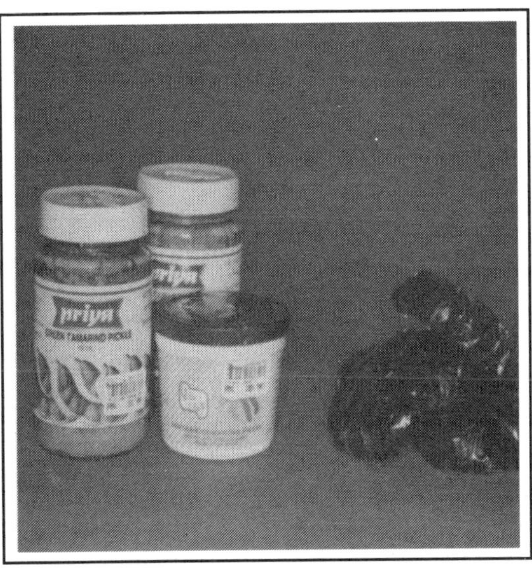

Plate 7.9. Value added products from Tamarind

Jam

In making tamarind jam, ripe fruits are shelled and the pulp is boiled for 10 minutes. The pulp is then drained and separated from the seeds. For every cup of pulp, two cups of brown sugar are added. The mixture is then cooked and constantly stirred while boiling until it becomes thick in consistency. The resulting jam is cooled, packed in dry, sterilized jars and sealed.

Syrup

Tamarind syrup is made by boiling immature fruit pulp until it is soft and then strained through cheesecloth. Forevery cup of juice, a half-teaspoon of baking soda is added. The mixture is boiled down to one-half the original quantity, removing the rising scum in the process. The juice is again strained and for every cup obtained, a quarter cup of sugar is added. The mixture is boiled again for 20 minutes. The cooled syrup is poured into sterilized bottles and sealed.

Candy

Sweetened tamarind fruit is made by peeling whole ripe fruits and pouring boiling (62^{0}C) sugar syrup over the fruits and placing in a deep enamel basin. Boiling syrup is prepared by mixing three parts sugar with one part water. After soaking for three days, the fruits are drained of the old syrup and again covered with freshly prepared syrup. The process is repeated until the fruits are sweet enough. The sweetened fruits are arranged on bamboo racks, dried in sum and covered with a wire screen to keep off flies. Final drying is done in an oven at a low temperature. After cooling, the dry fruits are wrapped in cellophane.

Future Research Needs

Although tamarind has commercial demand and being used widely, there has been little effort by scientists to improve the tree as a crop plant. The research undertaken by different institutions in the tamarind growing countries are scattered. Therefore, there is a strong need for a clear and prioritized research agenda for the development of tamarind. The priority research needs of tamarind are:

1. Management of genetic resources for various traits. Conservation of tamarind germplasm is pre-requisite to prevent the loss of genetic diversity of the species. Some collection has been made however, the number of accessions collected appears to be small and does not cover the entire spectrum of natural variability. Countrywide collections should be undertaken and maintained *in situ* and *ex situ* for use in future. Moreover, the collected germplasms in case of tamarind has not been fully exploited for tree improvement. Under the sponsorship of the ICRAF, Nairobi, the African Tamarind Network has been formed to study genetic variation in African tamarind. However, tamarind is extensively grown in other tropical regions, such as South and Southeast

Asia and tropical America therefore these countries should also be made partner of network programme. Although, several types and cultivars have been developed over the years through utilization and domestication in different regions but these needs intensification through international collaboration. Detailed information on genetic variation of the species is pre-requisite for its effective utilization. However, very little is known about the genetic variation of tamarind. Therefore, it is essential to initiate studies on genetic improvement such as genetic variation, genetic structure and effective size of breeding populations, both in natural and naturalized areas and development of cultivar by combining desirable traits.

2. Simple and easily practicable methods for rapid propagation are need of the time to produce seedlings and making them freely available to growers.
3. Post Harvest Management and Value addition to develop acceptable products should be taken at priority.
4. Efforts should be made for industrialization and marketing of tamarind pulp and its product.

REFERENCES

Allen, O.N. and Allen, E.K. (1936) Plants of the sub family Caesalpiniodeae observed to be lacking nodules. *Soil Sciences*, 42(2): 87-91.

Allen, O.N. and Allen, E.K. (1981) The *Leguminous : A Source Book of Characteristics, Uses and Nodulation.* Madison, WS, University of Wisconsin Press: p 812.

Amin, R.S. (1978) *Soft* wood grafting' a new technique for hardwood plants. *Current Science*, 47: 468-469.

Anon. (1976) *Tamarindus indica* L. In *The Wealth of India* (Raw Materials Series) Vol. X: 144-122. Council of Scientific and Industrial Research, New Delhi.

Anon. (1982a) Some recent developments. Central Food Technological Research Industry, 1: 43-44.

Anon. (1982b) Tamarind juice concentrate plant starts in Mysore. *Indian Food Industry*, I: 43-44.

Anon. (1991) *Revised Handbook of the Flora of Ceylon*. Vol. VII. *Fabaceae, Tamarindus indica* L. eds. M.D.Dassanayake and F..Fosberg. Oxford & IBH Publishing Co. Pvt. Ltd. New Delhi: 96-97.

Athar, M. and Mahmood, A. (1982) New Records of Legume Nodulation from *Pakistan. Pakistan Journal of Botany*, 14:36.

Baragano de Mosquida, M. (1980) Technology of clarified tamarind juice. In *Tropical and Subtropical Fruits.* Nagy, S. and Shaw, P.E. (Eds.), AVI Publishing, Westport, Connecticut: p 375.

Benero, J.R., Collazo de Rivera, A.L. and Georage, L.M.I. (1974) Studies on the preparation and shelf life of soursop, tamarind and blended soursop-tamarind soft drinks. *Journal of the Agricultural University, Puerto Rico,* 58-99.

Benero, J.R., Rodriguez, A.J. and Collazo de Rivera, A. (1972) Tamarind. *Journal of the Agricultural University, Puerto Rico,* 56(2): 185-186.

Benthall, A.P. (1933) *The Trees of Calcutta and its Neighbourhood.* Thacker Spink & Co.: p 513.

Bhattacharya, P.K., Bal, S. and Mukherji, R. K. (1994) Studies on the characteristics of some products from tamarind (*Tamarindus indica* L.) kernels. *Journal of Food Science and Technology,* 31(5): 372-376.

Bhutani, D.K. (1978) Insects pests of tamarind and their control. *Pesticides,* 12(11): 34-41.

Bueso, C.E. (1980) Soursop, Tamarind and Chironka. In *Tropical and Subtropical Fruits.* Nagy, S. and Shaw, P.E., (Eds.), AVI Publishing, Westport, Connecticut: p 375.

Chant, S.R. (1993) Fables. In *Flowering Plants of the World.* Editor V.H. Heywood. B.T. Batsford Ltd. London.

Chapman, K.R. (1984) Tamarind. In *Tropical Tree Fruits for Asutralia.* Compiled by P.E. Page. Queensland Department of Primary Industries. Information series Q 183018, Brisbane: 83-86.

Chattopadhyay, P.K. and Mohanta, S.K. (1988) Influence of the propagation medium on the germination and development of tamarind seedlings. *South Indian Horticulture,* 36(6):324.

Chaturvedi, A.N. (1985) Firewood farming on the degraded lands of the Gangetic plain. U.P. *Forest Bulletin* No.50. Lucknow, India. Government of India Press (1): p 286.

Chaturvedi, A.N., Bhatt, D.N., Mishra, C.M. and Singh, S.L. (1986) Root development in some tree species on usar soils. *Journal of Tropical Forestry,* 2(2): 119-120.

Chiemsombat, P., Naritoom, K., Jiwa Jinda, S., Klinkong, Krai-rod, A. and Feungchan, S. (1996) Microscopic studies on association of micro-organisms with tamarind roots and their rhizosphere. *Thai Journal of Agricultural Science.* Special Issue No. (1): 68-92.

Chundawat, B.S. (1990) *Arid Fruit Culture.* Oxford IBH Publishers, New Delhi: 158-163.

Chuntanaparb, L. and Ranganathan, R. (1990) Defining Tree Breeding Objectives for Multipurpose Trees in Asia. MPTS Research Series Report Number 10. Bangkok : Winrock International F/FRED. Eds. J.B. Raintree and David Taylor : p 88 and 105.

Coates-Palgrave, K. (1988) Trees of Southern Africa. 10. *Tamarindus indica* L. C. S. Striuk Publishers, Cape Town : 278-279.

Coronel, R.E. (1986) *Processing Fruits of the Philippines*. 2nd Ed., University of the Philippines, Los Banos.

Coronel, R.E. (1991) *Tamarindus indica* L. *In Plant Resources of South East Asia*, Wageningen : Pudoc, No.2. Edible fruits and nuts. Eds. E.W. M. Verheij and R.E. Coronel, PROSEA Foundation, Bogor, Indonesia: 298-301.

Dalziel, J.M. (1937) *The Useful Plants of West Tropical Africa. London.* Crown Agents for Overseas Governments and Administrations : p 612.

Ding, M.M., Yi, W.M. and Liao, L. Y., (1986) A survey of the (nitrogen) ase activities of tree legumes, including *Tamarindus indica* L., a species not widely known to nodulate, in artificial forest in Dainbai, Guangdong, China. *Nitrogen Fixing Tree Research Reports*, 4: 9-10.

Duke, J.A., (1981). *Handbook of Legumes of World Economic Importance.* Plenum Press, New York: 228-230.

Dwivedi, S.K., Ali Wahid and Pathak, R.K. (1996) Effect of sodacity on growth and mineral content of tamarind, *Tamarind indica* L. *Annals of Agricultural Research*, 17(4): 447-449.

Eggeling, W.J. and Dale, I.R. (1951) The *Indigenous Trees of the Uganda Protectorate Entabbe, Uganda.* The Government Printer: p 491.

FAO. (1988) Fruit Bearing Trees. Technical notes. FAO-SIDA Forestry Paper 34: 165-167.

Feungchan, S., Yimsawat, T., Chidaprasert, S. and Kitpowsong, P. (1996a) Tamarind (*Tamarindus indica* L.) Plant genetic resources in Thailand. *Thai Journal of Agricultural Science*, Special Issue No. (1): 1-11.

Feungchan, S., Yimsawat, T., Chindprasert, S. and Kitpowsong, P. (1996d) Effect of plant regulators in fruit setting. *Thai Journal of Agricultural Science, Special Issue*, No. (1): 48-51.

Feungchan, S., Yimsawat, T., Chindaprasert, S. and Kitpowsong, P. (1996f) Studies on the colour change of tamarind pulp in storage. *Thai Journal of Agricultural Science*, Special issue, No. (1): 52-56.

Galang, F.G. (1955) *Fruits and Nuts Grown in the Phillippines.* Malabon, Rizal AIA Printing Press.

Gowda, N. (1983) Studies on vegetative propagation of tamarind (*Tamarindus indica* L.) by air layering. *Thesis Abstract. Haryana Agricultural University*, 9(3): 276-277.

Grisebach, A.H.R. (1864) *Flora of the British West Indian Islands.* Lovell Reeve & CO, Henrietta Street, Covent Garden. London: 212-213.

Gunasena, H.P.M. and Hughes, A. (2000) Tamarind. *ICUC,* Southampton, U.K.: 1-170.

Gupta, G.N. and Mohan, S. (1990) Response of several tree species to management on degraded soils of semi arid region. *Indian Forester,* 116(8): 622-630.

Hasan, S.K. and Ijaz, S. (1972) Tamarind- A review. Science Industry (Karachi), 9(3&4): 131-137.

Hocking, D. (1993) Trees for Drylands. *Tamrindus indica* L. *Family leguminosae : Caesalpinoideae.* Oxford & IBM Publishing Co, New Delhi: 305-307.

Hong, T.D., Linnington, S. and Ellis R.H. (1996) *Seed Storage Behaviour : A Compendium.* Handbook for Genebanks No. 4. International Plant Genetic Resources Institute, Rome.

Irvine, F.R. (1961) *Woody Plants of Ghana.* Oxford University Press, London : p 868.

Ishola, M.M., Agbaji, E.B. and Agbaji, A.S. (1990) A chemical study of *Tamarindus indica* (Tsamiya) fruits grown in Nigeria. *Journal of Science, Food and Agriculture,* 51: 141-143.

Jaiwal, P.K. and Gulati, A. (1992) Micropropagation of *Tamarindus indica* L. from shoot tip and nodal explants. *National Academy Science Letters-India,* 15(3): 63-67.

Jambulingam, R. and Fernandes, E.C.M. (1986) Multipurpose trees and shrubs in Tamil Nadu State (India). *Agroforestry Systems,* 4(1): 17-32.

Jayaweera, D. M. A. (1981) *Medicinal Plants (Indigenous and Exotic) Used in Ceylon.* Part III. *Flacourtiaceae-Lytharaceae.* A publication of the National Science Council of Sri Lanka: 244-246.

Joseph, K.V. and Oommen, P. (1960) Notes on some insect pests infesting dry tamarind fruits in Kerala State. *Indian Journal of Entomology,* 22(3): 172-180.

Khan, S.N., Misra, B.M. and Rehill, B.S. (1988) Two new species *Cerospora. Indian Journal of Forestry,* 11(3): 188-190.

Knight, T., Jr, (1980) Origin and world importance of tropical and subtropical fruit crops. *In Tropical and Subtropical Fruits,* Nagy. S. and Shaw, P.E., Eds., AVI Publishing, Westport, Connecticut.

Kopp, M.S. and Nadaraja, K. (1990) *In vitro* plantlet regeneration from shoot tip cultures of *Tamarindus indica* L. *Indian Journal of Forestry,* 13(1): 30-33.

Kulkarni, R.S., Ganaprasad, S. and Swamy, G.S.K. (1993) Tamarind : Economically an important minor forest produce. *Minor Forest Products News,* 3(3): 6.

Lakshiminarayan Rao, M.V., Subramanian, N., and Srinivasan, N. (1954) Free amino acids Tamarind pulp. *Journal of Science and Industrial Research*, 13B: 377.

Lefevre, J.C. (1971) Review of the literature on the tamarind. *Fruits* (English Summary), 26: 687.

Leonard, J. (1957) Genera des Cynometrae et des Amherstieae africaines (Leguminosae-Caesalpinioideae). *Memoire Acad. Royale Belgique*, 30(2): 1-314.

Leung, W.T. and Flores, M. (1961) *Food Composition*. Tables for Use in Latin America, National Institute of Health, Bethesda, Md.

Lewis, Y.S., Neelakantan, S. and Bhatia, D.S. (1961) Organic acid metabolism in tamarind leaves. *Current Science*, 30: 381.

Lewis, Y.S. and Neelakantan, S. (1964a) The chemistry, biochemistry and technology of tamarind. *Journal of Science and Industrial Research*, 23: p 204.

Little, E.L. and Wadswoth, F.W. (1964) *Common Trees of Puerto Rico and the Virgin Islands, Agriculture*. Handbook 249, Washington DC. US Department of Agriculture: p 548.

Lokesha, S. and Shetty, H.S. (1991) A *Pestalotia* species causing stony fruit disease in tamarind. *International Journal of Tropical Plant Diseases*, 9(2): 179-181.

Mahadevan, N.P. (1991) Phenological observations of some forest trees as an aid to seed collection. *Journal of Tropical Forestry*, 7(3): 243-247.

Maksoud, M.A., Haggag, L.F., Azzay, M.A. and Saad, R.N. (1994) Effect of VAM inoculation and phosphorus application on growth and nutrient content (P and K) of *Tamarindus indica* L. (Tamarind) seedlings. *Annals of Agricultural Science* (Cairo), 39(1): 355-363.

Manjunath, M.N., Sattigeri, V.D., Rama, Rao S.N., Udah Rani, M. and Nagaraja, K.V. (1991) Physico-chemical composition of commercial tamarind powder. *Indian Food Packer*, 45: 39-42.

Meillon, S. (1974) Process for making drinks, syrups, juice and liquor and solid extracts based on tamarind products. French Patent No. 2231322 (*English Summary*).

Mital, V.P. and Khanna, H.C.L. (1978) Efficiency of some fumigants against adults of tamarind bruchid, *Caryedon serratus*. *Bulletin of Grain Technology*, 16(2): 147-148.

Morad, M.M., El Magoli, S.B. and Sedky, K.A. (1978) Physico-chemical properties of Egyptian tamarind seed oil. *Fette Seifen Anstrichmittel*, 80: 357-359.

Morton, J. (1987) Fruits of Warm Climates. Miami FL: 115-121.

Mukherji, K.G. and Bashin, J. (1986) *Plant Diseases of India* : A *Source Book.* New Delhi. Tata McGraw Hill: p 468.

Murashige, T. and Skoog, F. (1962) A revised medium for rapid growth and bioassays with tobacco tissue culture. *Physiologia Plantarum,* 15: 473-497.

Nagaraja, K.V., Manjunath, M.N. and Nalini, M.L. (1975) Chemical composition of commercial tamarind juice concentrate. *Indian Food Packer.* 29: 17.

NAS, (1979) *Tropical Legumes : Resources for the Future.* Washington DC : 117-121.

Navaneetha, N., Palniswamy, K.P., Abdul-Khader, Md and Kumar, N. (1990) Studies on rooting of cuttings of tamarind. *South Indian Horticulture,* 38(4): 220-224.

Navaneetha, N., Palniswamy, K.P., Abdul-Khader, Md and Kumar, N. (1991) Studies on rooting tamarind air layers. *South Indian Horticulture,* 39: 102-105.

Panigrahi, S., Blend, B. and Carlaw, P.M. (1989) Nutritive value of tamarind for broiler chicks. *Journal of Animal Feed Science and Technology,* 22(4): 285-293.

Parrotta, J.A. (1990) *Tamarindus indica* L. Tamarind. SO-ITF-SM-30 June, USDA Forestry Service. Rio Piedras, Peurto Rico : 1-5.

Pathak, R.K., Ojha C.M. and Dwivedi, R. (1992) Adopt patch budding for quicker multiplication in tamarind. *Indian Horticulture,* 36(2): 17.

Pathak, K.K. (1974) Seasonal pattern of tartaric acid metabolism underlying the phasic development in *Tamarindus indica. Biol. Plant,* 16: 1.

Postgate, J. (1979) *Nitrogen Fixation,* Southampton. The Camelot Press Ltd.

Prins, H. and Megehembe, J.A. (1994) Germination studies on seed of fruit trees indigenous to Malawi. *Forest Ecology and Management,* 64(2-3): 111-125.

Purohit, M., Jamaluddin, M. and Mishra, G.P. (1996) Testing seeds of some tropical tree species for germination and mycoflora. *Indian Forester,* 112: 492-495.

Purseglove, J.W., (1987) *Tropical Crops. Dicotyledons,* Longman Science and Technology: 204-206.

Purushotham, K. and Narasimharao, S.B.S. (1990) Propagation of tamarind by veneer and soft wood grafting. *South Indian Horticulture,* 38(4): 225.

Quiniones, S.S. (1983) Particularly fast growing tree species used in reforestation in the Philippines. *Nitrogen Fixing Tree Research Reports.* 1: 12.

Raizada, U. (1965) Life history *Scirtothrips dorsalis* Hood with detailed external morphology of the immature stage. *Bulletin of Entomology,* 6:30.

Rama Rao, M. (1975) *Flowering Plants of Travancore.* Bishen Singh Mahendra Pal Singh Publications Dehra Dun India, p 484.

Ramkumar, M.V., Babu, C.K., Subramanya, S., Ranganna, B. and Krishnamurthy, K.C. (1997) Development of a tamarind dehuller and short-term storage of pulp. Proceedings of National Symposium on *Tamarindus indica* L., Tirupati (A.P.) Organized by the Forest Department of A.P. 27-28 June 1997: p 145-150.

Rao, P.S. (1948) Tamarind seed (jellose pectin) and its jellying properties. *Journal of Science and Industrial Research*, 68: 89-90.

Rao, Saideswara. Y. (1995) Tamarind Economics. *Spice in India*, May Issue: 10-11.

Reena, J. and Bagyaraj, D.J. (1990) Growth stimulation of *Tamarindus indica* L. by selected VA micorrhizal fungi. *World Journal of Microbiology and Biotechnology*, 6(1): 59-63.

Relwani, L.L. (1993) Tamarind: A multipurpose tree for diverse environments. *The Bharathia Agricultural and Industrial Foundation Journal*, 13(2): 10.

Salim, A., Simons, A., Waruhin, A. and Orwa, C. (1998) Agro forestry Tree Database : A tree species reference and selection guide and tree seed suppliers directory. International Council for Research in Agroforestry, PO Box. 30677, Nairobi, Kenya.

Savur, G.R. (1948) Isolation and Characterization of Tamarind seed polysaccharide. *Journal of Biochemistry*, 199: 501-509.

Shankaracharya, N.B. (1998) Tamarind-Chemistry, Technology and Uses-a critical appraisal. *Journal of Food Technology*, 35(3): 193-208.

Siddaramaiah, A.L. and Kulkarni, S. (1982). Control of powdery mildew disease of tamarind seedlings. *Indian Forester*, 108(5): 361-364.

Simmonds, N. W. (1984) *Evolution of Crop Plants*. Longmans, London.

Singh, P.P. (1973) Oxalic acid content of Indian foods. Plant Materials Vegetables, Chemical Abstracts, 79: 135506w.

Singha, R.K. (1995) Biodiversity conservation through faith and tradition in India. Some case studies. *International Journal of Sustainable Development and World Ecology* (UK), 2(4): 278-284.

Sivarama Reddy, G., Jaganmohan Rao, S., Achyutaramayya, D., Azeemoddin, G. and Tirumala Rao, S.D. (1979) Extraction, Characteristics and fatty acid composition of tamarind kernel oil. *Journal of Oil Technology Association* (India), 11: 91-93.

Sone, Y. and Sato, K. (1994) Measurement of oligasaccharides derived from tamarind xyloglucan by competitive ELISA assay. *Bioscience Biotechnology and Biochemistry*, 58 : 2295-2296.

Sozolnoki, T.W. (1985) Food and Fruit Trees of Gambia. Hamburg, Federal Republic of Germany: 132p Tiftung Walderhatung in Africa.

Splittstoesser, W.E. and Mohamed Yaseen, Y. (1991) *In vitro* shoot regeneration of tamarind (*Tamarindus indica* L.) and carob (*Cerotonia siliqua*) with thidiazuron. Proceedings of the Inter-American Society for Tropical Horticulture. 35-37th Ann. Meeting, Vina-del-mar, Chile. 7-12 Oct: 6-8.

Srivasuki, K.P., Reddy, R.D. and Reddy, K.K. (1990) Rooting of terminal cuttings of *Tamarindus indica Linn. Indian Forester,* 116: 984-985.

Storrs, A.E.G. (1995) *Know Your Trees. Some common trees found in Zambia.* Regional Soil conservation Unit (RSCU). Zambia.

Swaminath, M.H. and Ravindran, D.S. (1989) Vegetaive propagation of fruit yielding tree species. *Myforest,* 25(4): 57-360.

Swaminath, M.H., Ravindra, D.S., and Mumtaz, J. (1990) Propagation of tamarind through stem cuttings. *Myforest,* 26(2): 207-208.

Tarber, W.C. (1915) Tamarind syrup. *Journal of Industrial Engineering Chemistry,* 7: 607.

Thimmaraju, K.R., Narayana Reddy, M.A., Swamy, N. and Sulladmath, U.V. (1977) Studies on the floral biology of tamarind (*Tamarindus indica* L.) *Mysore Journal of Agricultural Science,* 11: 293-298.

Timyan, J. (1996) BWA. Yo: Important Trees in Haiti. Southeast Consortium for International Development, 1634, I Street N.W. Suite 702, Washington D.C. 2006.

Troup, R.S. (1909) Indian Forest Memoirs, Indian Woods and their uses. Economic Product Series. 1 (1): 253. Superintendent of Government Printing, Calcutta.

Troup, R.S. (1921) The Silviculture of Indian Trees Vol.11. Leguminosae (Caesalpinieae) to Verbenaceae. 7. *Tamarindus indica* L.: 263-363 p. Oxford, Clarendon Press.

Ulrich, R. (1970) Organic acids. In *The Biochemistry of fruits and their products* Vol. 1, Hulme, A.C., Ed., Academic Press, New York.

Usha, K. and Singh, B. (1996) Influence of open and cross pollination on fruit set and retention in tamarind (*Tamarindus Indica* L.). *Recent Horticulture,* 3(1): 60-61.

Vogt, K. (1995) *A field workers, guide to identification propagation and uses of Common Trees and Shrubs of Dry Land Sudan.* 52p. SOS Sahel International, UK 1, Tolpuddle Street, London NI OXT.

Von Carlowitz, P.G. (1986) *Multipurpose Trees and Shrubs Seed Directory.* International Council for Research Agroforestry, Nairobi : p 265.

Von Maydell, H.J. (1986) *Trees and shrubs of the Sahel, their characteristics and uses.* Rossdorf L TZ-Verlagsgesellschaft: p 382-387.

Wenkam, N.S. and Miller, C.D. (1965) Composition of Hawaii Fruits. Hawaii Agriculture Experimental Station, University of Hawaii, Bulletin, 135.

Yaacob, O. and Subhadrabhandu, S. (1995) The Production of Economic Fruits in South East Asia. Oxford University Press: 133-142.

8

JAMUN

The Jamun (*Syzygium cuminii* Skeels), a member of family Myrtaceae is one of the important under utilized fruit crop, distributed through out the tropics and sub-tropic as an avenue plantation. It is one of the most hardy fruit crop and can easily be grown in neglected and marshy areas where other fruit plants can not be grown successfully. Its fruit is nutritious and are rich in minerals, fibre, carbohydrates, vitamin A, vitamin C, phenols, tannins and flamanoids. Mature fruit contains 1.0g iron, 0.02g calcium, 0.01g phosphorus, 0.7g protein, 0.1g fat, 0.9g fibre and 19.7g carbohydrates in 100g pulp. Colour in jamun fruit is due to anthocyanin pigments.

All parts of jamun tree are used for various purposes. The leaves have served as fodder for livestock and as food for tassar silkworms in India. The leaves contain crude protein (9.1%), fat (4.3%), crude fiber (17.0%), ash (6.0%), calcium (1.3%) and phosphorus (0.19%). They are rich in tannin and contain the enzymes esterase and galloyl carboxylase which are presumed to be active in the biosynthesis of the tannins. The essential oil distilled from the leaves is used to scent soap and is blended with other materials in making inexpensive perfume. Its chemical composition of jamun fruit has been presented in Table.8.1. It consists mainly of mono- or sesqui-terpene hydrocarbons which are "very common in essential oils." The leaves are antibacterial and are used for strengthening the teeth and gums. The fruits and seeds are sweet, acrid, sour tonic and cooling and are used in diabities, diarrhoea and ring worm .The bark is astringent, sweet sour, diuretic digestive anthelmintic used for curing diabetes. Jamun trees are mainly planted as windbreak and on roadsides. It is also experience in drier parts of Israel, Algeria, Malagassy, West Indies and southern Africa where the soil conditions are poor where other fruits plants cannot be grown successfully.

Jamun bark contains 8 to 19% tannin and is much useful to leather and preserving fishing nets industry. It is also used for brown dyes of various shades depending on the mordant and the strength of the extract. Bark also has many medicinal properties.

The wood is used for making beams and rafters, posts, bridges, boats, oars, masts, troughs, well-lining, agricultural implements, carts, solid cart wheels, railway sleepers and the bottoms of railroad cars. It is a fairly satisfactory fuel.

Table 8.1: Food value per 100 g of edible portion of jamun*

Nutrient	Food value	Nutrient	Food value
Moisture	83.7-85.8 g	Potassium	55 mg
Protein	0.7-0.129 g	Copper	0.23 mg
Fat	0.15-0.3 g	Sulfur	13 mg
Crude Fiber	0.3-0.9 g	Chlorine	8 mg
Carbohydrates	14.0 g	Vitamin A	80 I.U.
Ash	0.32-0.4g	Thiamine	0.008-0.03 mg
Calcium	8.3-15 mg	Riboflavin	0.009-0.01 mg
Magnesium	35 mg	Niacin	0.2-0.29 mg
Phosphorus	15-16.2 mg	Ascorbic Acid	5.7-18 mg
Iron	1.2-1.62 mg	Choline	7 mg
Sodium	26.2 mg	Folic Acid	3 mcg

*Values reported from Asian and tropical American analyses.

Fresh jamun fruits had 708 mg/100 g AEAC units of antioxidant capacity. There was significant increase from 708-1300 mg/100 g AEAC unit of antioxidant activity on storage at room temperature , mainly because of continued biosynthesis of anthocyanins.Total phenolic content of fruits increased from 360mg/100 g in fresh fruit to 570 mg /100 g in room temperature stored fruits reported by Rao *et al.*, (2006). The fruits are a good source of sugar (15.6%) and vitamin C (14.3 mg/100 g). The fruits act as effective medicine against diabetes, heart and liver diseases. The seed powder reduces sugar from urine very quickly. The fruit are eaten raw and can be made into tarts, sauces and jam. Astringent fruits are improved in palatability by soaking them in salt water or pricking or rubbing them with little salt and letting them stands for an hour. The juice is similar to grape juice. White fleshed jamun has adequate pectin and makes a very stiff jelly. The purple fleshed fruit yield rich coloured jelly but is deficient in pectin and requires the addition of a jelling agent. Good quality juice is excellent for sharbat and squash. Jamun vinegar is an attractive product in India with pleasant aroma and mild flavour.

Origin and Taxonomy

The jamun is a native of India. It is also found growing in Thailand, Philippines, and Myanmar, which is considered to be a secondary place of its origin. The jamun has been successfully introduced in many other tropical countries like West Indies, East and West Africa and also sub-tropical region including Florida, California, Algeria and Israel. The jambolan is native in India, Burma, Ceylon and the Andaman Islands. It was long ago introduced into and became naturalized in Malaya region. In southern Asia, the tree is venerated by Buddhists, and it is commonly planted near Hindu temples because it is considered to be a sacred tree to Krishna. The leaves and fruits are employed in worshipping the elephant-headed god, Ganesha or Vinayaka, the personification of "Pravana" or "Om", the apex of Hindu religion and philosophy.

Jamun belongs to genus *Syzygium* which has many wild and cultivated species. The important among are *S. cuminii* Skeels, *S. jambolana*. It belongs to family Myrtaceae. Earlier it was known as *Eugenia jambolana*. There are more than 1000 species of genus *Eugenia*. The genus *Syzygium* has 8 species which grows wild in Nilgiris and Western Ghats area. These species produce small, acrid fruits with thin pulp and large seed. Astringency in these fruits is mainly due to the tannins in the pulp. These species have high timber value. Some important species are *S. juvanicum* (water apple), *S. jambos* (rose apple) and *S uniflora* (Surinam cherry). Dry atmosphere favours good fruiting in jamun and fruit ripening is hastened in hot humid climate.

Jamun plant is an evergreen, multi branched tree of about 25 m height. The stem is grayish white and lower bark coarse and discolored. Leaves opposite, simple, entire, elliptic to broadly oblong, smooth, glossy, somewhat leathery, 5-10 cm (2-5 in) long, short pointed at tips; petioles 3 cm (1.2 in) long; leaf midrib prominent, yellowish; blades with many lateral veins closely parallel. Flowers white to pinkish, about

1 cm (0.5 in) across, in branched clusters (plate 8.1) at stem tips; calyx cuplike; 4 petals, fused into a cap; many stamens. Flower panicles emerge from the axils of leaves (Plate 8.2) which take 15-20 days to open. It is a cross-pollinated tree and pollination is done by honey bees. Only 32.6-36.0 per cent flowers set the fruits. Stigma remains receptive for only 24 hours and about 60 per cent flowers are dropped. Three distinct phases of fruit growth are observed, first between 15-52 days of fruit set when the growth rate remains slow, second about 6 days after the completion of first phase when development is very rapid and the third within 3 days when growth rate is slow with little increase weight. The whole process of fruit growth and development is completed within 62 days after fruit set in north India and, thus fruits mature in 3-4 months after full bloom. Spraying of 25 ppm 2, 4-D at fruit set induces about 83 per cent seedless ness in fruits. Fruit an ovoid, 1-seeded berry to 2 cm (0.8 in) long, dark purplish red, shiny, with white to lavender flesh.

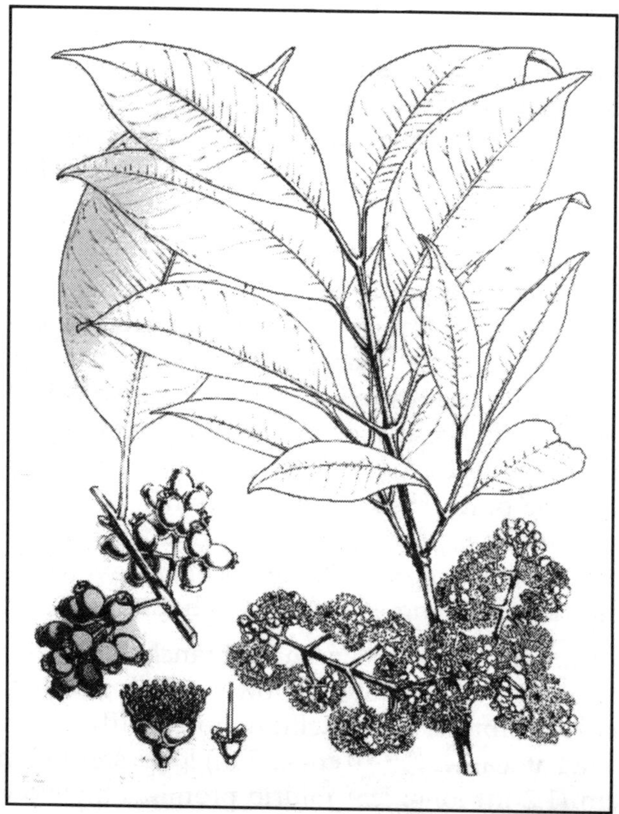

Plate 8.1. Sketch of Jamun leaf, inflorescence, flowers and fruits

Plate 8.2. Flowering and fruiting in Jamun

The jambolan (*Eugenia jambolana*) is a fast growing species, reaching full size in 40 years. It ranges up to 100 ft (30 m) in India and Oceania; up to 40 or 50 ft (12-15 m) in Florida; and it may attain a spread of 36 ft (11 m) and a trunk diameter of 2 or 3 ft (0.6-0.9 m). It usually forks into multiple trunks a short distance from the ground. The bark on the lower part of the tree is rough, cracked, flaking and discolored; further, it is smooth and light-gray. The turpentine-scented evergreen leaves are opposite, 2 to 10 in 5-25 cm long inch 2.5-10 cm wide; oblong-oval or elliptic, blunt or tapering to a point at the apex; pinkish when young; when mature, leathery, glossy, dark-green above, lighter beneath, with conspicuous, yellowish midrib. The fragrant flowers are in clusters 1-to 4 inch, which are ½ inch (1.25 cm) wide, 1 inch (2.5 cm) or more in length; have a funnel-shaped calyx and 4 to 5 united petals, white at first, then rose-pink, quickly shed leaving only the numerous stamens. The fruits are borne in clusters of just a few to10-40 which are round or oblong, often curved; ½ to 2 inch (1.25-5 m) long and usually turns from green to light-magenta, then dark-purple or nearly black as it ripens. A white-fruited types has been reported in Indonesia. The skin is thin, smooth, glossy, and adherent. The pulp is purple or white, very juicy, and normally encloses a single, oblong, green or brown seed, up to 1½ inch (4 cm) in length, though some fruits have 2 to 5 seeds tightly compressed within a leathery coat, and some are seedless. The fruit is usually astringent, sometimes unpalatable and the flavor varies from acid to fairly sweet.

Area and Distribution

Authentic information are not available in respect to area and production of jamun. In India, the maximum number of jamun trees is found scattered throughout the tropical and sub-tropical regions. It is also found in the lower range of the Himalayas up to an elevation of 1300 m and in Kumaon hills up to 1600 m. It is widely grown in the large part of India from the Indo-Gangetic plains in the North to Tamil Nadu in the South. Jamun is native of India to Myanmar and East Indies, but widely cultivated in tropics (Bailey and Bailey 1976). Jamun trees are found in peninsular Florida, mostly in southernmost counties; documented by herbarium specimens from Palm Beach, Collier, and Lee counties (Wunderlin *et al.*, 1995). It is also naturalized in Hawaii (Wagner *et al.*, 1990), Australia (Oommachan 1977, Morton 1987), the Philippines, Zanzibar, Pemba, Mombassa, and Kenya (Morton 1987).The tree is thought to be of prehistoric introduction into the Philippines where it is widely planted and naturalized. It is also naturalized in Java, East Indies, Queensland and New South Wales including the islands of Zanzibar and Pemba, Mombasa and adjacent coast of Kenya. In Ghana, it is found only in gardens. Introduced into Israel perhaps about 1940, it grows vigorously there but bears scantily, the fruit is considered valueless but the tree is valued as an ornamental and for forestry in humid zones. It is grown to some extent in Algiers. Jamun was introduced through seeds from Philippines to Florida for ornamental purpose in 1911 and to U.S.A. for edible fruit in 1920 and from Java and Zanzibar in 1912 (Popenoe 1939). Long recommended as a shade tree but considered messy (Broschat and Meerow 1991). The tree flourishes in California, especially in the vicinity of Santa Barbara, though the climate is not congenial for production or ripening of fruit. In southern Florida, the tree was rather commonly planted in the past. Here, in Hawaii, fruiting is heavy, only a small amount of the crop has been utilized in home consumption. Wunderlin, (1982) reported that jamun is so naturalized in southern Florida that now it is being treated as one of the more rapidly spreading non-native species, forming dense canopies that shade out young native trees in wet pinelands, hammocks, and well drained uplands (Duever *et al.* 1986). It is also reported from 17 natural areas in Pinellas, Lee, Collier, Dade, Palm Beach, and Martin counties (EPPC 1996) and naturalized on all islands of Hawaii, where it excludes native as well as introduced plants, such as *Schinustere binthifolius* (Wagner *et al.*, 1990). By 1870, it had become established in Hawaii and, because of seed dispersal by mynah birds, it occurs in a semi wild state on all the Hawaiian islands in moist areas below 600 m. There are vigorous efforts to exterminate it with herbicides because it shades out desirable forage plants. It is planted in most of the inhabited valleys in the Marquesas. It was in cultivation in Bermuda, Cuba, Haiti, Jamaica, the French Islands of the Lesser Antilles and Trinidad in the early 20th Century; was introduced into Puerto Rico in 1920; but still has remained little-known in the Caribbean region. At the Lancetilla Experimental Garden at Tela, Honduras, it grows and fruits well. It is seldom planted elsewhere in tropical America but is occasionally seen in Guatemala, Belize, Surinam, Venezuela and Brazil.

Soil and Climate

Jamun can be grown on wide range of soils. Deep loam and well drained soils are ideal for its cultivation. Such soils are also retaining optimum moisture for proper growth and development of jamun tree. It can also grow well under salinity and water logged conditions. Despite its ability to thrive in low lying wet areas, the tree does well on higher, well-drained land whether it is loam, sand or limestone. The jambolan tree grows well up to 1,800 m mean sea-level but above 600 m, it does not bear fruits and is being grown for its timber value. It develops most luxuriant growth in regions of heavy rainfall, as much as 1,000 cm annually. It prospers on river banks and has been known to withstand prolonged flooding. Yet it is tolerant of drought after it has made some growth. Dry weather is desirable during the flowering and fruiting periods. It is sensitive to frost when young but mature trees have been undamaged by brief below-freezing temperatures in southern Florida.

Jamun is successfully grown under tropical and sub-tropical climate .It is also found growing in lower hills of the Himalaya up to an altitude of 1300m.Young plants are susceptible to cold (Plate 8.3) and drought condition. The jamun tree requires dry weather at the time of flowering and fruit setting. In sub-tropical areas, early rain is considered to be beneficial for ripening and development of better fruit quality.

Plate 8.3. Young plant of Jamun damaged by frost

Plant Genetic Resource Management and Varieties

The jamun is one of the most popular fruit trees of India, grown throughout the length and breath of the country excepting high hills. A wide range of variability exists with regards to fruit size and quality. However, regular orchards of this fruit tree are very rare. Patel *et al.*, (2005) conducted survey in Uttar Pradesh (Lucknow, Varanasi and adjoining area) and Jharkhand (Ranchi and adjoining area) to find out the existing natural variability among the jamun seedling tree and identified superior genotypes with good fruit qualities. All the genotypes (32) showed considerable variations with respect to physico-chemical characters. The genotypes, RNC-26 and RNC-11 were found promising and had higher fruit number and pulp weight with sweet fruits. Highest pulp content (97.71%) was recorded in V-8 followed by V-7 (93.81%) genotypes collected from Varanasi region .Thin seed with almost negligible seed weight (0.12 g) was observed in V-8 followed by V-6 (0.16 g) and V-7 (0.31 g). Seedling progenies of jamun are wide spread all over India. There is no improved variety presently under cultivation. Some seedling selections have been made on the basis of fruit shape (oblong, obovate), size (big, small) colour (purple, deep purple, red, bluish, and black) and sweetness (sweet, slightly acrid). A local selection prevalent in North India is called Ra Jamun. It has big size oblong fruits of deep purple to bluish black colour. The fruits are juicy having small stones and ripens during June-July. EJ-4 is a selection made Aruppukottai (TN). It is prolific bearer (24.1 kg/plant). A seedless type has been identified at Periyakulam. Salvi *et al.*, (2006) surveyed and identified superior genotype for release under the name "Kankan Bahadoli", it is having bold fruits,small seeds, heavy and cluster bearing habit,high pulp to seed ratio, better table and processing quality at Regional Fruit Research Station, Vengurla. Whereas, Patil *et al.*, (2006) surveyed Karnataka, Goa and Maharashtra to find out the existing natural variability among the jamun seedlings. Observations on physical parameters of 137 genotypes were taken. All genotypes showed considerable variation with respect to fruit weight,seed weight and pulp weight. Other selections made at a number of Research stations have also shown promise.

The common types of jambolan in India are: 1) *Ra Jaman,* with large, oblong fruits, dark-purple or bluish, with pink, sweet pulp and small seeds; 2) *Kaatha,* with small, acid fruits. The named promising cultivars of big fruiting type jamun are, Early Wild, Late Wild, Pharenda; and the small jamun are Dabka (Dubaka). In Java, the small form is called *Djoowet kreekil;* and the seedless form is *Djoowet booten.* In southern Malaya, the trees are small-leaved with small flower clusters. Farther more, the variety called Krian Duat has larger, thicker leaves and red inner bark in which fruits are with purple flesh and more astringency than the white-fleshed types.

Plant Propagation and Nursery Management

Jamun is generally propagated by seeds. Seed germination is not a problem in jamun. Freshly harvested seeds germinate satisfactorily during rainy season. However, seed loose their viability, if stored for longer period. The jamun seed are extracted from ripened fruit of locally available genotypes during May-June. The seeds are sown in raised nursery beds at 2-3 cm depth and watered. Shanmugavelu (1970) reported that seeds treated with GA_3 improve seed germination and seedling growth. Seedling to be shifted to 15x25 cm perforated polythene bags (200 gauge) at 3-5 leaf stage. The polythene bags should be filled by mixing soil and FYM in the ratio of 2:1.After transferring into polythene bags, irrigation should be given at regular intervals. The seedling will be ready for planting in field after one year. Seedlings plants have long gestation period and bears fruit of variable size and quality. Seedlings are also used as root stocks for grafting true to type plant.

Jamun can also be propagated easily on their own roots by cutting and air layering. Bose and Sandhu (1974) reported that tip cutting (herbaceous cutting) of *Syzygium javanicum* treated with 5000 ppm IBA rooted well (100%) under intermittent mist. Air layering is also successful method of propagation in jamun. Singh (1968) reported 86% rooting when 1% IBA in lanolin paste was applied at the upper end of the ring and tied with white polythene by covering with moist *Sphaganum* moss grass. Saha (1969) found only 60% success in air-layers when treated with 500ppm IBA during spring season.

Use of rootstock has also been found successful in jamun propagation through inarching and veneer grafting and patch budding. Inarching is an old age practice for propagation of jamun. The inarching is performed in June-July on one to two-year old rootstocks with the matching thickness of scion. Rootstocks are watered if necessary, till the grafts are separated from the parent tree. The time taken for union formation is about six weeks. Veneer grafting is also successful in jamun. Saha (1970) recorded 31 per cent success in veneer grafting when one-year old seedlings were used as rootstock. The shoots were taken from spring flush and the operation was done in the month of July. Budding is generally employed for propagating jamun in Indonesia. It is an advantageous method of propagation over inarching because of the difficulty in finding suitable scion shoots in jamun near the ground. This method also reduces the expenditure in comparison with inarching. One-year old, 10-12 mm thick rootstocks budded in July-August showed better success (Singh *et al.*, 1967). The scion wood or bud wood is taken from one-to two year old vigorously growing shoots. In India, however, shield, patch and Forkert methods of budding have been tried. Singh *et al.* (1979) recorded 70 per cent success by patch budding in the month of March at Basti in Uttar Pradesh. Forkert method was better than shield or T budding (Singh *et al.*, 1967). Recently, Singh and Singh (2006) reported that plant budded through patch method in July took 16 days for sprouting. However, highest percentage of graft success was noted during March (40%) in patch budding. Soft wood grafting during August recorded 36% per cent success with higher number of leaves than patch budded plants. Irrespective of scion and rootstock, the maximum accumulation of nitrogen and carbohydrate content was recorded in March while it was found lower concentration during September to January. Thus, it can be recommended that patch budding in March and soft wood grafting in July-August should be followed for multiplication of elite jamun plants. Similarly Reddy and Singh(2006) also reported soft wood grafting with 60% success under north Indian conditions. The old, unproductive trees or damaged young plants can be rejuvenated easily by top working, using scion wood or bud from superior clone.

Planting

Seedling plants of jamun are planted in square system at 10x10 m spacing. Budded plants can be planted at 8x8 m spacing under semi-arid conditions. In poor soil condition, however, jamun can be planted at closer spacing of 6 m x 6 m. Rainy season is the best time for planting of jamun in field for better establishment. Planting should be done in the already prepared and filled pits of 1 x 1 x 1 m size. Pit should be filled before onset of rains and soil should be mixed with 20 kg well rotten FYM and anti-termite powder. In North India, planting of jamun can be done during the spring with assured irrigation facilities.

Training and Pruning

Jamun is an evergreen plant which do not require regular pruning however training of young plants is essential to provide better frame work. The tree bole should be kept straight and clean up to at least 1m height and then side branches should be allowed. Sprouts emerging from the rootstock portion should be periodically removed. Jamun branches are brittle and thus tree develops narrow crotch angle which leads to easy breaking of fruiting branches. Therefore wider crotch angles should be developed at least in case of main branches which help to provide better frame to the tree. During later part of plant growth, dried and diseased twigs and criss-cross branches should be removed regularly.

Orchard management

Although no systematic orcharding in jamun is being done, proper plant management help improved yield and quality of fruits in the improved genotypes.

Nutrient Management

Jamun trees are seldom manured. No systematic work has been done on its nutritional requirement for optimum fruit production. However, annual application of FYM @ 20 kg/tree during pre-bearing stage and @ 40 kg/ tree at bearing stage is suggested. Plants should also be provided inorganic fertilizers during bearing stage. Application of 1000g N, 500g P_2O_5 and 500g K_2O to 10-15 years old plants is considered useful under irrigated condition. February-March is the best time for application of manures and fertilizer in jamun.

Water Management

Jamun is a deep rooted plant and draws moisture from lower strata of the soil. To establish young plants, irrigation at 15-20 days interval is essential during summer months. Once established, the trees can withstand much adverse weather conditions under rain fed arid ecosystem. Under arid environment, young plants of jamun should be protected from frost during winters by light irrigations before the expected frost occurrences. Mulching with local weeds and crop residues helps in improving the moisture and moderation of temperature during summer month. In a bearing tree, irrigation is essential during April to June which helps in the fruit retention, development and ripening of fruits.

Orchard Floor Management

During the initial years of orchard establishment, intercropping with suitable leguminous crops and vegetable should be done to get additional income in the well planted jamun orchard. Under arid and semi-arid environment intercrops also improve the soil fertility conditions including its water holding capacity. The jamun tree is useful in silvi-horti cropping system and for integrated development of wastelands.

Disease Management

Being under utilized crop, much attention has not been paid to identify and control the diseases of jamun plant. Pathak (1980) reported that Anthracnose (*Glomerella cingulata*) is a common disease of jamun which affects the leaves and young fruits. Affected leaves show small water soaked circular and depressed lesions. Severely affected fruit become unmarketable. Spraying of mancozeb (0.2%) or 1% Bordeaux mixture checks the incidence of disease.

Pest Management

A number of harmful insects have been found to damage jamun plant both during establishment as well as during fruiting stage. Leaf damaging pests and termite are the main problem in the establishment phase whereas Bark eating caterpillar and Fruit fly are the major pests in the grownup plants.

Leaf eating caterpillar (Carea subtilis)

The caterpillar is a serious pest which defoliates the tender leaves of young growing plant. It can be controlled by two sprays 0.2 per cent carbaryl or 0.04 per cent monocrotophos at 15-20 days intervals.

Bark eating caterpillar (Inderbela sp.)

This pest feed and bore on the bark of main trunk, stem and branches of young as well as fruiting plants. The pest is widely distributed through out the country. Generally the incidence of this pest is more common in neglected and poorly managed orchards. It can be controlled by injecting the kerosene oil or any fumigating pesticide into the hole and plugging them with clay soil.

Fruit fly (Dialeurodes eugeniae)

Fruit fly attacks almost all part of tree in India. Some times the fruits of jamun get wormy due to attack of fruit fly. It can be control by maintaining sanitary situation in the orchard, which consist of picking up the affected fruits and burying them deep in soil and the area under the tree should be dug, so that maggot in the affected fruit and the pupae hibernating in the soil may be destroyed.

Jamun fruits are also damage by squirrels and birds such as parrot and crows. For keeping them away, beating of drum is commonly practiced.

Flowering, Fruiting and Yield

A seedling plant starts bearing after 8-10 years while the vegetatively propagated plants start bearing after 5-6 years. Flower panicles emerge in the axils of leaves during March-April. Flower buds take 15-20 days to open. It is a cross-pollinated tree and pollination is done by honey bees. Stigma remains receptive for only 24 hours and about 60 per cent flowers are dropped which can be reduced by 2 sprays of 60 ppm GA_3 first at full bloom and second after 15 days of initial fruit set. Under normal conditions, however only 32.6-36.0 per cent flowers set the fruits. Fruit development pattern in jamun is single sigmoid. Three distinct phases of fruit growth are observed, first between 15-52 days of fruit set when the growth rate remains slow, second about 6 days after the completion of first phase when development is very rapid and the third within 3 days when growth rate is slow with little increase weight. The whole process of fruit growth and development is completed within 62 days of fruit set in north India and, thus fruits mature in 3-4 months after full bloom (Plate 8.4). Spraying of 25 ppm 2, 4-D at fruit set induces about 83 per cent seedless ness in fruits. A full grown tree produces 70-80 kg fruits every year. The average yield of a full grown seedling tree of improved type is about 80-100 kg whereas grafted plant produces 60-70 kg fruits/ tree every year.

Plate 8.4. Bunch of Jamun with mature and immature fruits

Harvesting and Post harvest Management

Jamun is a non climacteric fruit botanically the berry with delicate skin. All fruits on a tree do not ripen at a time. Fruits ripen after 105-110 days after flowering in South India. The jamun fruit should be picked immediately when it is ripe because it can not be retained on the tree in ripe stage. The ripe fruits are picked singly by hand and in all cases care should be taken to avoid possible damage to fruits. For harvesting, the picker climbs the tree with bags of cotton slung on the shoulder. When the bag is full, the picker comes down from the tree and empties the bag in basket or with the help of rope, hangs the bag down to the ground and another person standing below the tree catches the bag and empties in the baskets. The fruits of jamun are generally harvested daily and send to market on same day. Fruits are harvested by 5-7 such picking in a season. These have to be picked individually to avoid damage by bruising. Harvesting can also be done by gently shaking the branches keeping a cloth or canvas sheet under the tree. The main characteristics of ripe fruits at full size is deep purple or black colour however, change in colour from green to purple, softness of pulp and their detachment from stem are some indications of fruit ripening. Harvesting season varies according to climatic conditions. Fruits are harvested during June-July in north India, May in Karnataka and Andhra Pradesh and during July-August in Tamil Nadu.

The jamun fruit are highly perishable and can only be kept in good condition for about 2-3 days under ordinary condition. The storage life of jamun fruit is 6 days at room temperature and three weeks at low temperature when precooled fruits are kept in perforated polythene bags. Precooled fruits are packed in polythene bags for storage for 3 weeks at 8-10^{0}C and 85-90 per cent relative humidity. Fruits are packed in baskets or paper cartons using local grasses or newspapers as cushioning materials. Fresh fruits are eaten after sprinkling salt.

Processing and Value Addition

Jambolans of good size and quality, having a sweet or sub acid flavour and a minimum of astringency, are eaten raw. Fruits can also be processed into various value added products such as tarts, sauces jellies, jam, squash, syrup, nectar, pickle, vinegar and wine. Natural dye is prepared from the fruit. Astringent fruits are improved in palatability by soaking them in salt water or pricking them, rubbing them with a little salt, and letting them stand for an hour. All kind of fruits preferably the inferior fruits can be utilized for juice which is much like grape juice. When extracting juice from cooked jambolans, it is recommended that it be allowed to drain out without squeezing the fruit and it will thus be less astringent. The white-fleshed jambolan has adequate pectin and makes a very stiff jelly unless cooking is brief. The more common purple-fleshed yields richly colored jelly but are deficient in pectin and require the addition of a commercial jelling agent or must be combined with pectin rich fruits such as unripe or sour guavas. The jamun fruit can be processed into different beverages. Fruit juice of jamun is excellent for R.T.S., sherbet, syrup and "squash". In India, the latter is a bottled drink prepared by cooking the crushed fruits, pressing out the juice, combining it with sugar and water and adding citric acid and sodium benzoate as a preservative.

RTS

This is a type of fruit beverage which contains at least 10 per cent fruit juice and 10 per cent total soluble solids besides about 0.3% acid. Before preparation of beverages, it is necessary to know the techniques for extraction of pulp/juice from jamun fruit. Srivastava and Kumar (2002) mentioned the technique for preparation of jamun juice (Fig 8.1)

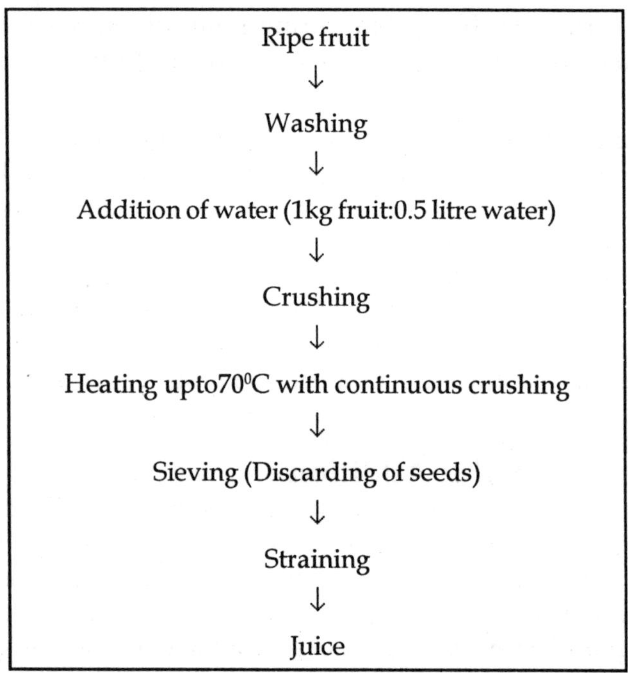

Fig. 8.1. Flow sheet for extraction of jamun juice

Nectar

Techniques for preparation of jamun nectar (13% TSS and 0.3% acidity) has been standerdised at NDUAT, Faizabad for home as well as for processing industry (Srivastava and Kumar, 2002). For commercial preparation of 100 liters nectar, 67 lit. water should be properly mixed with 20 lit. fruit juice, 13 kg sugar and 150 g citric acid. At home scale however, for preparation of nectar from 1 lit. fruit juice, 700g sugar and 7 g citric acid should be mixed with 3.3 lit. water. The juice should be properly mixed with sugar solution and packed in bottles after mixing 600 ppm sodium benzoate. The corked bottles should be pasteurized properly.

Squash

Jamun squash can be prepared by maintaining 25 % juice, 40-50 % TSS, 1% acidity. 600 ppm sodium benzoate should be mixed in prepared jamun squash. The recipes for jamun squash as suggested by Srivastava and Kumar (2002) is 1.8 kg sugar, 1.0 lit. water, 15 g citric acid and 3.0 g sodium benzoate.

Syrup

Jamun syrup should contain 25% fruit juice, 65 % TSS and 1.5 % acidity. At commercial level 50 kg sugar is dissolved in 13 liter water and 25 liter juice and 300 g citric acid is mixed with sugar syrup to prepare 100 liter jamun syrup. At home scale however, 1 lit juice should be mixed with sugar syrup prepared by mixing 2 kg sugar and 15 g citric acid in 0.5 liter of water. Prepared syrup should be bottled and stored.

Jamun Vinegar

Owing to better sugar : acid blend in fruit juice, jamun has been proven as better fruit with high antioxidant properties for preparation of brewed vinegar. Shrivastava and Kumar (2002) described the steps involved in vinegar production from jamun fruit.These are 1. Conversion of available sugar of juice in to alcohol by the yeast (*Saccharomyces* spp.) and the process is called as alcoholic fermentation. 2. Conversion of alcohol in to vinegar by acetic acid bacteria (*Acetobactor* spp. Which is aerobic but activity reduced in light, therefore the process should be done in dark) and the process is known as acitification. Although there are many methods to prepare vinegar, but Orlean Slow Process is very much suitable for jamun vinegar. The flow sheet for preparation of vinegar by Orleans Slow Process has been given in Fig 8.2.).

Jamun Cider

High quality cider can be prepared from jamun juice. In Goa and the Philippines, jamun fruits are important source of wine, somewhat like Port and the distilled liquors. Brandy and *Jambava* have also been made from the fermented fruit. Jamun vinegar is a common value added product throughout India. It is an attractive clear purple product with a pleasant aroma and mild flavour. Srivastava and Kumar (2002) described the procedure for preparation of jamun cider (Fig.8.3).

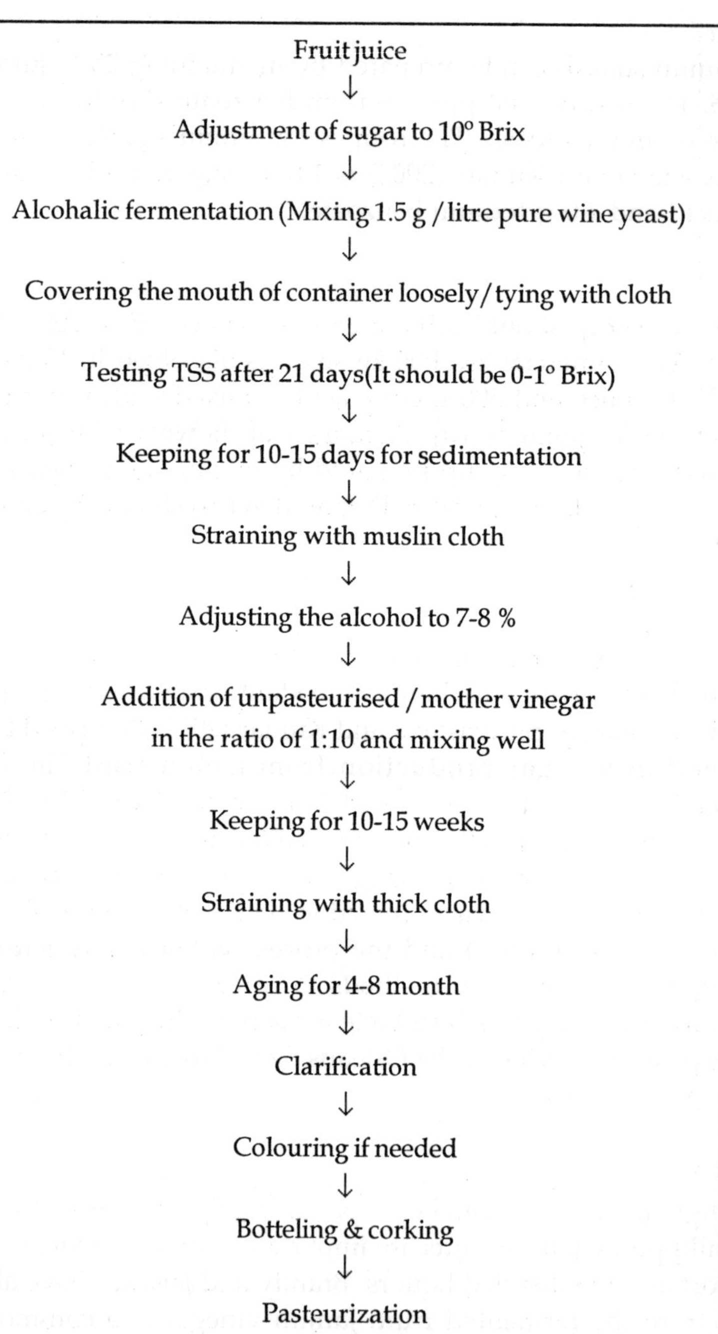

Fig. 8.2. Flow sheet for preparation of jamun vinegar
(Note-An ideal jamun vinegar should not contain more than 0.3% sugar)

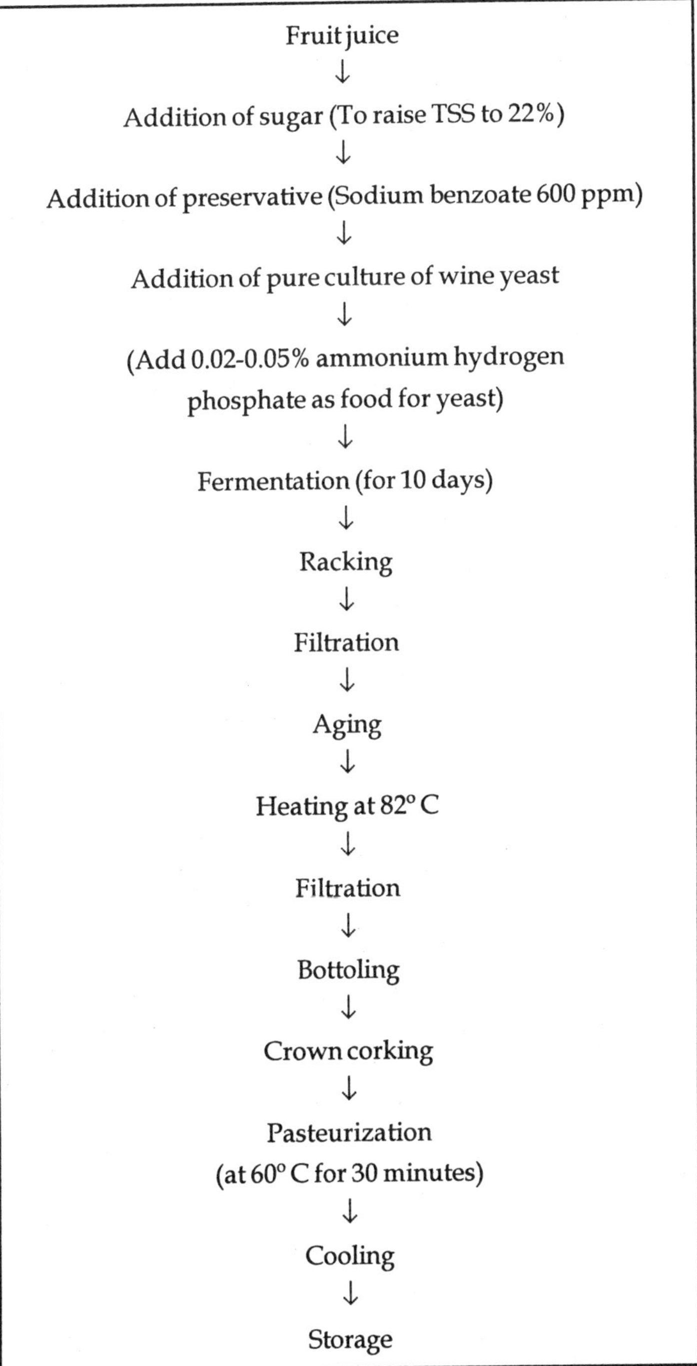

Fig. 8.3. Flow sheet for preparation of jamun cider

Future Research Needs

Jamun is an important fruit crop of marginal and wasteland. It is long lasting tree crop which yields for more than 50 years. This crop can be real crop for future provided the following reseach gaps are addressed.

1. Identification of drought hardy high yielding variety having high pulp ratio
2. Development of Dwarf statured variety with larger fruit which can be adopted for commercial cultivation
3. Development of variety suitable for processing and value addition
4. Development of complete production technology for utilization of various waste lands.

REFERENCES

Devi, S. P., Thangam, M., Desai, A.R. and Adsule, P.G. (2002) Studies on variability in physico-chemical characters of different jamun (*Syzygium cumini*) accession from Goa. *Indian Journal Horticulture*, 59:153-156.

Patel, V.B., Pandey, S.N., Singh, S. K. and Bikas Das, (2005) Variability in jamun(*Syzygium cumini* Skeels) accessions from Uttar Pradesh and Jharhkand. *Indian Journal Horticulture*, 62(3):244-247.

Patil, P.B., Swami, G.S.K., Anil I.Sabrad and Duragannavar, M.P. (2006) Variability in jamun (*Syzygium cumini* Skeels) accessions from port of western Ghat adjoining to Belgaum district of Karnataka. National symposium on underutilized horticultural crops. IIHR, Bangalore, June 8-9, p.50.

Pathak, V.N.(1980). Disease of fruit crops. Oxford IBH Publishing Co. New Delhi

Rao, V.K., Shivashankara, K.S. and Prakash, G.S. (2006) Antioxidant properties of some under utilized fruits. National symposium on underutilized horticultural crops. IIHR, Bangalore, June 8-9, p.54.

Reddy, B.M.C. and Singh, A.K. (2006) Propagation and production technology of under utilized fruits. National symposium on underutilized horticultural crops. IIHR, Bangalore, June 8-9, p.10.

Salvi, B.R., Malse, K.V., Bendole, V.W., Kshirsagar, P.J. and Jalgaonkar, V.N. (2006) Survey, collection,maintenance and evaluation of germ plasm of jamun(*Syzygium cumini*). National symposium on underutilized horticultural crops. IIHR, Bangalore, June 8-9, p.31.

Singh, S. and Singh, A.K. (2006) Standardization of method and time of propagation in jamun(*Syzygium cumini*) under semi arid environment of western India. *Indian Journal Agricultural Science*, 76(4):242-245.

Saha, A.K. (1969) *Allahabad Farmer.*, 43:187-189.

Singh, S., Krishnamurthy, S. and Katyal, S.L. (1967) Fruit culture in India. ICAR, New Delhi.

Singh, U.R., Pandey, I.C. and Prasad, R.S. (1979) *Punjab Horticulture Journal,* 19:74-75

Srivastava, R.P. and Kumar, S. (2002) Fruit and vegetable principal and practices. International Book Distribution Co., Lucknow.

—x—

9

CUSTARD APPLE

Custard apple (*Annona squamosa* L.) is one of the important fruit plant of tropical and sub-tropical areas and belongs to family Annonaceae. It is a drought hardy, grows well even on shallow soil without much care. Plants shed their leaves during stress period to avoid moisture loss from plant. Plant escape from the animal damage due to its hardy nature. The fruit of custard apple is very sweet and delicious. The ripe fruits contains about 73% moisture, 14.5% sugar, 1.6% protein, 0.02% calcium, 0.02% phosphorus,1.0% iron and mineral matter 0.7%.The ripe fruits are used as table fruit. It is also used for making custard milk, ice cream, jam, etc. The pleasant flavour and mild aroma is the typical taste of custard apple fruit.The food value of custard apple fruit pulp is presented in Table-9.1.

Table 9.1: Food value per 100 g of edible portion of custard apple pulp*

Constituent	Content	Constituent	Content
Calories	80-101	Phosphorus	14.7-32.1 mg
Moisture	68.3-80.1 g	Iron	0.42-1.14 mg
Protein	1.17-2.47 g	Carotene	0.007-0.018 mg
Fat	0.5-0.6 g	Thiamine	0.075-0.119 mg
Carbohydrates	20-25.2 g	Riboflavin	0.086-0.175 mg
Crude Fibre	0.9-6.6 g	Niacin	0.528-1.190 m
Ash	0.5-1.11 g	Ascorbic Acid	15.0-44.4 mg
Calcium	17.6-27 mg	Nicotinic Acid	0.5 mg

*Minimun and maximum levels of constituent from analyses made in Central America, Phillippines elsewhere

In India, the fruits are eaten mainly by the lower and medium class people. The fruit is appreciated by all in Central America, Mexico and the West Indies. When fully ripe, it is soft to touch and the stem and attached core can be easily pulled out. The flesh may be scooped from the skin and eaten as such as served with light cream and a sprinkling of sugar. Often it is pressed through a sieve and added to milk shakes, custards or ice cream. I have made a delicious sauce for cake and puddings by blending the seeded flesh with mashed banana and a little cream.

The seeds are so hard that they may be swallowed whole with no ill effects but the kernels are very toxic. The seeds, leaves and young fruits are having insecticidal properties. The leaf juice kills lice. The bark contains 0.12% anonaine. Injection of an extract from the bark caused paralysis in a rear limb of an experimental toad. Sap from cut branches is acrid and irritant and can severely injure the eyes. The root bark has yielded 3 alkaloids: anonaine, liriodenine and reticuline (muricinine).

The leaves have been employed in tanning and they yield a blue or black dye. A fiber derived from the young twigs is superior to the bark fiber from *Annona squamosa*. Custard apple wood is yellow, rather soft, fibrous but durable, moderately close-grained, with a specific gravity of 0.650. It has been used to make yokes for oxen.

The leaf decoction is given as a vermifuge. Crushed leaves or a paste of the flesh may be poulticed on boils, abscesses and ulcers. The unripe fruit is rich in tannin; is dried, pulverized and employed against diarrhoea and dysentery. The bark is very astringent and the decoction is taken as a tonic and also as a remedy for diarrhoea and dysentery. In severe cases, the leaves, bark and green fruits are all boiled together for 5 minutes in a liter of water to make an exceedingly potent decocetion. Fragments of the root bark are packed around the gums to relieve toothache. The root decoction is taken as a fabrifuge. The seeds of fruit contain 30% oil which can be used for making soap and paint. The oil have insecticidal properties. When mixed with neem oil it has demonstrated efficacy against Nephotettix virus and its transmission of rice tungro virus. The oil cake containing 4% nitrogen is used as manure and beneficial to the crop.

Origin and Taxonomy

The custard apple is believed to be a native of the West Indies but it was carried in early times through Central America to southern

Mexico. It has long been cultivated and naturalized so far in south as well as Peru and Brazil. It is commonly grown in the Bahamas and occasionally in Bermuda and southern Florida.

Apparently it was introduced into tropical Africa early in the 17th century and it is grown in South Africa as a dooryard fruit tree. In India the tree is cultivated, especially around cities, and also grows wild in many areas. It has become fairly common on the east coast of Malaya, and more or less throughout south east Asia and the Philippines. Eighty years ago it was reported as thoroughly naturalized in Guava. In Hawaii it is not well known.

The order Ranales is a large group containing 19 plant families under four sub-orders-Nymphacineae, Trochodendrineae, Ranunculineae and Magnolineae. The family Annonaceae belongs to sub-order Magnolineae to which some other important plant families also belong like Magnoliaceae, Myristicaceae and Lauraceae. Custard apples belong to the family Annonace and are native to South America. Among several related species of custard apple, *A reticulata* (Ramphal), *A. Cherimola* (Laxmanphal), *A. atemoya* (Hanumanphal), *A. glabra* (Pond apple) are the important ones.

Majority of the members of family Annonaceae are trees or shrubs, often scandant or lianas. Leaves simple, exstipulate, entire, and alternate. Flowers cauliflorous, i.e., borne on the older woody part of the branches, solitary or often fascicled; regular and spirocyclic, usually bisexual. Perianth in three whorls, each with three segments; the outer whorl is sepal like with much smaller segments than the inner whorl, valvate, free or connate below; the inner whorls with petal like segments, similar in the same whorl, often differing in size in different whorls; valvate or imbricate , free. The receptacle is convex or conical on which numerous stamens and carpels are borne. The stamens are spirally arranged; filament short, thick; anthers long, extrose with a prominent truncate connective. Gynoecium superior, of many free carpels, spirally arranged on receptacle, each carpel with one to many basal or marginal ovules, stipitate; style short or absent; stigma simple. Fruit an etaerio of drupes or berries, rarely follicles, sometimes fused to form a composite fruit. Seeds often arillate, with a small embryo and copious ruminated endosperm. The family has about 80 genera and 880 species distributed in the tropical and sub-tropical parts of the world. Some of the important members of this family are *Polyalthia longifolia* and *Artabotrys odoratissimus*.

Custard apple tree is not especially attractive. It is erect, with a rounded or spreading crown and trunk 10 to 14 inches (25-35 cm) thick. Height ranges from 15 to 35 ft (4.5-10 m). The ill-smelling leaves are deciduous, alternate, oblong or narrow-lanceolate, 4 to 8 inches (10-20 cm) long, 3/4 to 2 inches (2 5 cm) wide, with conspicuous veins. Flowers, in drooping clusters, are fragrant, slender, with 3 outer fleshy, narrow petals, 3/4 to 1 1/4 inches (2 3 cm) long; light-green externally and pale-yellow with a dark-red or purple spot on the inside at the base. The flowers never fully open.

The compound fruit, is 8-16 cm in diameter, may be symmetrically heart-shaped, lopsided, or irregular; or nearly round, or oblate, with a deep or shallow depression at the base. The skin is thin but tough, may be yellow or brownish when ripe, with a pink, reddish or brownish-red blush, and faintly, moderately, or distinctly reticulated. There is a thick, cream-white layer of custard like, somewhat granular flesh beneath the skin surrounding the concolorous moderately juicy segments, in many of which there is a single, hard, dark-brown or black, glossy seed, oblong, smooth, less than 1/2 in (1.25 cm) long. Actual seed counts have been 55, 60 and 76. A pointed, fibrous, central core, attached to the thick stem, extends more than halfway through the fruit. The flavour is sweet and agreeable though without the distinct character of the cherimoya, sugar apple, or atemoya.

Area and Distribution

It has adopted well in India where a considerable variability is found in Aravali hills and in southern India. In Andhra Pradesh, custard apple plants are found wild in Sangareddy and other areas. Custard apple plants can also be seen wild in Uttar Pradesh, Rajasthan, Karnataka, Madhya Pradesh, Maharashtra Tamil Nadu and Orissa. It is cultivated in Mexico, Phillippines, New Guinea, Malaysia and South American countries in the world.

The Spaniards probably carried seeds from the New World to the Philippines and the Portuguese are assumed to have introduced the sugar apple to southern India before 1590. It was growing in Indonesia early in the 17th century and has been widely adopted in southern China, Queensland, Australia, Polynesia, Hawaii, tropical Africa, Egypt and the lowlands of Palestine. Its cultivation is most extensive in India where the tree is also very common and the fruit exceedingly popular and abundant in markets. The sugar apple is one of the most important fruits in the interior of Brazil and is conspicuous in the markets of Bahia.

Soil and Climate

Well drained loam soil with normal pH value is suitable for cultivation of custard apple. Light soils having gravels and small pebbles, is also suitable for its cultivation. In Andhra Pradesh and in Deccan plateau regions, custard apple trees are found in wild form. Heavy soil with high clay content is not suitable for its cultivation.

Custard apple is a subtropical plant, requires hot and dry climate and average annual rainfall of 500-750 mm for better growth and fruiting. It is a deciduous plant and sheds leaves during winter. New growth flush comes during spring with the initiation of flowers. High temperature (40^{0}C) during flowering causes flower drop while rainfall restricts the pollination process. Fruit development takes place during monsoon. Drought spell during this period causes poor fruit set. Frost causes severe damage to young plants of custard apple.

Plant Genetic Resource Management and Varieties

A wide range of genetic diversity exists in custard apple from which some promising cultivars have been identified i.e. Balanagar, Red Sitaphal and Local Sitaphal. It is a cross pollinated crop and seedling plants produce variable size fruit and quality and offers ample scope for selection. The heterozygosity and nature of cross pollination bring wide variations in the seedling progenies. Such naturally induced variation can be used for selection of superior genotypes with desirable traits. Hiremath and Halappannavar (2006) surveyed and found higher fruit yield (9.69 kg/plant) in progeny No-3, which was mainly due to more number of flowers (69.2/plant), number of fruit (74/plant) with higher pulp weight (52.19g). In stark contrast, the Progeny No.2 recorded the highest fruit weight (142g).The introductions of custard apple and inter-specific hybrids such as Washington, Mammoth, Island Gem, Bullock's Heart, Barbadose, British Guinea and Pink's Mammoth were introduced in India. Out of these Washington, Island Gem and Bullok's Heart have proved promising under indian conditions. A promising hybrid, 'Arka Sahan' has been developed at IIHR, Bangalore. It has sweet, fragrant fruits which have low seed content and long shelf life.

The **'Seedless Cuban'** sugar apple was introduced into Florida in 1955, but produced scanty crop of slightly malformed fruits with mere vestiges of undeveloped seeds. The flavor was less appealing than that of normal fruits but it was vegetatively propagated and distributed as a novelty. Another seedless type was introduced from Brazil. Named cultivars growing at the Sabahia Experiment Station, Alexandria, Egypt, include: **'Beni Mazar'**—nearly round, large weighing 150-180 g; 56-60% flesh; 15 30 seeds. **'Abd El Razik'**—light-green or reddish rind; nearly round, large, maximum 8 1/3 oz (236.3 g); 69.5% flesh; 14 seeds.

Indian horticulturists have studied the diverse wild and cultivated sugar apples of that country and recognize ten different types: **'Red'** (A. *squamosa* var. *Sangareddyiz*)—red-tinted foliage and flowers, deep-pink rind, mostly non-reducing sugars, insipid, with small, blackish-pink seeds; poor quality; comes true from seed. **'Red-speckled'**—having red spots on green rind. **'Crimson'**—conspicuous red-toned foliage and flowers, deep-pink rind, pink flesh. **'Yellow'**; **'White-stemmed'**; **'Mammoth'** (A. *squamosa* var. *mammoth*)—pale yellow petals, smooth, broad, thick, round rind segments that are light russet green; fruits lopsided, pulp soft, white, very sweet; comes true from seed. **'Balangar'**—large, with green rind having rough, warty [tuberculate], fairly thick rind segments with creamy margins; sweet; high yielding (Plate 9.1). **'Kakarlapahad'**—very high yielding. **'Washington'**—acute tuberculate rind segments, orange-yellow

Plate. 9.1. Attractive fruit size of Custard apple cultivar Balanagar

margins (Plate 9.2); high yielding; late in season, 20 days after others. **'Barbados'** and **'British Guiana'**—having green rind, orange-yellow

Plate. 9.2. Fruits of Custard apple variety Washington (Left)

margins; high-yielding; late. Recently developed hybrid **Arka Sahan** (*Annona atemoya* x *A. squamosa*) at IIHR, Bangalore is vigorous and has a spreading habit (Plate 9.3). The fruit mature in September-October, the skin of mature fruit has a waxy bloom and light green in colour. The mature fruit weight is 210g and take about 7 days to ripe .The edible pulp is remarkable for its sweetness with 22.8% sugar and measures more than 30^0 Brix Total Soluble Solids as against 24^0B in Mammoth. On an average 12 t fruit can be harvested from a hectare and less infestation of mealy bug due to shallow crevices in between the eyes, they are prone to fruit borer.

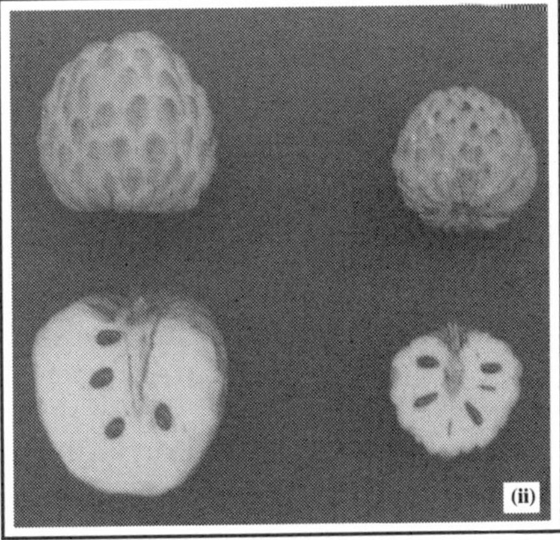

Plate. 9.3. The promising custard apple Hybrid Arka Sahan
(i) Vigourous plant in fruiting (ii) Close view of the fruits and its T.S.
in comparison to Balanagar

Plant Propagation and Nursery Management

In India, custard apple has been commonly propagated by seeds. However, it is advisable to adopt vegetative method of propagation to raise true-to-type plants in the orchard. Seeds are sown in polythene tubes filled with mixture of FYM, sand and clay in 1:1:1 proportion during June-July. Seedlings become ready for planting in field or for budding/grafting after one year. Custard apple can be vegetatively propagated by shield budding and side grafting. Shield budding during January and May gives better success. Healthy buds from previous season's growth are selected for budding on one year old rootstocks when they attain pencil thickness. Softwood grafting technique has now been found better and gives higher field establishment. For this, rootstocks of more than one year age can also be utilized. Softwood grafting with suitable scion is done during March-April which gives about 90 per cent success. Similarly, Giri and Lenka (2006) reported that grafting in the month of April gave >90% success under coastal Orissa condition. Veneer grafting is also done on one year old rootstocks which gives nearly 40 per cent success. Seedlings of Laxmanphal, Sitaphal, Ramphal and Pond apple can be used as rootstock. Custard apple can also be propagated by stool layering using 20,000 ppm IBA during September-October. Similarly more number of root formation in custard apple was also recorded by Rathore (1976). It gives more than 90 per cent rooting and 80 per cent field survival. Application of 10,000 ppm IBA for air-layering in custard apple gave better survival (77%) under Gujrat condition reported by Chovatia and Singh (2000).

Sugar apple seeds have a relatively long life and can be kept well for 3 to 4 years. They germinate better a week after removal from the fruit than when perfectly fresh. Germination may take 30 days or more but can be hastened by soaking for 3 days or by scarifying. The percentage of germination is said to be better in unsoaked seeds. While the tree is generally grown from seed, vegetative propagation is better the crop is important and early fruiting is a distinct advantage for getting uniform fruit quality and precocious bearing.

Seedlings may be budded or grafted when one-year old. In India, selected clones grafted on *A. reticulata* seedlings have flowered within 4 months and fruited in 8 months after planting, compared with 2 to 4 years in seedlings. The grafted trees are vigorous, the fruits less seedy and more uniform in size. *A. senegalensis* is employed as a rootstock in Egypt. *A. glabra* is suitable but less hardy. The sugar apple itself ranks next after *A. reticulata* as a rootstock. In India, budding is best done in January, March and June. Results are poor if done in July, August, November or December unless the scions are defoliated and debudded in advance and cut only after the petioles have dehisced. Side-grafting can be done only from December to May, requires much skill and the rate of success has not exceeded 58.33%.

Experiments in Mexico on utilizing cherimoya, llama, soursop, custard apple, *Annona* sp. A. *lutescens* and *Rollinia jimenezii* Schlecht. as rootstocks showed best results when custard apple scions were side-grafted onto self-rootstock, followed by soursop, A. sp. or A. *lutescens*. Custard apple seedlings are frequently used as rootstocks for the soursop, sugar apple and atemoya.

Planting

Custard apple is a medium stature tree with upright growth habit. Planting is done at 5x5 m spacing (Plate 9.4). Pits of 60x60x60 cm size should be prepared and filled with FYM+ top soil + fertilizer (50gN:50g P:25g K) mixture along with methyl parathion dust (50g/pit) before the onset of monsoon. During July-August, after 1 or 2 rains ,the saplings are planted in the centre of the pit. Small basin around each plant should be prepared. Orchard establishment *in situ* is beneficial under drought conditions. (Pandey and Sridhar, 2004).

Plate. 9.4. Plantation of custard apple

In Egypt, sugar apple trees are spaced at 3x3 m in order to increase atmospheric humidity and improve pollination. In Palestine Plant spacing of 5 m x 3 m has been found more feasible than 5 m x 5 m.

Training and Pruning

Strong tree frame work in custard apple is obtained by training the young plants during the initial stages of growth. Keeping single stem up to 50-70 cm height and then inducing scaffold branches in all the directions makes good framework of tree. To avoid crowding, proper space among the branches is maintained by removing criss-cross and overlapping branches. Custard apple generally bears on new growth but sometimes fruiting is also comes on old branches. Therefore, suitable combination of young and old branches should be maintained by judicious pruning. At Hyderabad, light pruning of branches (removing 25% top of branches) during March-April has been found beneficial. Judicious pruning to improve shape and strength of tree must be done only in spring when the sap is rising, otherwise pruning may kill the tree.

Orchard Management

Custard apple is grown commercially in southern states of India where orchard management techniques have been standardized for sole cropping.

Nutrient Management

Application of nutrients to custard apple depends upon the soil and climatic conditions. In light sandy and gravelly soils, an annual dose of 50:50:25 g NPK along with 10 kg FYM is found sufficient. The doses should be increased up to 5 years of age, thereafter a fixed dose should be given every year. Full amount of FYM, phosphorus, potash and half of nitrogen should be applied during monsoon season and remaining amount of nitrogen should be given during spring. Besides this, application of organic manures such as castor cake and bone meal has also been found beneficial at Sangareddy and in Rayalseema region of Andhra Pradesh. Green manuring with sunhemp, *moong*, *urd*, and *dhaincha* during monsoon period has been found effective for fertility improvement in custard apple orchards. On light soils, they apply 60-80 kg manure per tree annually and they recommend the addition of nitrogen. Application of chemical fertilizer significantly increases flowering, fruit set and yield.

Water Management

During the initial stage of orchard establishment, frequent irrigations at 15-20 days interval should be given particularly during the summer months. Once plant is established, life saving irrigations will be needed during summer month. In light soils of arid and semi arid region, provision of water harvesting structure by providing 5 per cent inward slope to the basin around the plant and moisture conservation by mulching in the basin with black polythene or locally available organic waste should be done. If water is available, irrigation by modified basin or drip irrigation system improves its production and quality of fruits.

Orchard Floor Management

During *kharif* and *rabi* seasons, inter-row spaces in custard apple orchard can be utilized to raise intercrops such as *moong, urd, clusterbean, moth, gram,* under rainfed conditions. Vegetables such as pea, chillies, okra, brinjal, etc. can be grown in areas having assured irrigation facilities. Proper hoeing and weeding are essential for good growth and fruiting. Custard apple has been found as very good filler crop in litchi and mango orchard because of its dwarf stature.

Disease and Pest Management

Custard apple is generally free from pests and diseases. Sometimes, due to improper nutrition and water management, flowering and fruit setting is very late. Fruits which set very late do not mature due to onset of winter and commencement of dormoney of plants. Such fruits turn hard and brown and remain hanging on the tree for long. Such fruits are called mummified fruits. The problem can be overcome by proper water and nutrient management.

Alternaria leaf spot causes some damage to custard apple which can be controlled by 3 sprays of 0.5 per cent benzimidazol or carbendazim or 0.2 per cent mancozeb at 15 days interval starting from the initiation of the symptoms. Wilting of trees have been widely noticed in acid soil conditions of plateau region. These wilted plants do not regenerate, hence, should preferably be replaced by new planting.

The mealy bugs infect the fruit surface and branches. They reduce the market value of fruits. The pest can effectively be managed by 2-3 sprays of 0.1% monocrotophos at 15 days interval after the appearance of nymph. Wrapping polythene sheets on tree trunk checks the upward crawling of the pest from the ground. Hoeing of tree basins during summer kills the eggs and nymphs.

Flowering, Fruiting and Yield

Under Orissa conditions, custard apple bears flower on the current season growth and very rarely on old shoots. The flowering period is very long, commencing from March-April and continue to July-August. The peak flowering period is April-May when temperature goes beyond 40°C, humidity is extremely low, desiccating wind blows during this month. In this situation, pollen production is scanty and pollen tube growth is restricted which causes poor fruit set. Low fruit set in custard apple is due to poor pollination , govern both by external and internal factor such as high and low humidity prevailing at the time of flowering, soil moisture stress, hypogyny and dichogamy condition, poor pollen germination and dearth of insect pollinator. It can be improved by hand pollination. Sahoo *et al.,* (2000) observed that in custard apple the flowers are borne mostly in new flushes after shedding of old leaves. Anthesis occurred from 2.30 AM to 6.30 PM with a peak period between 5.30 AM to 6.30 PM. Dehiscence of pollen began 10 hours after opening of flower and continued for a period of 2 hours only. Stigma was found to be receptive one day prior to anthesis and continued till 2-3 days after anthesis indicating a protogynous condition. The pollen grains were round to ovate in shape. However, very little variation was recorded on flowering behaviour of green and red fruited type.

Sahoo *et al.,* (2000) Also reported that pollen viability in custard apple ranges from 42.30 % to 93.33% in green type and 45.10 to 93.75% in red type. Maximum viability was observed during June to August in both varieties of A. squamosa. More or less similar results of pollen viability were obtained by Nalwadi *et al.,* (1975) and Thakur and Singh (1965).

Under moderate climatic conditions of Bangalore and Hyderabad, custard apple trees remain dormant before they put forth new growth during April. Flowering in custard apple takes place on one year old shoots during June-July and fruits mature by October under eastern Indian conditions. Under north Indian conditions having humid climate, however, the flowering in custard apple starts during the spring and fruits become ready for harvest during August-September. In dry areas, fruit matures during October-November (Plate 9.5). In

Plate. 9.5. Fruits affected by Mealy bug

order to increase the span of fruit harvest and thereby to increase fruit yield, early induction of new growth by defoliation of plants has been found effective. At Bangalore, manual as well as chemical defoliation during January gave goods response. Chemical defoliation by two sprays of 0.5 per cent potassium iodide at weekly interval during December-January has been found to increase number and weight of fruits in cultivar Balanagar. Grafted plants start fruiting at the age of 3-4 years but the commercial yield is obtained only after 5-6 years. A well managed grown up custard apple plant can produce 80-100 fruits /tree. Under Bangalore condition, Arka Sahan hybrid yielded 25 t fruits /ha.

Seedlings of 5 years old may yield 50 fruits per tree in late summer and fall. Older trees rarely exceed 100 fruits per tree unless hand-pollinated. With age, the fruits become smaller and it is considered best to replace the trees after 15 to 20 years. The fruits will not ripen but just turn black and dry if picked before the white, yellowish or redting appears between the rind segments, the first signs of separation. If allowed to ripen on the tree, the fruit falls apart. Moreover, tree ripe fruits have very short keeping quality and unfit for long distance transportation.

Harvesting and Post Harvest Management

Custard apple is a climacteric fruit which ripens after harvesting. Therefore, the fruits are harvested at proper maturity when they are hard. When the colour of skin and interspaces between the protuberances become yellow, the fruit is considered to has attained full maturity. The fruit has very poor shelf life. Therefore for distant markets, harvesting comparatively harder fruits is advised. Fruits harvested before proper maturity do not ripe well and remain less sweet. Newly developed hybrid Arka Sahan has better shelf life. Apart from fresh consumption, its pulp can be processed to prepare products like pastries, chocolates, ice creams, mixed fruit jam, etc.

Processing and Value addition

Custard apple fruits can be processed into various ways. The freeze drying is one of the important methods for processing of custard apple pulp. In this method, the pulp is first frozen at -18^{0}C on tray in the lower chamber of a freeze drier and the frozen materials dried (initially at 30^{0}C for 24 hours and then at 20^{0}C) under high vacuum (0.1mm Hg) in the upper chamber. Direct sublimation of the ice takes place without passing through the intermediate liquid stage (Srivastava and Kumar, 2002). Various value added products can be prepared from custard apple. Besides jam and beverages, fruit concentrates can be prepared from pulp.

Future Research Needs

Custard apple is an important fruit crop for rainfed area. Limited work has been done on production technological aspect. There is urgent need to take up work on following aspect in future:

1. Development of high yielding variety with more pulp content and keeping quality suitable for rainfed area.
2. Variety should be tolerant to major pest and diseases.
3. Development of variety suitable for processing purpose.
4. Development of suitable agro-techniques for higher production under rainfed condition.
5. Development of techniques and tools for management of tree wilting particularly under acid soil conditions.

REFERENCES

Chovatia, R.S. and Singh, S.P. (2000) Success of air layering in custard apple (*Annona squamosa* L) as influenced by ringing of shoots and growth regulators. *Orissa Journal Horticulture*, 28 (2):61-65.

Hiremath, S.M. and Halappannavar, A.B.(2006) Performance of custard apple seedling progenies for yield and its attributes. National Symposium on underutilized Horticultural crops., IIHR, Bangalore, June 8-9 p.49.

Giri, B.and Lenka, P.C. (2006) Studies on vegetative propagation of custard apple (*Annona squamosa*). National Symposium on underutilized Horticultural crops., IIHR, Bangalore, June 8-9 p.73.

Chundawat, B.S. (1990). Arid fruit culture. Oxford & IBH, New Delhi. pp.1-201.

Nalwadi,V.C., Sulikeri, G.S.S. and Singh, C.D. (1975) Floral biology studies of *Annona squomosa* L. under Dharwar condition. *Progressive Horticulture*, 7(1) :15-24.

Pandey, V. and Sridhar, V. (2004) Establishment of custord apple orchard *in situ* in drought prone areas of Orissa. *Indian Horticulture*, 45 (3): 21-23.

Rathore, D.S. (1976) Preliminary studies on the stooling of custard apple (*Annona squamosa*). *Indian Journal Horticulture*, 33(3-4):244-245.

Sahoo, S.C., Panda, J.M.and Mohanty,D.(2000).Studies on floral biology of custard apple (*Annona squomosa*) under Bhubaneswar condition.*Orissa Journal Horticulture*, 28(1):43-45.

Sahoo, S.C., Panda, J.M.and Mohanty, D.(2000A) A note on pollen morphology, viability, pollination and fruit set in custard apple (*Annona squomosa* L). *Orissa Journal Horticulture*, 28(2):109-110.

Srivastava, R.P. and Kumar, S.(2002) Fruit and vegetable Preservation: Principal and Practices., International Book Distributing Company, Lucknow. pp.1-474.

Thakur, D.R. and Singh, R.N.(1965) Studies on floral biology of *Annona*. *Indian Journal Horticulture*, 22:238-253.

—x—

10

FIG

Fig (*Ficus carica* L.) is an important acclimatized species under cultivation which is closely related to the *Ficus glomerata* (*gular*) commonly found in deciduous forests of subtropical region. It is one of the earliest fruit species cultivated by man. The Bible and some scholars believe the forbidden fruit picked by *Eve* was a fig rather than an apple, which would actually make more sense as it's supposed to be a fig leaf which *Adam* chose to 'protect his modesty'. Fig have been cultivated since ancient period (4000 BC.) and well known through out the world. Fruits are valued for their laxative and medicinal properties. The fruit contains 11.5g sugar, 1.5 mg iron, 270 IU vitamin A per 100g pulp. Fig fruits preserved in vinegar are used to cure liver and spleen enlargement in children and are effective in the treatment of chronic cough and piles. Fig fruit is used for their nutritive value in both fresh and dried form as part of the food in both Western Asia and Southern Europe. A sort of cake made by mashing up inferior quality figs are being serve in parts of the Greek Archipelago as a substitute for bread. Alcohol is obtained from fermented figs in some southern countries, and a kind of wine, still made from the ripe fruit, was known to the Ancients and is mentioned by Pliny under the name of *Sycites*. It is a nutritive and delicious fruit rich in calories (269), protein, and calcium (higher than milk), iron and fibre content. The fig fruit has great potential for domestic consumption, value addition and export.

Fig fruits are used fresh as well as for drying purpose. Ripened fruits are used for table purposes. The fruits are nutritious with high levels of calcium and iron. The total sugar content in fresh fig is 16% and that of dried fig is 52%. The chemical composition and flavour varies with the cultivars. Commercially, figs are peeled by immersion for 1 minute in boiling water or a boiling solution of sodium bicarbonate. In warm, humid climates, figs are generally eaten fresh and raw without peeling and they are often served with cream and sugar. Peeled or unpeeled fruits may be merely stewed or cooked in various ways, as in pies, puddings, cakes, bread or other bakery products, or added to ice cream mix. At household level the whole fruits are preserved in sugar syrup or processed into jam, marmalade, or paste. Fig paste (with added wheat and corn flour, whey, syrup, oils and other ingredients) forms the filling for the well-known bakery product, "Fig Newton". The fruits are sometimes candied whole commercially. Dried fruits are very delicious. In Europe; western Asia, northern Africa and California, commercial canning and drying of figs are industries of great importance.

The latex of the unripe fruits and of any part of the tree may be severely irritating to the skin if not removed promptly. It is an occupational hazard not only to fig harvesters and packers but also to workers in food industries and to those who employ the latex to treat skin diseases. The latex contains caoutchouc (2.4%), resin, albumin, cerin, sugar and malic acid, rennin, proteolytic enzymes, diastase, esterase, lipase, catalase, and peroxidase. It is collected at its peak of activity in early morning, dried and powdered for use in coagulating milk to make cheese and junket. From latex, protein-digesting enzyme '*ficin*' can be isolated which is used for tenderizing meat, rendering fat, and clarifying beverages. In tropical America, the latex is often used for washing dishes, pots and pans.

The latex has many medicinal uses. It is widely applied on warts, skin ulcers and sores and taken as a purgative and vermifuge, but with considerable risk. In Latin America, figs are much employed as folk remedies. A decoction of the fruits is gargled to relieve sore throat; figs boiled in milk are repeatedly packed against swollen gums; the fruits are much used as poultices on tumors and other abnormal growths. The leaf decoction is taken as a remedy for diabetes and calcifications in the kidneys and liver. Fresh and dried figs have long been appreciated for their laxative action. Fig fruits are used as medicine after preserving in vinegar for treatment of liver and spleen enlargement in children. It is also effective in chronic cough and piles (Chandra *et al*, 1994). Fig fruits valued for its laxative properties.

Fig leaves are used for fodder in India. They are plucked after the fruit harvest. Analysis show that leaves contain moisture, 67.6%;

protein, 4.3%; fat, 1.7%; crude fiber, 4.7%; ash, 5.3%; N-free extract, 16.4%; pentosans, 3.6%; carotene 0.002% on a dry weight basis contents also present are bergaptene, stigmasterol, sitosterol and tyrosine.

Dried seeds contain 30% of a fixed oil containing the fatty acids: oleic, 18.99%; linoleic, 33.72%; linolenic, 32.95%; palmitic, 5.23%; stearic, 2.1 8%; arachidic, 1.05%. It is an edible oil and can be used as a lubricant.

Origin and Taxonomy

The fig is believed to be indigenous to Western Asia and to have been distributed by man throughout the Mediterranean region since ancient times because of which it is thought to be a native of the southern Arabian Peninsula, Italy, Balkan Peninsula and Russia. Fig is also considered native of India may be the secondary centre of origin where it is grown in arid and semi-arid tracts since antiquity. In India, fig plantations are localised around Pune in Maharashtra and on the west coast in southern Gujrat. In north India, it is grown in kitchen gardens. It has been under cultivation for thousands of years as remnants of figs have been found in excavations of Neolithic sites traced to at least 5,000 B.C. As time went on, the fig-growing territory stretched from Afghanistan to southern Germany and the Canary Islands. Pliny was aware of 29 types. Figs were introduced into England some time between 1525 and 1548. It is not clear when the common fig entered China but by 1550 it was reliably reported to be in Chinese gardens. European types were taken to China, Japan, India, South Africa and Australia.

Fig (*Ficus carica* L.) belongs to family Moraceae and is closely related to the *Ficus glomerata* (*gular*) an ancient fruit yielding species. The fig is a tree of 3-9 meters hieght, with numerous spreading branches and a trunk rarely more than 20 cm diameter. It contains copious milky latex. The root system is typically shallow and spreading, sometimes covering 15 m of ground, but in permeable soil some of the roots may descend to 6 m. The deciduous leaves are palmate, deeply divided into 3 to 7 main lobes, these are more shallowly lobed and irregularly toothed on the margins. The blade is up to 25 cm in length and width, fairly thick, rough on the upper surface, softly hairy on the underside. What is commonly accepted as a "fruit" is technically a synconium (Plate 10.1), that is, a fleshy, hollow receptacle with a small opening at the apex partly closed by small scales. It may be obovoid, turbinate,

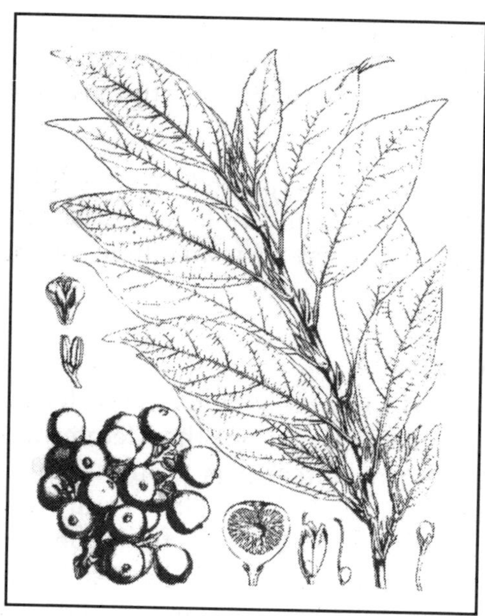

Plate 10.1. Flowering twig of fig

or pear-shaped, 1 to 4 in (2.5-10 cm) long, and varies in color from yellowish-green to coppery, bronze, or dark-purple (Plate 10.2). Tiny flowers are massed on the inside wall. In the case of the common fig

Plate 10.2. Fruits of fig

discussed here, the flowers are all female and need no pollination. There are 3 other types, the "Caprifig" which has male and female flowers requiring visits by a tiny wasp, *Blastophaga grossorum*; the "Smyrna" fig, needs cross pollination by Caprifigs in order to develop normal fruit; and the "San Pedro" fig which is intermediate, its first crop independent like the common fig, its second crop dependent on pollination. The skin of the fig is thin and tender, the fleshy wall is whitish, pale-yellow, or amber, or more or less pink, rose, red or purple; juicy and sweet when ripe, gummy with latex when unripe. Seeds may be large, medium, small or minute and range in number from 30 to 1,600 per fruit.

Area and Distribution

The plants of fig are grown in different parts of the country. However, the commercial area in India, is confined only in Maharashtra, Bangalore, Bellary (Karnataka), Saharanpur (U.P.), Coimbatore (Tamil Nadu), Gujarat, and Andhra Pradesh (Pareek and Vishal Nath,1996). The common fig plants are cultivated in sub-tropical to dry regions of the country. It is commercially grown in Turkey, Spain, Italy, Portugal and Greece. It is also grown in some parts of Australia, California, New Zeland and Argentina. Turkey is major fig producing country in the world with its 26 % production share. The United States, Turkey, Greece and Spain are the primary producing nations of dried fig. India is the second largest producer of fruits after China, with a production of 44.04 million tonnes of fruits from an area of 3.72 million hectares. At present, fig cultivation is being done in about 1,000 ha area in Maharashtra state. In dry land area, the cultivation of fig is possible since it requires dry climate and steady irrigation. In arid parts of Rajasthan, few growers have planted fig but the size of fruit is small and the performance is yet to be ascertained.

Soil and Climate

The fig can be grown on a wide range of soils. Light sandy, heavy clay or limestone having sufficient soil depth and good drainage for safely disposal of excess rain water are considered to be the best. Sandy soil that is medium-dry and contains a good deal of lime is preferred when the crop is intended for drying. Highly acid soils are unsuitable. The pH should be between 6.0 and 6.5. The tree is fairly tolerant of moderate salinity.

Fig is very sensitive to climatic conditions. It is generally grow in warm, relatively dry climate. It requires small winter chilling. In tropical areas generally, fig thrive between 2,600 and 5,900 ft MSL. In southern India, 'Marseilles' flourishes on hills above 5,000 ft MSL. The tree can tolerate 10° to 20° C of temperature in favorable sites. It should have a dry climate with light early spring rains if it is intended for the production of fresh fruit. Rains during fruit development and ripening are detrimental to the crop, causing the fruits to split. For best growth and fruiting, fig trees need full sunlight and freedom from competition of trees and shrubs particularly in the mixed plantation. The semi arid tropical and subtropical regions of the world are ideal for fig growing if assured irrigation facilities are available. Very hot weather and dry spell causes fruit drop even if the trees are irrigated.

Plant Genetic Resource Management and Varieties

The fig is one of the predominant crops in Tunisian oasis after date-palm. Fig culture consists of ecotypes and cultivars locally called varieties. These are clonally propagated and widely extended through out the areas. However, for several decades Tunisian fig groves are seriously menaced by drastic genetic erosion due to biotic and abiotic stresses. In order to preserve local genotypes and to improve fig culture in oasis groves, it is important to analysis the genetic diversity of this important crop. Hence, a study has been carried out using a set of varieties and a large number of morphological characters. The isozymic analysis has been also performed and markers suitable in the evaluation of fig genetic diversity and as criteria in the variety identification delineated. These traits were also useful for the genetic preservation in this crop. Review of the current situation of fig culture and genetic resources in different countries revealed that the crop has wide genetic variability. Diversity among 22 cultivars of southern arid regions was analyzed using morphometric studies. Cultivar characterization and multivariate analysis were made on the basis of 18 physical and chemical characters of the main crop fruits. Results showed that the fruit parameters of high discriminatory value were fruit size and shape, ostiol opening, flesh and skin thickness, titrable acidity and total soluble solids. The study suggested that fig germplasm is diverse. It was also possible to differentiate three cultivar groups and two distinct cultivars.

For selection of fruit crops and its varieties for dry land horticulture, one of the basic requirements is that the crops should complete their vegetative growth and reproductive phase during the period of maximum moisture availability. The crops must have xeric characters, e.g., deep root system, summer dormancy, high 'bound water' in the tissues, reduced leaf area, leaf surface having sunken stomata, thick cuticle, wax coating and pubescence and ability to adopt to shallow soils, rocky, gravelly and wastelands. There are more than 700 varieties of fig available in the world. Out of which some introductions have been made at Bangalore from USA, China and Greece. Based on the sex of flowers and pollination, fig has been grouped into four types. Large sized figs belong to common fig group. The varieties are divided into green and yellow figs and dark-colored figs. The descriptions of some of the important varieties are as follows.

Green and Yellow fig cultivar

Alma

A small- to medium-sized golden brown, pear-shaped (pyriform) fig with amber pulp. Sweet in taste and delicate flavor. Bred by Texas A&M and released in 1975. Well adapted in the Southeast. Highly resistant to fruit rots diseases. Wood is very hardy.

Blanche

An old, reliable variety usually called Lemon in the South. A medium to large greenish yellow fig with white pulp and many seeds. Top-shaped (turbinate) without a neck. Sweet and delicate flavor with a nutty texture from the soft seeds. Well adopted in the South. Fairly hardy plant type.

Brunswick

A medium to large fig with bronzy yellow skin and rich flavor. Oblique- turbinate. Well adapted in the Southwest and drier areas of the South. The fruit is ruined by excessive rain since it has open eyes. Fairly hardy.

Flanders

A greenish yellow medium-sized fig with violet stripes and amber pulp. Bred and released by Ira Condit in 1975. Pyriform with a long, slender neck. Fine flavor. Plants are vigorous but not particularly hardy. Good on the West Coast.

Green Ischia

A small grass-green fig with dark strawberry pulp. Oblate to spherical shape. Small, fairly well-closed eye. Good quality but not as good as Brown Turkey. Matures late. Fairly hardy.

Gulbun

It is also known as Jewel. A large light green to pale yellow fig with a translucent pulp tinged with pink. Good flavour. Seems moderately hardy.

Ischia

A small- to medium-sized yellow fig with fair flavor. Oblate to spherical with or without a short neck. Well adapted in coastal California. Its quality is poor in South part of country.

Jurupa

A very large green fig with pink pulp and a medium-sized closed eye. Pyriform. A Condit hybrid selected by Julius Enderud. Good flavor. Under trial in the South. Somewhat hardy but tends to leaf out early which is a negative point in areas with late frost.

Kadota

A greenish white small to medium sized fig, Pyriform, Vigorous, Delicious fresh or dried. It is a rich, sweet, all-purpose fig and the most common canned fig. Well adapted in the Southwest and drier areas of the South. Fairly hardy.

King

A medium-sized greenish yellow fig with strawberry pulp. Pyriform to oblique in shape. It ripens between late June and August. A San Pedro type, it sometimes sets main-crop figs without pollination. Sweet and rich in nutrients. Well adapted in the Northwest and cooler areas of the South. Fairly hardy.

Dark coloured fig cultivar

Beall

A medium to large purplish black fig with amber pulp. Brebas are pyriform with prominent necks; main-crop figs are oblate to pyriform with short, thick necks. Very good flavor. Well adapted in California and hardy in the South. A chance seedling identified in California in the 1920s.

Brown Turkey

A small- to medium-sized light brown to violet fig with strawberry pulp. Turbinate to oblique, mostly without a neck. Small eye has a reddish color from very early stage (unlike Celeste). Cold hardy. It fruits on new growth if winter killed. Often bears two crops in year. Its fruit are very sweet but not rich. The tree is prolific. The main crop, beginning in mid-July, is large; the early crop is small. The cultivar is well adopted to warm climates. It is grown on all the islands of Hawaii. Moreover, this variety is used as rootstock in India (Anon, 2002).

Black Ischia

A small purplish black fig with strawberry pulp. Turbinate. Fairly sweet and rich in flavor. Well adopted in coastal California but not very productive in the South. Susceptible to frost therefore replaced by Celeste. It is grown in near by Lucknow, Saharanpur in U.P. Fruits are become black after ripening (Plate 10.3).

Plate 10.3. Cultivar Black Ischia

Conadria

A medium to large yellow-green fig with light strawberry pulp and rich flavor. Bred by Ira Condit and released in 1957. Pyriform in shape. Well adopted in California and the Southeast. Hardy with good rebound from freezes. Its performance has been found better under Indian conditions.

Excel

A medium-sized yellow fig with amber pulp. Also bred by Ira Condit and released in 1975. Oblate to spherical in shape. Well adapted in California. Early trials in the Southeast are very promising. Seems to be very hardy and has superb flavour.

Deanna

It is an exotic variety. Medium size yellowish, bigger fruit size and weight (55g) and TSS 22.6 %. It is performing better under Maharashtra and Karnataka conditions (Plate 10.4).

Plate 10.4. Fruiting in fig cultivar Deanna

The exotic varieties viz. Conadria, Deanna and Excel are performing better under Bangalore conditions and these are good for drying, canning and table purposes. The performance evaluation showed that fig cultivars Deanna, Excel and Conadria are suitable for dry areas. Excel and Conadria develop small canopy and are suitable for high density planting. Poona fig and Dinkar cultivars are being grown in drier parts of Maharashtra (Anon, 2002).

Indian cultivars

Poona fig

Fruits are bell-shaped, of medium size, weighing about 20-30 g; thin-skinned; light purple with red flesh, of sweet, good flavor. It is most popular variety and it is commercially cultivated near by Poona. It is suitable for fresh consumption and drying purpose.

Dinkar

It is an improvement over Daulatabad for yield and fruit quality. This variety has been released by Marathawada Krishi Vidyapeeth, Parbhani and it is gaining commercial cultivation in Maharashtra state. It is medium type variety in yield. It produces about 200-300 fruits per tree.

Plant Propagation and Nursery Management

Fig trees can be raised from seeds extracted from commercially dried fruits however the seed propagation is employed only to produce new varieties. Seeds are used for breeding/hybridization purpose. Female trees may also produce apomictic seeds i.e. without pollination and subsequent fertilization. The seed germination is about 60 per cent under warm greenhouse condition however the seed propagation is not a common method.

Fig trees are readily propagated by dormant hardwood cuttings. Fig propagation using hardwood stem cuttings under field conditions is, therefore, of great importance, as it helps produce vigorous and potential saleable plants. Therefore, hard wood cutting is considered to be the commercial method of propagation of fig. Planting of fig cuttings is generally done from June to January. Cuttings are placed in bundles with with right polar ends in callusing pits for about 1.5 to 2 months before being planted in main nursery or poly bags containing the rooting medium. Single bud cuttings between the third and sixth nodes take minimum period for rooting. The development of root and leaf has been found the best in 30-35 cm long cuttings. Cuttings having less than 20 cm length resulted in only 30 per cent rooting. Maturity of scion wood selected for taking cuttings in fig play crucial role in rooting and successful plant regeneration. Faster and prolific rooting has been observed in the basal nodes having comparatively mature wood. Hard wood cuttings from 3-year-old scion wood resulted 90 per cent rooting whereas in 1-year-old wood only 55 per cent rooting was noticed. Cuttings prepared in autumn and planted in spring rooted better than those prepared and planted in spring. Autumn-planted cuttings gave poor results. Application of 30 kg N and 22 kg P along with 10 t/ha well rotten FYM in the nursery bed increases the production of standard quality transplants. Various concentrations of IBA on rooting of hard wood cuttings gave promising result.

Experiments conducted at the University of California to determine the extent and degree of root and shoot regeneration in tip cuttings of fig and their response to various concentrations of IBA revealed that treatment of cuttings with indole butyric acid (IBA) at 2,000-10,000 ppm resulted in better rooting. Treatment with IBA and NAA showed higher percentage of rooting and larger root number compared to IAA. The effective growth substance also promoted shoot growth. In another study treatment of 10-15 cm long cuttings having 3-4 leaves with NAA and IBA at 20 ppm proved very effective in inducing rooting in a medium containing soil and 10 kg FYM/m^2. Thus, it is recommended that as a general practice, take cuttings which are 15 to 30 cm long and pencil to finger thickness during autumn season and treat with IBA or Rootone at the basal portion. The best cuttings should have some of last year's wood on them. Plant the cuttings in 10 cm deep plastic pots and pack a half sheet of newspaper tightly into the bottom of the pot. Put a little rooting mixture in the bottom to hold the cuttings upright in each pot. Water the pots thoroughly and put them in a very bright, but not in sunny day. The water should be applied only when they are very dry. After a few days, it will be time to shift the cuttage in to the pot/nursery bag. There will be 4 or 5 leaves and few roots on each cutting. The vigorous growth takes place after establishment.

Pareek and Sharma, (1991) reported that for commercial production, the plant is commonly propagated by cuttings of mature wood. Leafless, fully mature, dormant hardwood cuttings of 1.5 to 2 cm thickness and 25-30 cm length are severed from parent plant. Under hot and dry condition of Southern Ukraine, propagation of fig was more successful than from the seed. Cuttings root quickly and plants so raised may produce fruits before they are one year old. One year old layered plants also produce the first crop in the first year itself and a fair crop can be obtained from third year onwards. Select foot-long pieces of dormant wood, less than 1-inch diameter, with two-year-old wood at base. One-year old twigs with a heel of two-year branch at the base may also be used. Dip the cuttings in a rooting hormone & allow them to callus one week in a moist place at 10-15°C. Summer cuttings may also be made, but they do best if defoliated and winterized in refrigeration for 2-3 weeks before potting. Leafy shoots require a mist bed. Antunes *et al.*, (1996) suggested the treatment of cuttings with 100 ppm IBA and then planting them in sand: soil (1:1) mixture without stratification for better rooting and shoot development. The cuttings of Fig (EC 538098) Brazilian varieties were introduced from Nigeria at CIAH, Bikaner during the year 2004, but did not sprouted and dried after some time due to small size cuttings and low temperature during the month of January. Thus, appropriate size of cuttings should be introduced for success. Planting of cuttings must be done within 24 hours but, first, the upper, slanting end of the cutting should be treated with a sealant to protect it from disease and the lower flat end with a root-promoting hormone.

Fig can be easily propagated through air layering. Air layering is very popular in South India and is widely practiced. For layering, one year old two cm thick branches are preferred. Air layers are done during early spring; they will root well by mid summer (Hartmann, *et al.*, 1997). Plants are later transferred to pots then planted in the main field. Ground or air layering can be done satisfactorily and rapid mass multiplication by tissue culture has been achieved in Greece.

Budding or grafting also has been used to top work a tree to a better variety. Trees of unsatisfactory varieties can be top worked by shield or patch budding, or cleft or bark grafting. Several *Ficus carica* cultivars have been successfully budded or grafted onto *F. glomerata, F. cocculifolia, F. Graphalocarpa* and *F. palmata* rootstock. No incompatibility was noted after a year or more. Under Bangalore conditions, the plants of fig varieties are propagated on Brown Turkey rootstock which is found suitable root stock under Indian condition. Chip budding was best for the economical production of large numbers of container grown trees. Ring budding and side grafting can also be performed for multiplication of propagules. Figs can also be grafted on nematode resistant rootstock such as *Ficus cocculifolia* and *F. glomerata* but these rootstocks are cold sensitive. Results obtained at Kerala Agricultural University, Vellanikkara revealed that ring and patch budding were the successful methods in *Ficus* species. Study suggests that the rare cultivars of fig can be propagated on local rootstocks or on the older trees by top working using whip, cleft or crown grafting and chip or patch budding techniques.

Planting

The plant density mainly depend upon the plant type, soil fertility status and management practices while planting system to be adopted in fig depends largely upon the topography of the land and soil type. In the plains, planting is generally done in square or rectangular system at a distance of 5- 6 meter. However, Excel and Conadria can be planted at 2.5-3.0 m spacing. Plant fig trees when they are dormant. Spring is considered to be the best time for planting of container grown plants. In warm areas, bare-root trees can be set out in fall or early winter, but where late spring frosts are common, it is best to set them out in spring after the danger of hard winter freezes has passed. In arid region, the best time of planting is onset of monsoon preferably during August-September. Pareek and Vishal Nath (2006) reported that one year old plants should be planted at 5x5m spacing in well prepared pits of 60x60x60 cm size which are filled with a mixture of top soil and FYM in equal proportion along with 50g Phorate 10G. In light soils, application of 500g super phosphate per pit is advisable. Irrigation immediately after planting is needed to improve field establishment. Cultivars on vigorous rootstocks would require wider spacing.

Training and Pruning

Optimum and quality fruiting in fig can be ensured by proper training and pruning of the trees. Plants are usually trained to single stem. At locations having pest problems such as shot hole borer leading to death of the main stem, multi-stem training is advised. When plants are trained to single stem, growing tip of the plant is headed back to force a number of side shoots, to develop a balanced frame of the tree. In red soils of Andhra Pradesh, plants are trained to multi-stems.

Fig generally bears individually or in bunches in the axils of leaves on current season's growth. There are two seasons of flowering in fig in India i.e., October-November flowering and March-April ripening and February-March flowering and June-July ripening. October flowering comes on old branches whereas spring flowering comes on current season's growth. Pruning time and severity in fig is decided considering the growth habit and bearing of the plant (Pareek and Vishal Nath, 2006). In Pune, light pruning is done just after crop harvest while in north India, severe pruning is done leaving only 3-4 buds on previous season's shoots during December. In Karnataka however, plants are headed back every year in January-February leaving about 2 buds on each shoot. Little or no pruning is done in Tamil Nadu.

Orchard Management

Fig plants propagated in the nursery are generally used to raise orchards. Plants raised in containers develop coiled roots. For success in dry lands, plants must have root architecture with a strong tendency to penetrate deep into the soil. Standard orchard management practices (Plate 10.5) should be timely followed in fig orchards to ensure optimum productivity.

Plate 10.5. Healthy plantation of fig

Nutrient Management

Fig is seldom manured however, to obtain good yield, adequate nutrition must be ensured in the growing as well as in the grownup fig trees. At Bangalore, a fertilizer mixture containing 220g nitrogen, 450g phosphorus and 450g potassium per tree apart from 25-30 kg well rotten Farm yard manure has been recommended. Fig cultivar Black Ischia planted at normal spacing of 5x5m produced the maximum fruit yield by application of 125 kg nitrogen and phosphorus alone or in combination under semi-arid conditions of Udaipur. At Rahuri, however, application of 900 g nitrogen + 250 g phosphorus+250g potassium to 6-7 years old tree of Poona fig was adjudged optimum (Pareek and Vishal Nath, 1996). Doses of manures and fertilizers according to the age of plants has been given in table-10.1 which should be applied at appropriate time to get optimum fruit yield in fig.

Table 10.1. Manure and fertilizers for fig tree.

Age of plant	Organic manure (Kg)		Inorganic manures (g)		
	FYM	Oil cake	N	P	K
1-2	15.0	0.5	75	50	50
3-5	25.0	1.0	150	100	100
Above 5	40.0	2.0	300	200	200

In 6 to 7 years old fig trees planted at 5m x 5m spacing, fertilization with 900g N + 250g K improved fruit production under Bangalore conditions (Anon., 2002). Organic fig farming in Turkey has great importance among other fruit crops with the production of 2615 MT from 4053 ha area. Micronutrients are often found deficient in semi-arid and arid soils. Foliar feeding of nutrients such as nitrogen (0.5-2.0% urea), zinc (0.05 to 1.0% Zinc sulphate), and Boron (0.05 to 1.0% borax) has given beneficial results.

Water Management

Although fig plant can tolerate high degree of drought and moisture stress but to obtain optimum growth and fruiting irrigation is essential. *In situ* or *ex situ* system of water harvesting is the key factor in fig orchard under dry land condition. The excess moisture or stress condition has detrimental effects on fruit growth and development and yield. Mulching with organic materials (e.g., hay, straw, dry leaves and local weeds) has been found highly beneficial in reducing evaporation loss (Pareek and Sharma, 1991) in fig plantations. The practice also suppresses weed growth, prevents erosion and adds organic matter to the soil. Although local organic mulch materials are cheaper than polythene mulches but these require proper care to maintain effective cover thickness. Sugarcane trash mulch was found effective in Maharashtra. It is necessary to irrigate fig plants during summer months. In light soils of arid region, frequent irrigation is required at 15-20 days interval for plant establishment during summer. Bearing orchard must be frequently irrigated during fruit development. In heavy soils of semi-arid region especially in Maharashtra, Karnataka and Andhra Pradesh, irrigations are done only at the time of fruit development. Drip irrigation (10-15 lit/day/tree) can be practiced for fig production, which depends upon soils and climatic conditions of the region.

Water loss due to transpiration can be reduced by use of radiation reflectants, stomata closing chemicals, and plastic films. Spraying of 4-6 per cent Kaolin, 0.5-1.0 per cent liquid paraffin and 1.5 per cent power oil, after occasional rains in low rainfall areas and after the post-monsoon rains in high rainfall areas, considerably reduced plant water losses (Pareek and Vishal Nath, 1996). Shelterbelt and windbreaks can reduce evapo-transpiration by reducing the wind speed and stabilizing microclimate.

Orchard Floor Management

The control of weeds has special significance in rainfed orchards in reducing soil moisture losses. Timely weeding is essential to improve fruit quality and yield. Weeding is also necessary to control diseases and pest attack. Timely weeding and hoeing in the tree basin should be done for good hygine of the root rizosphere. In fig orchards, leguminous crops such as *urd, moong,* etc. can be grown as intercrop during *kharif* while during *rabi* season vegetable are grown with the application of additional nutrients and water. Green manuring crops are recommended in light soils.

Disease Management

There are number of diseases affecting various parts of fig plants. A few of these have been found economically important in India.

Leaf rust (Cerotelium fici)

Fig rust is a fairly serious fungal disease. It attacks young leaves, causing defoliation. It is also easy to recognize from the small yellow-green spots that appear on leaves. The spots will get bigger and turn yellowish brown and the leaves will soon become yellow and fall. It can be controlled by using sanitation methods or spray on the plant with a 4-4-50 Bordeaux mixture at 3 to 4 week intervals particularly during rainy season/wet weather condition.

Leaf spot (Cylindrocladium scoparium M)

It develops brownish angular spots on the leaves, becoming light ashy coloured later on spot appears on the active leaves which reduces the photosynthetic efficiency of leaves and ultimate production capacity of the tree. It can be control by spraying of 0.1% bavistin.

Physiological Disorders

Fig is susceptible to sun burn, fruit splitting and fruit drop. Sun-burn is found mostly in young plantations. The newly planted trees are very much affected by scorching heat. This situation arises when plants are under leaf fall stage. It may be due to excessive pruning. Developing a good plant canopy by proper pruning and pasting of the exposed limbs with lime protect the plants from sun-burn. Fruit splitting is occurred due to sudden change in humidity during fruit ripening. It deteriorates the fruit quality and exposed to fungal infection. Fruit drop may be result of excessive moisture stress, cold nights or frost. Lack of pollination also causes fruit drop in figs. There is no appropriate control measure of splitting and fruit drop. However, it can be checked by maintaining proper soil moisture in the field.

Pest Management

Fig plants are prone to number of pests and diseases which cause considerable loss to the young growing plants as well as to the bearing trees. Some of the important pests and their management schedule have been enumerated.

Stem borer (Batocera rufomaculata)

The insect feeds on the young branches and leaves and grubs bore into the main stem and make zigzag tunnels in the wood resulting in slow deaths of these parts and reduction in production capabilities. For control, insert cotton soaked swab in carbon disulphide or put few drop of kerosene in holes and plug them with mud.

Leaf eating caterpillar (Ocinaria varians)

This is wild silk moth harmful to fig. Caterpillar feed on young and tender leaf and causes damage. This can be control by 2-3 sprays of 0.05% Metasystox or any other systemic insecticide.

Leaf roller (Phycodes minor)

Moths lays eggs on tender leaves. On hatching the caterpillar feed on epidermis of fig leaves during spring; later they fold the leaves singly or one over another and feed within. Defoliation of leaves takes place which adversely affects the production. To control the pest, collection and destruction of rolled leaves is recommended. In case of severe infestation, spray with 0.05% Endosulphon at weekly intervals should be done.

Termites

Termite is also a serious pest of fig. It causes considerable damage to the establishing plants particularly in low rainfall areas. Application of 5% Methyl parathion dust @ 50 g /pit at the time of planting and drenching of basins with 5 ml/liter chlorpyriphos at 15 days interval during dry period is recommended for control of this pest. Subsequently, if the problem is permanent, water-soluble insecticides (chloropyriphos) should be applied with irrigation water.

Flowering, Fruiting and Yield

The flowering in fig starts after 3-4 years of planting. Fig is a gynodioecious species and some female types need pollination while other set fruit parthenocarpically. However, under North India, commercial fruiting takes place after 6-7 years. Three types of flowers are found in fig i.e. male, female and gall. The development of fruit takes place without pollination. Pollination is affected by wasp, which develop inside the syconium of a male fig. Treatment with GA has shown 10% increase in fruiting and such fruits ripen 15 days earlier (Chandra *et al*, 1994). Immature fruit drop during dry weather condition has also been observed in fig. In India, a fig tree bears 280 to 450 fruits every year in the main crop. The second crop is generally poor in quality of fruits. Under Rahuri condition, cultivar Conadria produced significantly higher fruit yield than other cultivars (Anon, 2002). The fruit yield of some fig cultivar grown under Indian condition has been presented in Table-10.2.

Table 10.2 : Fruit yield of some fig cultivars.

Varieties	Av. weight of fruit (g)	No of fruits/tree	Yield (kg/tree)
Poona fig	31.25	427.50	13.36
Dinkar	29.41	406.33	11.92
Excel	27.78	385.14	10.70
Conadria	36.71	465.40	16.62
Deanna	55.56	288.81	16.05

Harvesting and Post Harvest Management

The harvesting of figs depends on the uses. The fruits may be picked from the tree or gathered manually or by mechanical sweepers after they fall to the ground. When figs are grown for drying purpose, they are allowed to ripen and dry partially on the trees. Workers must wear gloves and protective clothing because of the latex oozing from the detached end of the fruit. Sticking of latex on the fruits spoils the appearance of produce. Harvested fruits are spread out in the shade for a day so that the latex will dry a little. Then they are transported to processing plants in wooden boxes of 10-15 kg capacity.

Fresh figs are very perishable. Fruits are graded according to fruits weights (50g, 40-50g, 30-40g and below 30g) and packed in cartons for marketing. At 40° to 43°F (4.44°- 6.11°C) and 75% relative humidity, figs remain in good condition for 8 days but have a shelf life of only 1 to 2 days when removed from storage environment. At 50°F (10°C) and 85% relative humidity, figs can be kept no longer than 21 days. They remain in good condition for 30 days when stored at 32° to 35° F (0°-1.67° C).

Processing and Value Addition

Fig can be processed in to various value added products. In world market, high quality dried product of fig are available. In India, however drying of fig is done in Poona at a very limited scale. Recently, the solar-drying of fig has been initiated in Guatemala. Fig is one of the first fruits to be preserved by drying. Apart from canning and drying, it is also processed in to paste and jelly. To develop a dehydrated product, usually the fruits are allowed to fully ripen and partially dehydrate on the tree itself. After harvesting, they are exposed to sulphur fumes for about half an hour, placed out in the sun and turned daily to achieve uniform drying and pressed flat during the 5 to 7 day process. 'Black Mission' and 'Kadota' figs are suitable for freezing whole in syrup or sliced and layered with sugar. Dried cull figs have been roasted and ground as a coffee substitute. In Mediterranean countries, low-grade figs are converted into alcohol. An alcoholic extract of dried figs has been used as a flavoring for liquors and tobacco.

Fruits are very delicates and easily spoiled during handling and transportation. Proper marketing facilities are very essential for the fig. The growers prefer to sell fruits in local market. Otherwise it should be dried to fetch high price in the market. At present, fig production is relatively less so marketing is not a major problem but new market should be developed. The quality standards, packaging, season of crop harvest, etc. should be worked out for International markets.

Future Research Needs

Improved fig varieties in India are mainly introduction from Mediterranean countries. In the country, the fig has not got the status of commercial crop. As a result a lot of work in almost all facets of fig production is essential. Some of the urgent future research needs for fig production in India has been mentioned.

1. Some more introductions particularly for disease resistance, high yield and better shelf life should be made from iso-climatic region of world to enrich genetic resources and study their performance under Indian condition.
2. The genotypes having better keeping quality should be identified on priority.
3. Identification of drought hardy rootstock is needed for arid regions
4. Organic farming package for fig production should be developed and popularized to fetch high price of the produce in the International market.
5. Integrated approach for diseases and pests management should be developed.
6. Development of agro-techniques for different agro-ecological regions and popularization of fig cultivation should be taken at top priority.
7. The market should be developed for quick disposal of produce.

REFERENCES

Anonymous (2002) Annual Report, AICRP on AZF, CIAH, Bikaner,

Antunes, L.E.C., Chalfun, N.N.J., Pasqual, M., Dutra, L.F. and Cavalcante-Alves, J.M. (2003) Factors affecting on rooting of figs (*Ficus carica* L.) cuttings. *Acta Horticulture,* 605:141-146.

Chandra, A., Chandra, Anju and Gupta, I.C. (1994) Arid Fruit Research, Scientific Publisher, Jodhpur pp.250-54.

Hartmann, H. T., Kester, D. E., Davies, F. T. and Goneve R. L., (1997) Plant propagation- Principle and practices. Prentice Hall of India Pvt. Ltd. New Delhi, pp. 635.

Pareek, O.P. and Sharma, S. (1991) Fruit trees for arid and semi-arid lands. *Indian Farming*. 41:25-30.

Pareek, O.P. and Vishal Nath (1996) Fig. In: Co-ordinated Fruit Research in Indian Arid Zone- Two Decades Profile. NRC for Arid Horticulture, Bikaner. pp. 1-105.

Pareek, O.P. and Vishal Nath (2006) Arid and Minor fruits. *In*: Text book of Horticulture (K. L. Chadha, Eds.): *In Press*.

—x—

11

KARONDA

Karonda (*Caissa carandas*) is an evergreen, thorny and bushy plant generally used as live fence. It is a hardy multipurpose horticultural bush flourishes well without much care in arid and semi-arid conditions. It can be grown as hedge for beautification of nursery or orchard which serves as one of the best bio-fencing material because of thorny in nature. Karonda plants are useful in many ways. The karonda fruit is a rich source of iron (39.1%) and also contains 2.2 per cent protein, 96 per cent fat, 2.8 per cent minerals, 0.10 per cent calcium, 0.06 per cent phosphorous and 201-555 mg/100g vitamin C. Its fruit are astringent, anti-scorbutic and used as a remedy for biliousness. Mature fruit contains high amount of pectin and therefore, besides being used for making pickles and chutney, it can be exploited for making jelly. Leaves are palatable and contain fair amount of nutrient which browses by sheep, goats and camels in arid and semi arid areas. The fallen leaves enrich the soil by adding organic carbon and other mineral nutrients. It reduces evaporation rate from the soil underneath due to dense foliage. Leaves tassers are used for feeding silk worm. The leaf decoction is used in cases of intermittent fever, diarrhoea, oral inflammation and earache. The roots of the plant are heavily branched which improves binding capacity of the soil. During rainy season, it minimizes the runoff and soil loss on sloppy land. The roots serve as stomachic, vermifuge, remedy for itches and insect repellent. The roots contain salicylic acid and cardiac glycosides causing a slight decrease in blood pressure. Report reveal the carissone; the D-glycoside of B-sitosterol; glucosides of oderside H; presence of carindone; lupeol; ursolic acid and its methyl ester; also carinol a phenolic lingan in the roots and karonda plants. The wood is hard and is used for making article like combs, toys, utensils and small agricultural implements. It is also used as a fuel wood in dry region.

Origin and Taxonomy

The *Carissa carandas* L. indigenous fruit of India which belongs to the family Apocynaceae. However, Natal plum (*C. grandiflora*) came to India from South Africa and *C. edulis* Vahl. has been introduced from the USA (Singh *et al.*, 1982, Pareek and Sharma, 1993). Its occurrence is also common in Mayanmar, Malacca and dry areas of Sri Lanka.

According to recent taxonomic records, the botanical name of karonda has been changed as *Carissa conjesta* Wight. (Syn. *C. carandas* Auct.), formerly known as *C. carandas* L. It is an important species of the family Apocynaceae and which has over 30 species originating in South Africa, Asia and Malaya (Singh *et al.*, 1982). Some of the important cultivated species of the family are *Carissa grandiflora* DC, *C. bispinosa* Desf., *C. spindrum* DC, *C. ovata, C. edulis* Vahl. *C. inermis* Vahl. (Syn. *C. macrophylla*), *C. paucinervia* D.C., *C. spinarum* L. (Syn. *C. diffusa*) (Pareek and Sharma, 1993, Singh and Arora, 1978).

Karonda grows as a small shrub or tree attaining a height of 2-3 meter. Botanically, it is large dichotomously branched evergreen shrub with short stem and strong thorns in pairs simple or forked up to 2" (5 cm) long in pair in the axils of leaves. Cohen and Arzee (1980) observed two phase rhythm of apical growth in *C. grandiflora*, the first being vegetative and second leading to thorn and flower formation.The leaves are simple, evergreen, oval or elliptic in shape, 1 to 3 cm long, dark green, leathery, glossy on the upper surface, lighter green and dull on the under surface. New emerging leaves are slightly reddish in colour. Cohen and Arzee (1980) reported that in *C. grandiflora*, individual branches of sympodial shoot usually bear 2-3 pairs of decussate leathery ovate foliage leaves and a pair of inconspicuous ovate leaves from whose axils, the branched thorny shoots grow. Flowers are small (3-5 cm in diameter), found in corymbose cymes are pedicellate, bracteate, hermaphrodite, complete, hypogynous and pentamerous. Cohen and Arzee (1980) observed that in *C. grandiflora*, the flowers could be borne terminally (Plate 11.1) between thorny branches or forked thorn or on the branches in place of a thorny tip. Flowering in *C. carandas* takes place in spring (Hayes, 1966). *C. bispinosa* also produces flower during February-March under Allahabad conditions (Shankar, 1962). Flower bud takes 19-21 days to develop

Plate 11.1. Flowering in Karonda

to full bloom stage in *C. conjesta* (Syn. *C. carandas*) (Srivastava, 1966), whereas in *C. edulis* it takes about 19 days (Bankar and Verma, 1987). In *C. congesta,* the flowers open mostly between 8-10 pm (Srivastava, 1966), whereas in *C. edulis,* the anthesis time was reported between 12.30-4.30 pm under Jodhpur condition and it took 3-4 hours for complete bud opening. Bankar and Verma (1987) observed, only 36 per cent bud transformed into fruits and rest dropped off at different stages of development.

Area and Distribution

The authentic information on area and production of kronda is not available in India as well as in other part of world. In India, karonda is a subsistance crop mostly meeting local demands. Although, it is grown in different parts of the country, but due to lack of systematic plantation, no official estimates are available. It is also difficult to make a realistic estimate of the area under this crop due to scattered nature of cultivation. In world, it is grown in Philippines, Thailand, Cambodia, South Vietnam and in East Africa.. In India, karonda plants are found in Madhya Pradesh, Maharashtra, Rajasthan, Bihar, Orissa, Andhra Pradesh, Karnataka and Uttar Pradesh in wild form. Mount Abu region of Rajasthan and Rajgarh region of MP have its natural variability (Vishal Nath, 1999). Another species *Carissa inernuis* Vahl. is native of south India and is found in abundance in the forest of western Ghats. Karonda is cultivated in parts of Rajasthan, UP, Bihar and Maharashtra and in lower hills of Shiwalik. It can be good alternative for resource poor farmers in saline, sodic soil and degraded uplands.

Soil and Climate

Karonda is a hardy plant which thrives well in a wide range of soils. Deep, fertile and well drained soils are ideal for its cultivation. It can also grow successfully on poor, sandy and rocky soils however in good soils its growth in luxuriant (Vishal Nath, 1999). In Florida, the plant grows vigorously on sand or limestone. It has been classified according to salt tolerance limit, karonda as medium salt tolerant crop which could be grown up to 30-40 Exchangeable Sodium Per cent (ESP) and Electrical conductivity (EC) 9-12dsm^{-1} (Pathak, 1991). Bernstein (1972) reported that C. *grandiflora* is affected little by salinity of 8 dsm^{-1}. There is direct relationship between transpiration and ESP levels in karonda as reported by Valia *et al.*, (1993).

The crop is well adapted to varying climatic condition from an altitude of 300 to 1800 m in the Himalayas. The crop is susceptible to frost if prolong for long period. Karonda has been classified as a dry land crop for scarcity zone (Tambe *et al.*, 1993).The plants posses drought avoidance organs like thorn on the stem and branch and less stomata on lower surface of the leaf which allow the plants to perform better under drought conditions of Rajsthan and Gujrat.

Plant Genetic Resource Management and Varieties

Genetic variability in karonda is available in the states of Bihar, West Bengal, South-Eastern part of Chattisgarh, parts of Orissa and South India. The variability is found in form of morphological character, qualitative and quantitative fruit characters and bearing habit. Diversity in *Carissa* also occurs in North-Western India, particularly Mt. Abu in Rajasthan and Khandala in Maharashtra (Mehra and Arora, 1982; Vishal Nath, 1999). Plants with spreading and climbing growth habit have been reported by Sawant *et al.*, (2003). Phenotypic variability with respect to plant growth, fruit size, shape and colour of fruits has been reported by Bankar *et al.*, (1992). Dhumal *et al.*, (1993) reported variability in physico-chemical characters of fruit. Based on colour, fruits have been grouped into four categories i.e. grape green, green with purple blush, white pink blush and maroon with wide variation in biochemical composition (Kumar and Singh, 1993; Singh *et al.*, 1999; Vishal Nath, 1999 and Singh and Singh, 2000). Fruits of two types of karonda (Red green and white yellow) were studied by Singh and Singh (1998). At maturation stage, fruits from the Red-green type were significantly larger (weight and size) than those from the White-yellow type.

Conservation of plant genetic resources in the areas of their concentration, where they continue to perpetuate has special significance in enhancing the bio-diversity *In-situ*. However, no systematic efforts have been made in case of karonda, except on farm conservation by the farmers on their farm land because of their multiple utility. Very few survey and selection studies have been made so far. About 212 genotypes of karonda were selected from Kolhapur district by Sawant *et al.*, (2003) and are being maintained at College of Agriculture, Kolhapur (Maharashtra). At Rahuri, some promising types No. 13, 16, 12 and 3 have been identified (Karale *et al.*, 1989). At Central Arid Zone Research Institute, Jodhpur (Rajsthan) 13 genotypes of karonda collected from Rajasthan and Maharashtra are being maintained at its Regional Station (Pareek and Vishal Nath, 1996). Under survey, collection and evaluation programme, 5 genotypes of karonda were collected from Bharatpur, Udaipur and Ajmer are being maintained at CIAH, Bikaner, while at its Regional station Godhara, 2 promising genotypes have been identified and conserved for further evaluation (Anon. 2001). At NDUAT, Kumarganj, Faizabad, two promising line of karonda are being maintained.

Earlier, there are no well established cultivars of karonda in our country. Recently, few selections based on variation in fruit colour, size and quality have been made. The existence of entomophily in addition to cross pollination has resulted in selection of some promising plant types from the widespread natural variability followed by their propagation and use by man. Most of the existing cultivars are local selection based on their shape, size, colour and quality of fruits. Some promising types have been observed at Dapoli, Maharashtra. Karale *et al.*, (1989) evaluated seedling population from MPKV, Rahuri and reported three promising selections. Five promising genotypes have been identified by Sawant *et al.*, (2003) and are being evaluated under Kolhapur condition. Two promising types have been identified based on colour and size of fruit at NDUAT, Faizabad (Kumar and Singh, 1993). However, most of these studies include a few genotype with limited characters, since these have been developed through single plant selections from the natural population. The need for improvement is realized particularly when commercial karonda has to be undertaken under different agro-ecologically environments. The main objectives for crop improvement in karonda have been focused on development of varieties resistant to drought, tolerant against saline and alkaline soil conditions, high yielding under low and high temperature conditions, high pulp: peel ratio, longer shelf life and suitable for processing industry. Work initiated during the 1980's on collection and evaluation of germplasm has led to the identification of few location specific cultivars (Plate 11.2). Quality characters of ripe fruits of karonda lines PK-3, PK-4, Pant Manohar, Pant Sudarshan and Pant Suvarna have been studied at Horticultural Research Station Patharchatta, Pant Nagar, (Misra and Jaiswal, 1998). The characteristics of some of the selected lines of karonda developed by different institutions have been enumerated.

Plate 11.2. Good fruiting in a promising genotype of Karonda

Pant Suvarna

This is a seedling selection from G.B. Pant University of Agriculture and Technology, Pantnagar. Plant grows upright and is bushy in nature. Fruits are green with light maroon blush. On maturity they attain deep maroon colour. Fruit size is 2.55x1.66 cm and average fruit weight is 3.62g. Pulp is 88.27 per cent, dry matter 12.39 per cent; Total soluble Solid 3.83 per cent, acidity 2.30 per cent and the average fruit yield is 25 kg / bush.

Pant Manohar

The variety Pant Manohar is developed by a selection from G.B. Pant University of Agriculture and Technology, Pantnagar. Fruits are attractive in colour having deep pinkish blush on white background. Fruit is medium size (2.13x1.69 cm) having an average fruit weight of 3.49 g. Pulp composition is 88.27 per cent, dry matter 12.77 per cent,

Total Soluble Solids 3.92 per cent and acidity 1.82 per cent. The average fruit yield is 35 kg/bush.

Pant Sudarshan

It is also evolved from G.B. Pant University of Agriculture and Technology, Pantnagar is a medium sized bush. Fruits are light pink in colour measuring 2.16x1.69 cm. Average fruit weight is 3.46 g and the number of seeds/fruit is 4.68. Pulp is 88.47 per cent; dry matter 11.8 per cent, Total soluble Solids 3.45 per cent, sourness 1.89 per cent and average fruit yield is 32 kg/bush.

Maroon Coloured

A promising type identified and developed from N.D. University of Agriculture and Technology, Kumarganj Faizabad, has an attractive maroon colour fruit. Average fruit weight is 3.2 g, T S S- 7.1 per cent, acidity 4.1 per cent, vitamin C 17.9 mg $100g^{-1}$, anthocynanin (OD) 0.7, Iron 3.8 mg $100g^{-1}$. Average fruit yield is 3.7 kg/bush.

White-Pink blush

This is also developed by NDUAT, Faizabad. Fruit weight is comparatively less than Maroon coloured 2.9 g. Average number of seed/fruit is 7.0, T S S - 6.6 per cent, acidity 4.7 per cent, vitamin C 15.4 mg $100g^{-1}$, anthocynin (OD) 0.2 and iron 4.15 mg $100g^{-1}$. The average yield is 2.8 kg/bush.

Besides these promising karonda types such as white and red fleshed karonda have been located in tribal tract of Chattishgarh, Bastar and are under evaluation at the College of Agriculture, Jagdalpur, Bastar and Chattisgarh.

Plant Propagation and Nursery Management

Karonda can be propagated through sexual and asexual means. Seed has short viability, hence should be sown just after extraction from ripe fruits. The ideal time for sowing is rainy season. Vishal Nath (1999) reported that freshly extracted seeds from ripe fruits should be sown directly in nursery beds or in polythene bags filled with proper nursery mixture during September. Seeds treated with GA_3 at 25 ppm improve seed germination (67%) after 58 days after treatment (Bankar, 1987). One year old seedlings are used for planting preferably after first shower of monsoon. For promoting early growth, gibberlic acid 1000 ppm have been reported effective by Misra and Jaiswal (1998).

Sexually propagated plant starts bearing quite late. At the same time the quality of fruit from the seedling population can not be assured. Therefore, the vegetative propagation techniques need to be explored. Not much work has been done in respect to vegetative propagation of karonda. Hard wood cuttings from trimmed hedge resulted (20%) rooting during November. Cuttings from well established mother plants under nursery gave best results. In a study conducted at Allahabad revealed that pretreated cuttings with Indole-butyric acid at 500 ppm in 50 per cent alcohol produced 10 per cent rooting in late September, 20 per cent in early October, 30 per cent late October and 50 per cent in early November. Shankar (1962) reported air layering for propagation of 3 species of *Carissa* and reported rooting in *C. carandas, C. grandiflora* and *C. bisponia* by treating with IAA and IBA at higher concentrations under Allahabad conditions. Vishal Nath (1999) reported very good success in air layering during monsoon period with 1000 ppm IBA treatment. Moist Mass grass produced better roots than soil based rooting medium. Soft wood grafting is also successful and very useful technique for *in situ* propagation of karonda in arid zone (Chundawat, 1990). July- August was found to be the best time for soft wood grafting. For grafting purpose, Shankar (1966) reported that *C. carandus* , *C. grandiflora* or *C. bispinosa* could be successfully used as rootstock for *C. carandas* whereas *C. grandiflora* found to be better root stock for *C. grandiflora*. Some success was obtained with *C. grandiflora* on *C. carandas* root stock but grafts of *C. grandiflora* on *C. bispinosa* rootstock did not survive.

Planting

Planting should be done in square system at 2 m x 2 m distance or hedge row plantation along the boundary of orchard at a distance of 1 m x 1 m at prefilled pit (50 x 50 x 50 cm) with organic manure and soil (1:2).With the onset of monsoon, planting should be done as per the spacing. Total number of plant per hectare is 2500 when planted at a spacing of 2 m x 2 m.Whereas 400 plant /ha is required when planted along the boundary at a spacing of 1 m x 1 m. Planting should be done during the monsoon, particularly in dry regions. This will reduce the need of frequent watering until the plants are firmly established in the soil. Karonda needs minimal after care however, hoeing of the soil around the plants, once a year helps better plant growth.

Training and Pruning

Karonda is a bushy fruit plant. However, plants can be trained on single or double stem in regular plantations. For proper training, lateral shoots should be removed up to 15-20 cm height from the ground level. This will form the main stem of the plant. Suckers arising out from ground, diseased, insect infested and dried twigs should be removed at regular intervals. Plants grown as protective hedges around the other perennial orchards do not require any pruning except old hedges which may be headed back to induce new growth and form thick, impenetrable barrier for common animals. Bearing plant, usually do not require pruning. However, to give desired shape and keeping the stem clean in the young plants unwanted twigs should be pruned as and when required during preparation of frame.

Orchard Management

Karonda may be grown as pure crop or as a filler crop in agro-forestry system (Horti-silviculture) or as boundary plantation. There are several methods for establishing karonda such as *in situ* establishment through direct seeding, transplanting nursery grown plants into the field or vegetatively propagated plants can be raised in nursery beds and transplanted to the field.

Nutrient Management

Being an underutilized fruit crop, very meagre information is available on manure and fertilizer requirements of karonda. In practice, the dose of manure and fertilizer for optimum growth and bearing of karonda plants will depend upon the nutrient status of the orchard soil. However, as a matter of common experience it has been observed that the crop responds well to organic manuring. Without adding optimum quantity of manures and fertilizers, the orchard soils slowly get exhausted and ultimately the karonda plants show poor vegetative growth, produce fruits of poor quality and subsequently the die back symptoms appear. Well rotten Farm yard manure @10 kg/plant with the onset of monsoon will be beneficial in dry areas. Application of 50, 100 and 150 g /plant/ year of nitrogen, phosphorus and potash respectively should be done for better results (Pandey and Reddy, 1989). Srivastava (1984) has suggested application of 2-3 kg well rotten FYM and 100 g of fertilizer mixture of N P K to each of the bearing plant for better yield. Vishal Nath (1999) mentioned that 500 g urea along with 15-20 kg FYM per plant is sufficient for well grown up karonda plants. The manures and fertilizers should be applied when there is sufficient moisture in the basin of the plants.

Water Management

Karonda is a xerophytic plant and therefore needs very little irrigation after establishment. However, if there is no rain during the period of fruit development, single irrigation is sufficient for increasing the fruit size. Irrigation after new planting and manuring in grown up plant is beneficial. In arid regions, water harvesting structure should be created to collect run off water during rainy and post monsoon season near root zone for proper growth and development of plants.

Orchard Floor Management

Weeds are not a major problem in full grownup karonda orchard, however, field should be free from weeds at initial stage of growth. Elmore *et al.*, (1979) have reported that oxyflurofen 2 per cent granules at 1.12 or 2.24 kg ha^{-1} or oxadiazon 2 per cent granules at 2.24 kg ha^{-1} gave good control of *Oxalis corniculata, Euphorbia maculata* and *Senecio vulgaris* in karonda orchard. However, the Oxyfluorfen treatment caused necrosis of young growth of C. *grandiflora* hence; one has to be careful in using herbicides in young karonda orchards for weed management. Spray of 1.25 kg ha^{-1} oxadiazon and application of 0.6 kg ha^{-1} of 2 per cent Oxyflurofen granules have been found effective in weed management in karonda orchards (Brosh and Zilberstin, 1985). Considering the ecological impact of pesticides on soil properties, it is felt that till the agricultural labour is relatively cheaper and abundantly available, manual weeding should be given preference over chemical weed control practices. Exposure of soil to direct and strong sunlight by deep summer ploughing could be another alternative and most effective method of weed control, more so under fragile agro-eco system and with resource poor farmers.

Seasonal vegetables or medicinal plants particularly *Matricaria* can be taken in regular plantation of karonda. The crop can also be tried as under storey crops with aonla, bael, ber etc. at different crop geometry. Growing of aonla+karonda in association have been reported to improve physical and chemical properties of soil (Malviya and Singh, 1998). The work done at Narendra Deva University of Agriculture and Technology, Kumarganj at Faizabad in Uttar Pradesh has proven that aonla- guava- karonda and aonla-ber- karonda are ideal crop combinations for salt affected soils. This practice not only ensures income from the newly established orchards but also reduces the evaporation losses from soil hence helps in moisture conservation in water scarcity conditions (Pathak, 1991).

Disease Management

Several diseases have been reported to infect karonda plant. The major diseases reported are Anthracnose (*Collectotrichum inamdrii*) and Bacterial canker (*Xanthomonas carissa*). Anthracnose is a wide spread disease. It is very much common in U.P. but practically it is found in all areas wherever karonda is cultivated at varying virulence. The disease can be identified as black greyish spots on the leaves. Spray of copper oxychloride @ 2 g/liter water effectively manages this disease. The symptoms of Bacterial canker first appear as spots on lower surface of leaf. These spots are small, round and water soaked and later on turn dark brown. A yellowish green halo may surround the spots which may cause necrosis in sever cases. The disease can be checked by removing diseased leaves and twigs by pruning followed by spray of 200 ppm phytomycin. Leaf blotch of *karonda* caused by *Pseudocercospora* has also been observed by Singh and Mukherjee (1979) under Faizabad conditions.

Apart from foliar diseases affecting the standing crop and plants of karonda, quite good number of Post harvest diseases also occurs on karonda fruits during its storage, transportation and marketing. Fruit rot of karonda due to infection with *Myrothecium roridum* (Kumar and Tandon, 1979) and brown rot of fruits due to *Alternaria alternata* (Tandon and Kumar, 1974) have also been observed as common and serious disease in local orchards under Allahanbad conditions of Uttar Pradesh. These diseases rendered the fruits unfit for human consumption. Both immature and mature fruits are susceptible. Fruits of karonda are also attacked by fungus *Bartalinia robillardoides* Tassi (Sharma and Agrawal, 1976). Beacause of infection of fruits with fungi like *Pestalotiopsis versicolar*, *Myrothecium roridum* and *Alternaria alternata* the quantity of tartaric and malic acid decreases while fumaric acid disappears completely in infested tissues (Kumar and Tandon, 1979).

Pest Management

There is no serious pest to infesting karonda plants. However, during initial stage of crop growth, it is infected by larvae of *Digama hearseyana,* which results in stunted growth due to defoliation of leaves.

This can be controlled by spray of 0.2 per cent endosulphan or 0.1 per cent nuvocron. Agrawal (1984) reported karonda plants to be infected by *Simcronyx roridus,* a gall forming weevil almost throughout the year with its peak from March-May. The removal of infested twigs is the most practical approach of managing gall midges in karonda.

Flowering, Fruiting and Yield

Seedling plant comes into bearing after 4-5 years of planting whereas air layered karonda plant start bearing within two years. Plant comes in floriferous phase during Spring-Summer in north Indian condition. In southern India, however, plants produce two crops. Flowering in karonda takes place more than one year old shoots. Plant produces hermaphrodite flowers but due to lack of desired pollination almost 60-70 per cent flowers are dropped without fruit set. Flowering in karonda takes place in February- March and fruit attains physiological maturity by May- June. The fruit become ready for harvest after 100-110 days of fruit set (Plate 11.3). Change of colour with specific gravity above 1.0 is a good indicator of fruit maturity. Ripening of fruits in karonda coincides with rainy season under north Indian condition. Fruit yield in karonda varies from plantation to plantation according to the orchard management practices. In sandy soils, 5 kg fruit tree^{-1} year^{-1} can be harvested under normal conditions. Under

Plate 11.3. Fruiting in Karonda-fruits ready for harvest

good management conditions, Pareek (1990) reported that 10-12 kg fruit yield from a full grown tree/ year can be obtained. Singh *et al.*, (1982) reported a yield of 500-600 number of fruits /bush/year from *C. grandiflora* and 1.2-1.6 kg from *C. edulis* under Jodhpur conditions from 7-8 year old trees. The fruit yield varied from 1-3 kg /plant under hedge row planting system along the field boundary in Agra conditions (Kumar and Yadav, 2006).

Harvesting and Post Harvest Management

Karonda fruits do not have synchronous maturity therefore; Two to three pickings are generally required for a complete harvest of 4-5 kg physiologically mature fruits /plant. Both mature and ripe fruits can be harvested depending upon their use. There is no standard practice for grading and packing of fruits. Fruits after harvest are collected in shade, where blemished and undesirable fruits are sorted out. Fruits are kept in baskets and marketed. Generally, mature and freshly harvested fruits are marketed locally for fresh consumption. There is no organized marketing network of karonda, basically because of the fact that it is highly seasonal in production and perishable in nature. Its fresh consumption is also extremely area oriented. The fruit harvested at maturity can be stored for 5-7 days at room temperature before they begin to shrivel. If harvested at ripe stage, it has a very short life and can be stored only up to two days at normal temperature. Fruits can also be stored for long time in a solution of 2000 ppm concentration Potassium Metabisulphite. Three marketing channels i.e., Grower-consumer; Grower-retailer-consumer and Grower-wholesaler-processor-consumer is generally followed in India for karona fruit trade.

Processing and Value Addition

Karonda fruits do not have table use, however the fruits can be processed in to a number of products. Mature fruits are processed into pickle, chutney, candy and jelly. Fruits can be used for making jam, jelly, squash and syrup. Recipes and methods for some of the important preparation from 1 kg of karonda fruits have been described by Srivastava and Kumar, 1994.

Chutney

The ingredients required for making chutney from 1 kg fruit is 50 g chilli powder, 1 chopped ginger, salt 10 g, garlic chopped 10 g, hot spices 20 g and acetic acid 20 ml. The fruits after thoroughly washing are cut into slices and boiled with little water by adding sugar and salt. While cooking is continued, chopped ginger in a muslin bag are put in it and pressed with a spoon to obtain the extract. After cooking to a desired consistency, spice bag is again slightly heated and pressed and thereafter, it is filled in clean bottles, capped and stored in cool and dry place.

Pickle

For making pickle, after washing the fruits, salt 200 g is added and kept for 3 days. On the fourth day, all spices (chilli powder 25 g, mustard powder 50 g, turmeric 10 g, fenugreek 10 g) is added and mixed thoroughly. After mixing, 300 ml heated then cooled mustard oil is added and filled in jar followed by capping and storage in a dry place (Fig. 11.1).

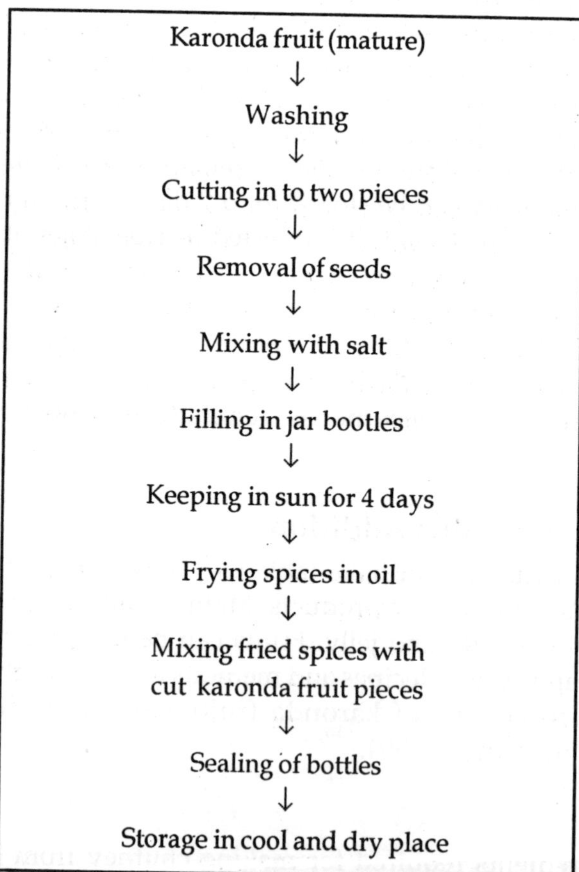

Fig. 11.1. Flow chart for preparation of karonda pickle

Jelly

Karonda jelly is made by boiling mature fruits. The fruits after washing in water are cut in the form of thin rings and equal amount of water is added. After adding water it is boiled for 20-30 minutes and 2 g citric acid is added. After boiling is completed, extract is strained and tested for pectin content. Thereafter, 1 kg sugar is added and cooked and during this cooking process, scum is removed. Finally, the ready product is filled into clean and sterilized bottle followed by sealing, capping and storage in cool and dry place.

Candy

For making candy, mature and healthy fruits of karonda are washed properly. The seed is removed by giving a slit with the help of a sharp knife. The ready fruits without seeds are steeped in 2 per cent salt solution containing 600 ppm Potassium Meta Bisulphite (KMS) for 24 hours to bleach the fruits followed by washing, pricking and blanching in water containing 0.05 per cent erythrosine and 0.25 per cent citric acid to fix colour. Thereafter, the fruits are steeped in sugar syrup of 50 per cent Total soluble solids for 24 hours and subsequently to higher concentration of sugar syrup in phased manner till it is steeped in 75% TSS sugar syrup. The adhered syrup is ultimately removed by dipping in boiling hot water for few seconds. The ready product is packed in suitable bottles, sealed and stored in cool and dry places.

Future Research Needs

Visualizing the adaptability of karonda plants under adverse soil and climatic conditions and tolerance against biotic and abiotic stresses, karonda crop has bright future in India. In the programme on increasing area and production of fruit crops in India, utilization of degraded and waste lands for fruit plants karonda crop has imperative future. The crop diversification programme will have to consider karonda as a component crop in multi-storied cropping system of fruit crops in waste lands. As such, the following priorities of research needs to be addressed:

1. Identification and development of high yielding varieties with desired fruit qualities
2. Standardization of agro-techniques especially nutrient and water management for higher production

3. Management of pests and diseases including post harvest diseases to minimise the losses
4. Crop diversification options under different agro-climatic conditions needs to be worked out
5. Development of crop production technologies for waste and degraded lands
6. Post harvest handling techniques and product diversification for high value addition.

REFERENCES

Anonymous (2001) Annual Report-2001. CIAH, Bikaner.

Agrawal, B.D. (1984) Life history of *Cuscuta* gall weevil Smicronyx roridus Mshl. (Coleoptera: Curculionidae). *Cecidologia International,* 5 (1-2): 41-50.

Bankar, G.J. (1987) A note on influence of gibberlic acid on seed germination and vigour of seedlings in karonda (*Carissa carandus* L.), *Progressive Horticulture,* 19:1-2.

Bankar, G.J. and Prasad, R.N. (1992) Shushk kshetron mein karonda ki Bagwani. *Udyan Lok,* 4 (2): 13-15.

Bankar, G.J. and Verma, S.K. (1987) *Progressive Horticulture,* 19: 163-166.

Bankar, G.J., Verma, S.K. and Prasad, R.N. (1992) Vegetative and Fruit Characterization of some *Carissa* species in the Thar Desert. *Progressive Horticulture,* 25 (3):7-9.

Bernstein, L. (1972) *Journal of American Society Horticultural Science,* 97: 550-556.

Brosh, S. and Zilberstein, J. (1985) *Phytyoparasitica,* 13:245.

Chundawat, B.S. (1990) Arid Fruit Culture, Oxford and IBH, New Delhi. pp. 171-176.

Cohen, L. and Arzee, T. (1980) *Botanical Gazette,* 141:3

Dhumal, S.A., Patil, D.S. and Jadhav, M.G. (1993) Studies on variability and correlation in fruit characteristic in karonda. *Journal of Maharashtra Agricultural University,* 18 (3): 498-499.

Elmore, C.L. (1979) Flower and nursery Report, *Summer*: 4-5

Hayes, W.B. (1966) The *Carissa,* In: Fruit growing in India. IIIrd Ed., Kitabistan, Allahabad, pp.416-17.

Karale, A.R., Keskar, B.G., Shete, M.B., Dhawale, B.C., Kale, P.N. and Chaudhari, K.G. (1989) Seedling selection in karonda (*Carissa carandus* L). *Mahrashtra Journal of Horticulture,* 4: 125-129.

Kumar, D. and Yadav, R.C. (2006) Karonda (*Carissa carandus* L.)-a dual purpose horticultural bush for protection and production in wastelands. National symposium on under utilized horticultural crops, IIHR, Bangalore, June 8-9, p-29

Kumar, S. and Singh, I.S. (1993) Variation in quality traits of karonda (*Carissa caradus* L.) germplasm. *South Indian Horticulture,* 41: 108-110.

Kumar, S. and Tandon, M.P. (1979) *Indian Phytopathology,* 32: 134-135.

Malviya, S.K. and Singh, I.S. (1998) Effect of fruit based cropping models on chemical properties of soil. *Indian Journal of Horticulture,* 29 (1-2): 18-21.

Marlatt, R.B. and Riddings, W.H. (1976) *Plant Disease Reporter,* 60: 842.

Mehra, K.L. and Arora, R.K. (1982) Plant Genetic Resources of India-Their Diversity and conservation, NBPGR, New Delhi

Misra, K.K. and Jaiswal, H.R. (1998) Effect of foliar sprays of gibberlic acid on the growth of karonda (*Carissa carandus* L.) seedling. *Indian Journal of Forestry,* 21 (1): 70-71.

Misra, K.K. and Jaiswal, H.R. (1999) Studies on quality characteristics of ripe fruits of some selected lines of karonda (*Carissa carandus* L.). *Scientific Horticulture, 6: 45-48.*

Misra, R.S. (1968) *Punjab Horticulture Journal,* 8:47-49.

Pandey, R.M. and Reddy, Y.N. (1989) *Intensive Agriculture,* 26: 6-15.

Pareek, O.P. (1990) *Agriculture Review,* 11: 203-220.

Pareek, O.P. and Sharma, Suneel, (1993) Genetic resources of under-exploited fruits. In: Advances in Horticulture, Vol. I (K.L.Chadha and O.P.Pareek; Eds.), Pub. Malhotra Publishing House, New Delhi, pp. 189-225.

Pareek, O. P. and Vishal Nath (1996) Coordinated fruit research in Indian Arid Zone- Two decades profile. NRCAH, Bikaner:1-106.

Pathak, R.K. (1991) Problems and Prospects of wasteland utilization for fruit production. In: Proceeding of Summer Institute on Fruit production and Utilization in wasteland June 24-July12, 1991, N.D. University of Agriculture and Technology, Kumarganj, Faizabad, U.P., pp161-167

Sawant, R.B., Desai, U.T., More, T.A., Ranpise, S.A. and Sawant, S.V. (2003). Performance of Promising selections of karonda from Hills of Kolhapur District. *Journal of Maharashtra Agriculture University,* 28 (3): 339-340.

Shankar, G. (1962). *Allahabad Farmer,* 36: 19-23.

Sharma, N.D. and Agrawal, G.P. (1976) *Indian Phytopathology,* 29: 193.

Singh, A.K. and Singh, P. (1998) Power of significance of difference among fruit and seed size parameters of karonda (*Carissa carandus* Linn.). *Annals of Agricultural Research,* 19 (1): 66-77.

Singh, B.B. and Mukherjee, P. (1979) *Indian Phytopathology*, 32: 468-470.

Singh, H.B. and Arora, R.K. (1978) Wild edible plants of India, ICAR, New Delhi.

Singh, I.S. and Singh, V. (2000) Karonda-Bulletin. Deptt. of Horticulture, NDUAT, Faizabad.

Singh, I.S., Srivastava, A.K. and Singh, V. (1999) Improvement of some underutilized fruits through selection. *Journal of applied Horticulture*, 1: 34-37.

Singh, R. (1982) *Indian Horticulture*, 27 (3): 10-12.

Singh, R. , Chopra, D.P. and Gupta, A.K. (1982) Some *Carissa* for western Rajasthan. *Indian Horticulture*, 27: 10-11.

Srivastava, R.P. (1984) Karonda. In: *Phal vrikshon ka Poshan*. ICAR, New Delhi, p. 258.

Srivastava, R.P and Kumar, Sanjeev. (1994) Fruit and Vegetable Preservation: Priciples and Practices. International Book Distributing Co., Lucknow, p.192.

Srivastava, V.K. (1966) *Indian Journal of Horticulture*, 23:151-155.

Tambe, T.B. Ranshe, D.G. and Walunjkar, R.B. (1993) Dryland fruit crops for scarcity zone. *Mahrashtra Journal of Horticulture*, 7 (1):105-106.

Tandon, M.P. and Kumar, S. (1974) *Indian Phytopathology*, 27:112-113.

Valia, R.Z., Patil, V.K., Patil, Z.N. and Kapadia, P.K. (1993) Effect of soil sodicity on growth and physiological parameters of karonda (*Carissa carandus* L.). *Annals of Plant Physiology*, 7(2):211-215.

Vishal Nath (1999). *Bagwani main karonda ugayen. Krishi evam Gramin*, 2 (11):10.

—x—

12

MULBERRY

Mulberry (*Morus alba*) is a very hardy plant belonging to the family Moraceae. There are more than 100 species under the genus *Morus* among which *Morus alba, Morus rubra* and *Morus nigra* are the main species. It is perennial, fast growing fruit tree, thrives well under tropical, sub- tropical and temperate region. The main constituents of ripe fruits are water (85-88%), carbohydrate (7.8-9.2%), protein: (0.4-1.5%), fat (0.4-0.5%), free acid (1.1-1.9%), fibre (0.9-1.4%) and minerals (0.7-0.9%).

The ripe fruit of mulberry is highly appreciated for its delicious fruit which is consumed fresh or for extraction of juice. The ripe fruits are processed throughout the Asia particularly in Central Asia and in the Near East. The dominant taste of the ripe fruit is sweet but usually with sub acidic blend due to the high water content and low level of other flavoring ingredients. The fruit is very testy and used for chutney preparation at immature stage while for table purpose on ripening. The flavour improves if the fruit is dried; it can then be used as a raisin substitute. The ripe fruit of mulberry contains 8-9 % sugar and 1% acidity (Sturrock, 1940). In recent years, with considerable work on cultivating the mulberry plants under various conditions, mulberry fruit juice has been commercially produced as a health beverage, and it has become very popular in China, Japan, and Korea. Singh (1952) reported good quality of jam prepared by mulberry fruit. Without adding preservatives, the original juice of mulberry fruit remains fresh under cold storage for 3 months, while the bottled beverage remains fresh at room temperature for 12 months.

The mulberry wood is used for making agricultural implements, table, chair, boat and sport items like hockey sticks and tennis rackets.

Pruned twigs are used for basket making in the Indian subcontinent. Bark is used for making good quality paper. The most important use of mulberry leaf is rearing of silk worm for production of silk. The silk producers have traditionally fed mulberry refusals and leftovers from silkworm feeding to farm animals and to herbivorous carp in poly culture fish ponds (Gongfu *et al.*, 1997).

A variety of medicinal properties have been attributed to the different parts of the mulberry plant (Datta, 2000). Mulberry fruit juice is used for curing throat infection, controlling fever and also for blood purification. Bark is used for making nerve tonic. Mulberry trees are also utilized for landscaping in Asia, Europe and Southern U.S.A. (Tipton, 1994). Because of low water requirements and amenability to pruning mulberry trees are suitable plants for urban conditions, house gardens, street shade and city embellishment.

Origin and Taxonomy

The genus *Morus* belongs to family Moraceae and Asia is considered to be center of origin of mulberry. There are three major species of mulberry; *Morus alba* L. (white mulberry), *Morus rubra* L. (red mulberry) and *Morus nigra* L.(black mulberry) originated in China, North America and Iran and India, respectively. According to Janaki Ammal (1960) only *M. rubra* is native to America. The systematic taxonomic position of mulberry is Kingdom-Plant, Division-Angiosperm (flowering plant), Class-Dicotyledon, Order -Urticales, Family-Moraceae, Genus-Morus and Species-*alba, nigra* and *rubra*. Wild species of *Morus* like *M. australis* Poir, *M. indica* Lin., *M. laevigata* wall. X Brandis (wild mulberry) and *M. serrata* Roxb. (Kimu, Himalayan mulberry) grow wild in the Himalayan region (Singh and Arora, 1978). In the Republics of the ex-USSR, the most common species are *M. multicaulis, M. tartarica* and *M. nigra* (Datta, 2000). In Indonesia (West Java) there are seven species: *M. alba* (varieties *tartarica* and *macrophyla*), *M. nigra, M. multicaulis; M. australis; M. cathyana* and *M. mierovra* (Katsumata, 1972). In Vietnam they have over 100 varieties mainly *M. alba, M. nigra* and *M. laevigata* (Katsumata, 1973). The red mulberry is native of moist bottom lands from Canada to Southern Texas and is supposed to be the latest species in the entire genus. However, the trees of *Morus alba* var. *tomentosa* are bushy and spreading in nature. In India, the main species are *M. indica, M. alba, M. serrata* and *M. laevigata,* which grow naturally in the northern part of the country (Ravindran *et al.*, 1997). Four species, *Morus australis* Poir (Tot), *Morus indica* Lin. (Tot), *Mores laevigata* Wall ex. Brandies (wild mulberry) and *Morus serrata* Rob (Kim, Himalayan mulberry) grow wild in the Himalayan region (Singh and Arora, 1978). There are about 68 species in the genus *Morus* and majority of them occur in Asia (Datta, 2000).

The mulberry trees are small to medium in size which typically reaches up to a height of 6 to 18m. The bark on mature trees is light brown with loose vertical plates. The leaves are heart-shaped, leaves 8 to 24 cm long and 8-15 cm wide; they roughly textured on the upper surface and have a long tapered tip. The sap in the leaves is milky. The flowers are pendulous catkins of yellowish green to reddish colour (Plate 12.1). The edible cylindrical fruit botanically, an aggregate fruit is 2 to 5 cm in length and turns white black or red at ripening stage depending on the species. Mulberry is a wind-pollinated crop; flowers emerge with the growth of new leaves in late spring. Individual trees may have both male and female catkins, or may bear all male catkins or all female catkins in one inflorescence stock. The mulberry plants are monoecious in sex expression. Birds are likely the most significant dispersers of mulberry seeds.

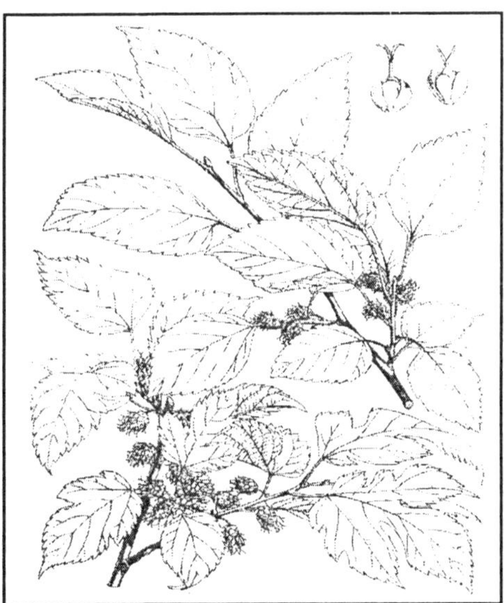

Plate. 12.1. Infloresence of mulberry with branches and leaves

Area and Distribution

Globally, the most important use of mulberry is in the production of silkworm, which feeds exclusively on its leaves. The country with the largest area of mulberry is China with approximately 6,26,000 ha, followed by India with nearly 2, 80,000 ha. Other countries where mulberry is being grown are Brazil (38,000 ha), Cuba, Dominican Republic, El Salvador, Honduras and Panama, Mexico and Saint Vincent with less than 1000 ha (Manuel, 2004).

In India, mulberry is being grown through out but its more extensive area in Karnataka particularly Mysore especially for sericulture. The other mulberry growing states in India are Tamil Nadu, Kerala, Uttar Predesh, Bihar, Madhya Predesh, West Bengal, Rajsthan, Himanchal Pradesh and Assam. The domestication of mulberry must have started several thousand years ago as a requirement for silk worm rearing (FAO, 1990). Considering that the silk trade has existed for a long time throughout the Old World and that mulberry is also cultivated for its fruit and for landscaping. Its germplasm has been taken to many countries, and now it has a very wide distribution range in Asia and Europe (from Korea to Spain including China, India, Central Asia and the Near East); in Africa (North and East Africa) and in the America (from the United States to Argentina, including Mexico, Central America, Colombia and Brazil).

Soil and Climate

The soil for mulberry cultivation should be deep, well drained, clay loam in texture, friable structure with good water holding capacity. Slightly acidic soils, free from injurious salt are ideal for mulberry cultivation. Saline and alkali soil are not suitable for mulberry cultivation. It is possible only after soil improvement by the application of gypsum / green manuring. *Dhaincha* and cow pea are recommended as green manure crops for soil improvement. Application of gypsum, incorporation of heavy dose of compost or farm yard manure with the provision of adequate drainage are necessary to improve the productivity of these saline and alkali soil. In the case of acid soils, it will be profitable to resort to application of lime.

Mulberry can be grown in a wide range of climatic conditions varying from temperate to tropical region. In the temperate climate mulberry does not sprout during winter season while in the tropics the growth is continuous. In India, the climatic condition is favorable for luxuriant growth of mulberry throughout the year. In Karnataka state, the temperature generally ranges between 21^{0}C and 31^{0}C while in West Bengal it varies between 15^{0}C in winter to 37^{0}C in summer. In both the states, mulberry exhibits good growth throughout the year. But at extremely low (below 13^{0}C) or high (above 38^{0}C) temperatures the sprouting of buds and growth of mulberry will be affected. Mulberry can be grown where the rain fall range from 600 to 2500 mm per year. In Karnataka and West Bengal the annual rainfall varies from 500-750 mm and 1500 to 3400 mm, respectively. Kashmir valley receives annually 2346 mm snowfall and 1039 mm rainfall. Under low rainfall conditions the growth of mulberry is limited due to shortage of moisture resulting in poor yield of mulberry as in Karnataka state. Mulberry growth is optimum at an altitude of 600-700 m from mean sea level (MSL).

The hot arid zone is characterized by the conditions of high aridity, with low average annual rainfall 180 mm in Jaiselmer to 360 mm in Jodhpur of Rajasthan. In these areas the rainfall is confined from July to September with 9-21 rainy days. The water holding capacity of light sandy soil is very low. The atmosphere is very dry with a vapour pressure deficit (V.P.D) of more than 24 mb in April and more than 30 mb in May. In such areas, mulberry can also be grown with proper water management.

Plant Genetic Resource Management and Varieties

Indian sub continent is bestowed with enormous diversity of several fruits in a wide range of agro-ecological situation through out the tropics and temperate region. Gene pool represents total genetic diversity accumulated over a period of evolution under domestication and natural selection. Principally the genetic resources in plant come from (i) Wild species and primitive forms in primary centre of diversity and (ii) Introduction/migration/trans domestication in the secondary centers of diversity. Under the present scenario, due to green revolution and different land use programmes, the traditional genetic resources which dominated the scene for thousand of years are gradually getting obsolete as cultigens resulting in erosion of indigenous plant wealth. Mulberry comprises of about 68 species mainly from Asian countries like Japan, India, China, Korea and Taiwan. Four species *M. serrata, M. indica, M. Alba and M. laevigat*a are from India. *M. serrata* is endemic to North-west India whereas, *M. laevigata* is widely distributed through out the country. Biodiversity particularly of plant species is the base for plant improvement. Conservation of bio diversity in general and plant genetic resource in particular assumes crucial importance for global food, health and livelihood security. Genetic stock conserved in specific ecological niches offers unexploited genes for productivity as well as resistance to biotic and abiotic stresses. Conservation of any genetic resource ensures easy assess to desired genes, a resource to revitalization, a guard file for disaster management, safe gourd against gene erosion and comparative profiling of gene / traits etc. The choice of conservation strategy depends mainly upon nature of the materials to be conserved i.e. life cycle, mode of reproduction, size and ecological status .The major approach for conservation are *in-situ* and *ex -situ.*

Mulberry being perennial and out breeding tree, exhibit high degree of heterozygosis and often produces recalcitrant seed. Hence, for conservation of mulberry outside its natural habitats, a field gene bank has been developed at a Central Seri plasm cultural Germ Resources Centre, Hour, Karnataka which maintains *M. austral,* M. *cathay Ana* Hems, *M. multicaulis* Poir, *M. rotundiloba* Koidz, *M. alba* L., *M. indica* L., *M. teliaefolia* Makino, *M. nigra* L., *M. serrata* Roxb., *M. laevigata* Wall., *M. rubra* L., *M. sinensis* Hort. and M. *bomycis* Koidz. Saraswat *et al.,* (2004) mentioned 1023 (761 indigenous and 262 exotic) mulberry accession in the field of gene bank.

Choice of mulberry varieties for planting is very important. As mulberry is a perennial crop and initial planting involves heavy investment, a great deal of care is necessary to ensure that a proper variety of mulberry is selected as planting materials. The varieties are recommended on the basis of favourable response to the existing environmental conditions. Many cultivars of mulberry have been introduced in India from France, Ramania and Japan. The good fruited variety includes EC 120294 (Ramania), Back English, Wellington mulberry, Tout Amer, Variegala Globosa (EC 200365 - 280368) from France. Many high yielding varieties of mulberry are currently available for cultivation in different states.

Karnataka, Tamil Nadu and Andhra Pradesh	:	Kanva-2, S-30 and S- 54.
West Bengal	:	S-1, S-799
Darjeeling district of West Bengal	:	Kosen (a Japanese variety)

These varieties easily yield 30-70 per cent more leaves per hectare and as such are preferred by the sericulturists. Kanva-2 strain has broad leaves and responds to fertilizer application. The leaves are succulent and retain moisture for longer period. It yields about 9000 kg of leaf in rain fed condition and 30000 kg in irrigated condition per hectare per annum as compared to 6000 kg and 15000 kg of leaf per ha of local variety in rain fed and irrigated condition respectively. Similarly, S-799 variety is preferred in West Bengal which yields about 30000 kg leaves / ha per annum under irrigated condition. In Jammu and Kashmir, Japanese grafted varieties -Gasheoerami, KNG etc. are popular. There are about 223 mulberry cultivars: 78 indigenous; 44 exotic; 21 unknown and 101 elite hybrids maintained at Central Sericulture Research and Training Institute at Mysore, India (Sastry, 1984). High yielding mulberry varieties suitable for plantation especially for hilly areas are K-2 and S-54. and for the plain areas of Andhra Pradesh and Karnataka, M-5 is recommended.

With regard to fruit yielding varieties, out of the several varieties of mulberry (*M. nigra*) developed in Europe only Black Persian has been most popular which has juicy, black plump fruit of 4x2.5 cm size with a rich, sub-acidic flavour. The white mulberry is native of China and is mainly grown for silkworm forage and its fruits are dry pale and insipid. The varieties grown for fruit in the USA are Trowbridge Thorburn, Ramsey White, Victoria Munson and Merit which produce large fruits with good flavour (Ottman, 1987). In Brazil, there are about 90 varieties all belonging to *M. alba* (De Almeida and Fonseca, 2000). Hybrid mulberries have also been developed in many countries. In China, there are over a thousand varieties under cultivation originated from four main species; *Morus alba, M. multicaulis, M. bombycis* and *M. atropurpurea* (Yongkang, 2000). In Vietnam, there are over 100 varieties, mainly *M. alba, M. nigra* and *M. laevigata* (Katsumata, 1973).

Plant Propagation and Nursery Management

Mulberry can be propagated either through seeds or vegetatively through stem cuttings. The stem cutting is the most common method of propagation in India because of the distinct advantages like speedy multiplication of parent materials and maintenance of the desired characters of the plant. In temperate climates like Japan, South Korea and Jammu and Kashmir State in India the improved varieties of mulberry do not root readily, therefore root grafting is popular for commercial cultivation. Layering is also another vegetative method for propagating mulberry.

Seeds of mulberry are orthodox and can be stored under dry cool conditions. Soaking of seeds in 100 ppm GA_3 for 24 hrs resulted in higher seed germination (Rao *et al.*, 1963). Treatment with gamma rays at 7.5 Kr gave 91 per cent germination as compared to 70 % in control, although some abnormalities in plant characteristics were observed (Das, 1970). In mulberry, the sexual propagation through seeds carries a varied population and utilized in selection and hybridization purposes. Seeds are sown during June in well prepared nursery beds for raising of seedlings. After attaining 15-20 cm height, seedlings are shifted to earthen pots or of polythene bags. For seed germination, certain prerequisites such as selection of quality seed, preparation of land needed to be fulfilled. This is possible only when the seed is subjected to suitable environmental conditions, embryo of seed is alive and healthy. The fresh seeds will have greater germination rate than the stored once. Seeds must be cleared from the flesh by repeated washing with fresh water and dried well. Minimum six per cent moisture in seed should be maintained for optimum germination. Seeds should be shown by broadcasting or in line in well prepared beds. After germination the seedlings are shifted to polythene bags pre-filled with a mixture of orchard soil, compost and sand in equal proportions. During rainy season plant should be transplanted in well prepared field. Seed propagation is mainly for raising hybrid populations.

Mulberry is by and large propagated by semi hard wood cuttings. The mature shoots of 6-8 month age with 15-20 cm length, 10-12 mm in diameter, 3-4 healthy buds are selected for cutting. The cut ends of the cuttings should be clean at an angle of 45^0 with a sharp knife without bark split. Cuttings are potted in sand beds or pots during winter which start sprouting with increase in temperature during February- March which can be shifted to polythene /earthen pots for growth. Irrigation is done regularly afterwards. By the onset of monsoon season saplings are ready for planting in the main field. Semi hard wood cuttings give better result in respect of sprouting and survival.

Other vegetative methods for propagation of mulberry are budding and tissue culture. For budding 8-10 month old seedlings are used as rootstocks. Shield or ring budding is done during spring season with swollen buds for optimum success. Micro- propagation is not very common in mulberry. However, recent experiments have shown the potential of artificial seed preparation for micro- propagation and long term storage of mulberry germplasm. Standard MS medium is used for germination of seed *in vitro*. Under *in vitro* condition, the artificial seeds are planted into small plastic pots containing artificial /natural soil moisture with either quarter strength MS medium devoid sucrose or distilled water and kept inside the environmentally tuned growth house covered with polythene sheets or green house. In green house high humidity is maintained. Under *in vitro* conditions (growth chamber or incubator), the germination rate is 50-60 % whereas, under glass house condition the germination is 20-30 % (Pattnaik and Chand, 2000).

Seedlings of mulberry are of common occurrence in arid and semi-arid regions of the country. These plants have very poor fruit yield and quality. Such plant may be rejuvenated to promising types by top working with improved cultivars.

Planting

Mulberry can be planted based on purpose. For fruit production, planting should be done in square system at a spacing of 6-8 m (Plate 12.2). For silk worm feeding, closure planting is recommended. Normally tree growth is restricted in arid and semi-arid region of the country. Hence planting distance need to be reduced by 25-30 % than the normal planting distance. Pit should be prepared as per the physical and chemical properties of soil. In sodic / rocky soils, the hard pan should be broken and add 5-10 kg gypsum or 4-8 kg pyrite or replacement of sodic soil with pan sediment along with 40-60 kg well rotten farm yard manure for better plant stands. Planting is done during rainy season. In India planting season varies in different parts of the country. For example in Kashmir valley, which has a severe

Plate. 12.2. Mulberry orchard

winter, planting in spring and late autumn is popular. In West Bengal also, cuttings are planted during October/November i.e., late autumn. In Karnataka, the most ideal season for plating is July-August i.e., the onset of south-west monsoon. July and August are favourable months for planting of mulberry in Andhra Pradesh and Tamil Nadu. In Rajasthan, monsoon season is the ideal time for planting. Thus in general early spring and late autumn seasons are best time for mulberry plantations. However, planting during extremes of winter and summer should be avoided. In temperate countries where typical winter is there, planting prior to sprouting of buds is the proper time during spring season. For leaf cutting, planting should be done at a distance of 60 cm row to row and 22 cm plant to plant under irrigated condition. In tropical countries like India where sunlight is available in plenty, mulberry rows can be aligned in any direction. Direction of planting is very important in temperate regions where sunlight plays a very crucial role in determining the growth rate of mulberry. In these regions, the mulberry rows should have a north-south alignment. Only in sloppy lands we should ensure that the mulberry rows run parallel to the contour lines.

The normal practice, both in Karnataka and West Bengal is to plant the cuttings directly in the field. Sometimes when the field is not ready for planting, it becomes necessary to preserve the cuttings in sand beds under shade for about a week before planting. This method of storage helps the formation of "callus" and such cuttings give better performance in respect to establishment. In some cases the cuttings are planted especially in prepared nursery beds; with a close spacing of 6 to 8 cm and transplanted after a couple of months. This method has an advantage over the system of direct planting in that it reduces the incidence of failed pits in the mulberry plantation. In arid regions, it is better to plant cuttings first in nursery for rooting and thereafter these should be planted in the well prepared field during monsoon season. Orchard establishment *in situ* in the arid regions is better. For this seeds should be sown in the prepared pits during monsoon season and subsequently budded thereafter. *In situ* established plants have deeper root system and grow better in low water availability conditions.

Training and Pruning

The plant canopy is the major site of metabolic activities which ultimately decide the fruiting of the tree. Therefore, it is essential to quantify and establish a relationship between total leaf area and ultimate productivity of the tree. The purpose of training is to give suitable shape to the plant to obtain the optimum production. This is achieved by adoption of suitable pruning schedules. If the plant is kept in a certain shape, routine pruning and management works become easy. In Karnataka and West Bengal the practice in vogue is to raise mulberry as bushes only. The periodical pruning help maintain the bush form for mulberry in these states. In Jammu and Kashmir however, high cut tree of 2 to 2.5 meter height at the branching levels are commonly used for raising block plantation and for planting along river and canal bunds.

Mulberry trees remain in the field for 15-20 years. Thereafter, it is desirable to replace old plants as productivity declines .However, during the productive phase, mulberry plants need to be periodically pruned to keep the production at the optimum level. Pruning is the removal of certain branches of the plants with for giving a convenient shape and size to plants, and to improve the fruit yield and quality. Under rain fed condition, the plants are pruned once in a year during July- August to 0-15 cm height from the ground level known as bottom pruning. After taking two successive flushes of leaf, middle pruning is some times resorted to for inducing the sprouting of buds to lower down the stem especially during winter months of December and January. In this, the branches are cut at a height of 45-60 cm above ground level. Plants are pruned by sickles, pruning saw and secateur. In Karnataka, two bottom pruning is suggested for pit system, first during June and second during November. This helps in taking five to six leaf harvests annually. Irrigated mulberry plants under the row system in Karnataka and also in the strip system of cultivation in West Bengal, the practice of cutting the whole shoot to the ground level at each harvest is practiced. In West Bengal during rainy season (August) however, plants are cut at the height of 0.20-0.25 m to avoid likely wastage of bottom leaf due to splashing of water. This is followed by drastic pruning at 3-5 cm below ground level in October. This method also provides cutting for subsequent planting. Thus, pruning and harvesting of leaves become a combined operation. As there are five crops in a year there will be five pruning. This system of drastic pruning calls for heavy fertilization and irrigation. In Kashmir where mulberry is grown as bushes (Plate 12.3), bottom pruning is practiced during spring and summer in alternate years. In other words during the first year the plant are pruned in summer after harvesting of leaves in spring

Plate. 12.3. Pruning of mulberry to obtain bush form for leaf production

season. This enables the plants to be in full flush for the autumn rearing. After harvesting of leaf in autumn, these plants gets a pruning in early spring in the subsequent year and provide leaf for the second or summer rearing. This system is comparable to what prevails in Japan and South Korea.

Orchard Management

Orchard establishment in arid and semi-arid region is difficult due to limitation of soil moisture. Moisture conservation is of utmost importance in these areas for the establishment of mulberry plant. Use of bentonite as a sub-soil barrier in controlling the moisture loss to facilitate the establishment of young mulberry plants in such areas are advisable. A hollow metallic cylindrical drum with a diameter of about 2-3 cm less than that of pits is placed in the pit. This creates a hollow space all around the drum placed in the center of the pit in which two to three centimeter thick layer of soil and bentonite mixture is filled at the bottom and interspaces of drum. After sometime when the soil and bentonite mixture becomes hard enough the drum is taken out thus, creating a bentonite lined layer at the outer periphery of the planting pit. This layer reduces the loss of precious irrigation water both through the side and bottom of the pit.

Nutrient Management

The cost of inputs like manures and fertilizers are very high, thus integrated nutrient management not only helps in increasing the present fruit production level but also sustains the production and protects the environment from different types of hazard occurring due to misuse of costly fertilizers. The soils of arid and semi-arid region are impoverished and hungry for plant nutrients. Considering economy, energy and environment, it is imperative that nutrient are used effectively by applying appropriate dose of nutrient at correct time to ensure optimum yield on sustained basis. Being perennial in nature, there is need to supply nutrients both for current vegetative growth and subsequent flowering and fruiting. The nutrient use efficiency in mulberry trees varies with the soil type, variety and age of trees.

Soil testing alone in perennial fruit crops has not been found very useful to make fertilizer recommendation. Due to their specific root distribution pattern and plant density, it is necessary to use soil and plant analysis techniques efficiently for recommending fertilizers for mulberry crop. Modern technology of nutrient management provides scope of tissue analysis for nutrient diagnosis. Because tissue is the principal site of plant metabolism therefore, changes in nutrient supply are reflected in composition of tissues. These changes are sharp at certain stages of development and nutrient concentration in tissue at specific growth stages are related to crop performance. Balanced fertilization ensures an optimum supply of all essential nutrients; promotes synergistic interactions and keeps antagonistic interactions out of the crop production system. It should take in to account physico-chemical properties of soil, nutrient requirement of crop and the nutrient uptake and soil fertility built up pattern of the concerned field.

The concept of balanced fertilization suggests that every time a crop is grown it is not necessary to apply all the nutrients in the same proportion. This may indicate the need for the application of different nutrients at specific times, in a particular order to derive the maximum benefits from the application of a given quantity of nutrient. Thus balanced fertilization should aim at: (a) increasing crop yield with improved quality (b) increasing farm income, (c) correction of inherent soil nutrient deficiencies, (d) avoiding damage to the environment, and (e) restoring fertility and productivity of the land that has been degraded by faulty land use practices. The common organic manures used for mulberry are farmyard manure and vermi- compost. Growing of green manure crops like Sun hemp, *Dhaincha*, cowpea in between the rows of mulberry and incorporating the same in soil before they flower.

Vermi composting is an alternative of farm yard manure for enhancing microbial population and improving physico- chemical properties of soil. The earth worm feed on organic matter viz, agriculture / farm/ kitchen waste, cattle dung, saw dust etc and convert into coarse granular aggregates as organic manure. Good quality vermi compost contains 0.9 -1.7% Nitrogen, 0.74% phosphorus and 1.08% potash which is better than cow dung manures(0.05% nitrogen,0.47% phosphorus and 1.4 % potash) on dry weight basis. Regular application of farm yard manure helps in improving the organic matter content of the soil which results in improving its microbial activity as well as water holding capacity. In Karnataka, under rainfed conditions application of about 10 tones of farmyard manure or compost per hectare per year is recommended as a basal dose. Under irrigated conditions the dosage should be 20 tones per hectare per year. In West Bengal also about 20 tones of farmyard manure/compost per hectare per year is recommended. Application of balanced dose of fertilizer however, improves yield and quality of mulberry. The uptake of nutrients depends on the availability of moisture in arid and semi-arid regions. The choice and dosage of fertilizers are related to the soil conditions and irrigation potential. The quantity of fertilizers to be applied depends upon the soil type, rainfall distribution pattern, spacing, pruning and harvesting methods and also seasonal conditions.

In India, where mulberry is in active growth phase all round the year, application of fertilizers depends to a great extent on monsoon rains. Care should be taken to apply fertilizers when there is sufficient moisture in soil. Moisture is the limiting factor in arid and semi-arid region; therefore, half the dose of nitrogen and full dose of phosphorus and potassium are applied during July/August after pruning and rest after first leaf harvest in October. In bush system of cultivation in Kashmir the schedule is 300 kg N: 150 kg P: 200 kg K per hectare per year. Nitrogenous fertilizers are applied in split dose during the growing season of spring and summer so as to meet the extra nitrogen needed during the maximum absorption period of the plant. First dose of nitrogen is applied just before the spring sprouting (March) and the second after the spring leaf harvest in June. Phosphorus and potash are applied in single doses along with the first application of nitrogen.

The recommended doses and manuring schedule for mulberry has been given in table 12.1 and 12.2 (Srivastava, 1984). First dose of fertilizers should be given after two and half month of planting at the rate of 40 kg nitrogen/ha. Plant will be ready for first leaf harvest after six month of planting. Jolly (1986) recommended fertilizer application at the rate of 300:120:120 kg NPK /ha/ year in split doses for better leaf harvest.

Water Management

Moisture stress is the main limiting factor for mulberry cultivation in arid and semi-arid regions of the country. If there is a 1000 mm rainfall and distributed through out the year uniformly, mulberry requires no supplementary irrigation. But such ideal conditions do not exist in the major mulberry growing countries of the worlds. In such situations, efforts should be made to work out proper irrigation scheduling for mulberry. Generally basin irrigation at fortnightly intervals between November / December to April/May is sufficient. The mean annual rainfall in the Indian arid region is very low and varies from 100 to 450 mm in Rajasthan, 300-500 mm in Gujarat and 200-400 mm in Haryana and Punjab and from 522 to 748 mm in peninsular India. Most of the precipitation in this region occurs during July-September in about 9-21 rainy days. The moisture storage capacity of the shallow and light sandy soil is very low. The atmosphere is very dry with a vapor pressure deficit (V.P.D) of more than 24 mb in April and more than 30 mb in May and June.

Table 12.1. Manure and fertilizer for mulberry trees

	Manure and fertilizers	Quantity
a. At the time of planting	1. Farm yard manure 2. Nitrogen 3. Phosphorus 4. Potash	10-20 kg 20 g 15 g 20 g
b. Up to 2-3 years	1. Farm yard manure 2. Nitrogen 3. Phosphorus	25 kg 40 g 30 g
c. Full grown trees	1. Farm yard manure 2. Nitrogen 3. Phosphorus 4. Potash	30 kg 50 g 40 g 40 g

Table 12.2. Fertilizer schedule for mulberry planted in different system of planting.

Application of fertilizer	Row system of planting	Pit system of planting
1st application	60kg N:60kg P:60kg K /ha	60kg N:60kg P:60kg K/ha
2nd application	60 kg N as straight fertilizer	40 kg N as straight fertilizer
3rd application	60kg N:60kg P:60kg K /ha	40 kg N as straight fertilizer
4th application	60 kg N as straight fertilizer	60kg N:60kg P:60kg K /ha
5th application	60 kg N as straight fertilizer	40 kg N as straight fertilizer
6th application	-	40 kg N as straight fertilizer
Total	300kg N:120 kg P:120 kg K	280kg N:120 kg P:120 kg K

Amongst the modern methods of irrigation, drip irrigation is most appropriate and input efficient system which is gaining popularity in the arid and semi arid regions of the country. In this method of irrigation, the plants watered at the rate equivalent to its consumptive use so that the plants would not experience any moisture stress from fruit setting to maturity during February to May. Drip irrigation ensures uniform distribution of water to plants, prevents water losses through conveyance and seepage, reduces weed growth and keeps the harmful salt below the root zone and save the labour. Daily drip irrigation at 40-60 per cent of pan evaporation through four to six drippers during fruit setting to maturity is a general rule for optimum growth and fruit yield of mulberry. Mulching the plant basin with organic waste materials or black polythene sheets reduces the evaporation by interfering with radiation on the soil surface and thus delays drying of soil and reduces soil thermal regime during day time. It also reduces the weed population and improves the microbial activity of soil by improving the environment in the root zone. Continuous use of organic mulches is helpful in improving the organic matter content which in turn improves water holding capacity of soil. Some green manure crops like sun hemp (*Crotolaria juncea*) and *Dhaincha* (*Sesbania aegyptiaca*) sown in between the row of mulberry plant can be conveniently pulled out later and used as mulching materials. Paddy straw, leaf mould, dry grasses, stubbles etc. also serve as good mulch materials.

Orchard Floor Management

The purpose of weeding is to remove unwanted plants and loosen the soil so as to allow rain water to percolate into soil and to ensure better aeration. Weeding is very essential as these compete with mulberry plants for nutrient and water. Inter-cultivation can be done either by manpower, bullock drawn or mechanized tillers. First weeding should be done after 45-50 days of planting and repeated periodically during first year of plantation. Thereafter ploughing of inter row spaces periodically to a depth of about 15 cm should be done to eradicate weeds. Power tiller can conveniently be used in wider spacing plantations for inter-cultivation from the second year onwards.

Mulberry fruit plants have long gestation period, it is beneficial to utilize vacant space for intercropping during initial year of orchard establishment. The basic objective of intercropping is to utilize vacant space, minimize soil erosion, add organic matter and realize additional income. Promotion of intercropping will depend upon the main plantations, their synergy regarding vegetative and reproductive growth. Growing of legume like cowpea in rainy season is very promising, while in winter season several vegetable crops such as chilli, brinjal, tomato, radish, carrot, cabbage, cauliflower and in summer season water melon, musk melon and bitter gourd and aloe can be grown if irrigation facilities are available. Medicinal plant like agave and aloe can successfully be grown in arid region.

Disease Management

There are a number of diseases which affect the normal growth of mulberry plant such as powdery mildew, leaf spot and root rot etc.

Powdery mildew is of common occurrence through out India. The lesions first appear as whitish patches on the lower surface of leaves and corresponding area on upper surface gradually turns yellow, subsequently brown and ultimately black. In severe cases it quickly covers the entire leaf surface. Finally the leaf dries and falls on the ground. The fruit yield is drastically reduced due to leaf shedding. The infected leaves are not suitable for silkworm rearing. Spraying of sulphur 50WP@ 2g/l fortnightly gives satisfactory control of powdery mildew in mulberry. But care should be taken not to feed these leaves to the silkworms. Picking and destroying the affected leaves is recommended. There are a few varieties of mulberry which are comparatively resistant to powdery mildew. Farmers should go in for these, especially in areas where this disease attains endemic form.

There are several leaf spot diseases affecting mulberry. The most common is caused by a fungal pathogen *Cercospora moricola* Cooke. The affected leaves develop circular or irregular brownish black spots of varying sizes. Usually the incidence is more during rainy season. The affected leaves are unfit for feeding silkworms. Spraying of suitable fungicides like difolatan 0.2 % on the young leaves at fortnightly intervals. Care has to be taken to see that the sprayed leaves are not used at least for a period of 15 days as feed for the silkworm.

The root knot nematode caused by *Meloidogyne incognita* is serious in mulberry. The parasite is soil borne and spreads primarily through contaminated soil, farm implements and runoff irrigation water from one field to another. The secondary source of infection is plantation of contaminated saplings. The disease affects underground systems which includes the formation of knots/galls on the roots. The root galls are spherical and vary in size; young galls are too small and yellowish white, while old galls are big and blackish brown in colour. The nematode disorganized the vascular tissues and cortex which affects the uptake of water and mineral from the soil to aerial parts of the plants. The death of grown up plants are very rare unless some root rot causing pathogenic fungi like *Fusarium solani* or *F. oxysporum* are involved. The nematode infested plants remain stunted with marginal necrosis and chlorosis of leaves. As mulberry is a perennial crop, it is difficult to control nematodes by single method. Sharma (1999) has suggested to follow an integrated approach for controlling the nematodes.

1. *Deep digging and soil solarization:* The nematode population in infected gardens can be reduced by deep ploughing at 30-40 cm depth during summer. This practice exposes the nematode egg and larvae to desiccation. Exposing them to direct solar heat helps in their destruction.

2. *Thermotherapy:* Hot water treatment of roots of mulberry saplings at 48^{0}C for 20 minutes may also be adopted as quarantine measure for elimination of nematode populations. The saplings should be soaked up to the collar zone and temperature does not exceed 48^{0}C as well as also make sure that the age of saplings does not exceed 3 months.

3. *Soil amendment with oil-cakes:* Application of neem/pongamia oil cake @ 2 tonnes/ha/year in four split doses at an interval of three months can be done during intercultural operations or

fertilizers application. Oil-cakes should be placed around basin of plants along with fertilizers and mixed well in the soil by ploughing or digging followed by irrigation. It reduces 65-70% root galls and egg mass production of nematodes and increases 16-18% leaf yield.

4. *Intercropping of nematicidal plants* : Intercropping of plants like marigold or seasame (*Sesamum indicum*) or sunhemp (*Crotalaria spectabilis*) at 30 cm distance in between mulberry rows are found to be effective and reduces 60-65% root galls and egg mass production of nematodes resulting in marginal increase of the leaf yield by 8-10%.

5. *Chemical Control:* Organophosphates and carbamates nematicides like application of Furadon 3 g @ 40 kg/ha/year @ 30 kg/ha/ year in 4 split doses at an interval of 3 months can effectively reduce 70-75% root galls and egg mass production of the nematode as well as improves the leaf yield to an extent of 12-16%.

6. *Biological control:* Since the residual toxicity of recommended nematicides persists in mulberry leaves for more than 40 days and also affects the beneficial micro fauna in soil, an eco-friendly method has been developed to control the disease which involves the use of talc-based bionematicide named as Bionema (a product of CSRTI, Mysore) produced by a bio-control agent *Verticillium chlamydosporium* along with neem oil cake. The formulation contains colony forming units (cfu) of 114.3 cm (4.5 x 10"/g) and its shelf life is 6 months under room temperature (25-30^{0}C) and 10 months under low temperature (4-6^{0}C). One kilogram Bionema should be mixed with 200 kg FYM and 24 kg neem. This should be kept under shade for a week for multiplication of fungus. This mixture is sufficient for application in 1000 plants @ approximately 200g / plant.

Pest Management

There are number of insect pest attack on mulberry crops in India. Leaf eating caterpillars (*Diacrisia obliqua* Walker) is a lepidopeterous insect and is commonly known as "Bihar hairy caterpillar". It is a polyphagous pest and causes maximum damage to mulberry plants. It also feeds on castor and it is a pest of jute. The caterpillars are voracious leaf eater and devour the leaf very fast causing extensive damage. In Karnataka their incidence is frequent between November and January. The life cycle of the insect is completed in 48 days with a larval duration of 30 days. It can be controlled by Collection and destruction of egg masses is a mechanical way to reduce the number of successive generation. At early stage larvae can easily be collected and destroyed. Deep ploughing and flood irrigation is also useful in killing of the hiding pupae in soil. Among chemical control measures spraying of 0.2 % Nuvan at fortnightly interval during emergence of new leaves for controlling of insect is useful. Jassids (*Empoasca flavescens* Fab.) attack mulberry plant extensively during October–May. The little greenish hoppers feed on the under side of the leaf, sucking the sap from the veins causing characteristic systems known as 'hopper burns'. It is reported that the cause of hopper burn is a toxic virus for which this insect is a vector. The first symptoms of this are the appearance of a triangular brown spot on the top of the leaf. Similar triangles appear at the end of each vein or the entire margin may roll up as though scorched by fire or drought. These brown margins increase in width until a narrow strip of the leaf along with the midrib remains green. Spraying of 0.1 % Rogor is quite effective for controlling the jassids. Mealy bug (*Maconellicoccus hirsutus*) feed on tender leaves and shoots. They are also supposed to be carriers of viruses which causes a disease locally known as 'Tukra' in west Bengal. The affected plants show curling of leave at the growing point and the plant growth is stunted. Spry of insecticide like Metasystox 0.05 % is useful in mealy bug management. Better results have been obtained with soil application of Furadon @7 kg/ha. Safety period for utilization of sprayed leaves for silk worm rearing is after 15 days. Different types of Scale insect like red, black etc. feed on leaves and twig of mulberry and when the attack is severe the branches dry and the leaves turn yellow and finally plants die. To manage the scale insects 1. Stem cutting for clonal propagation should be taken only from shoots which are free from the attack of scales. 2. Infected parts should be cut and burnt. 3. Release of predator *Coccinellid* (*Chilocherus kawanae*) is helpful as a measure of biological control. Spreading parasitic fungus *Attractum indicum* is also helpful. 4. Spraying lime sulpur solution is effective in reducing the incidence. Emulsion of soap and diesel oil is also helpful in controlling the scale insects.

Thrips attack on mulberry grown under rain fed conditions. There are several species of thrips attacking mulberry plants. They attack on the ventral side of the leaf and produce characteristic secretion. Spraying of 0.1 % Rogor is found beneficial in controlling the pest.

Stem girdler beetle are also common pest in India. This beetle girdles the young or green branches of mulberry plants. Damage is done during night time. The girdled branches dry up and die to enable the grubs to tunnel into the dry wood. This pest can be controlled by spraying endosulphon @ 0.07% at 15-20 days interval as per requirements.

Powder pest bettle (*Sinoxylon pubense*) and many wingless gross hoppers also cause considerable damage to mulberry crops at various stages. Bordeaux mixture combined with kerosene oil could be injected into the affected parts to force out the bettle which should be destroyed whereas wingless grasshopper can be managed by regular cultural operations like ploughing and weed control. Insecticides like endosulphan 0.07%. may be sprayed at fortnightly interval to control grasshoppers.

Flowering, Fruiting and Yield

Mulberry plants are monoecious in which flowers emerge with the initiation of new growth in late spring. The catkins are brone on current season growth and a tree can have both male and female catkins on the same twig (Plate 12.4). It is a wind pollinated crop in which whole catkins are the main edible material after maturity.

Plate. 12.4. Fruiting catkins of Mulberry

Mulberry is a non-climacteric fruit which consumed as fresh or after processing. Fruit and leaf yield depends upon variety and management practices. On an average a full grown tree can produce 30-40 kg fresh fruit / tree and 250 to 350 q/ha good quality fresh leaf with optimum management conditions.

Harvesting and Post Harvest Management

Timely harvesting of fruit is very important for value addition point of view. Stage and method of harvesting depends on the commodity however, morning harvesting is found better in all most all the fruit crops. Time of harvesting significantly affect the fruit quality and shelf life. Morning time is usually preferred because of more cool and turgid fruit which can be disposed or process during the day. Colour change is the main index of fruit maturity in mulberry. In general the fruits turn black, red or white respectively, in *M. nigra, M. rubra* and *M. alba* species at the time of maturity and attractive shining on surface and typical aroma.

Harvesting is done by shaking of individual secondary branches on tree at maturity stage or picking of individual fruits. There is severe damage to fruits if they fall directly on the ground. Therefore, fruits are allowed to fall on a cloth or plastic spread sheets held by two to three persons in suspended condition just below the tree. The berries are very delicate and therefore should be handled carefully. The mature fruits should be collected and disposed off quickly.

Mulberry fruits are perishable in nature and once harvested these should be protected from adverse weather conditions. Immediately after harvesting, the fruits are brought under shade for cleaning, grading and packing. Proper cleaning and grading into various classes (according to their size, appearance and uniformity) before packing increases the returns from the produce. Generally the fruits are graded into super, medium and small fruited categories. Since the fruits are extremely perishable in nature, therefore, these should be packed in aerated boxes, bamboo or mulberry baskets giving waste paper or dry grasses as cushioning materials. Generally mulberry fruits are disposed in local markets because of very short shelf life. Therefore, proper grading and packing is not common in India. But for distant markets proper post harvest handling like grading, packing and transportation have to be followed for getting optimum return. Fast transportation with minimum damage during shipment is very important in successful marketing of fruits. In India, the road transport is 3-4 times more expensive than the rail. Moreover, rail shipment is 8-10 times more efficient in the use of energy than road haulage for the movement of the same tonnage yet in India road transport is preferred for shipment of fresh fruits because of the faster movement of perishable commodities and advantages of door to door service (Arun, 1981). Both the road trucks and railway wagons should have controlled atmosphere storage (CAS) arrangement to reduce the spoilage of fruits during long distance transport. Storage is primarily practiced for extending the period of availability and to avoid immediate glut in the market. Proper temperature and humidity management are very effective tools in maintaining the produce in good condition throughout the storage period. Generally mulberry fruits are not stored in cold storage for long time under Indian conditions. But the ripe fruits are used for preparation of juices which is concentrated and can be stored for some time.

Processing and Value addition

Generally, the mulberry fruits are consumed as fresh however, ripe fruits can be processed into delicious juices and juice concentrates. The fruits can also be converted into chutney, mulberry rub, vinegar and many other value added products. In India, however, mulberry fruits needs to be processed into various diversified value added products.

Future Research Needs

Mulberry crop needs popularization for its use as fruit. For commercial cultivation of mulberry in India and other countries, following aspects needs to be addressed on priority basis.

1. Development of high yielding, deep rooted and drought hardy varieties suitable for arid and semi arid region.
2. Standardization of agro-technique for sustainable production.
3. Development of variety which is resistance to pest and diseases.
4. Development of suitable post harvest handling and value addition techniques

REFERENCES

Armand, D. and Meuret, M. (1995) *Culture en sec et utilisation en élevage de* Morus alba *"Kokuso 21" en Provence*. Rapport final (1991-94). Avignon, France.

Chand, P. K.; Sahoo, Y. and Pattnaik, S. K. (1994) Artificial seeds: A novel approach for mulberry propagation. *Indian silk,* (Sep.) p. 33-38.

Das, B. C. (1970) Effect of gama radiation on germination and seedling development of mulberry. *Science culture,* 36:60-61.

Datta, R.K. (2000) Mulberry cultivation and utilization in India In: *Proceedings of the electronic conference.*

De Almeida, J.E. & Fonseca, T.C. (2000) Mulberry germplasm and cultivation in Brazil In :*Proceedings of the electronic conference.*

De Almeida, J.E. and Fonseca, T.C. (2000) Revision of mulberry (*Morus* spp) studies in Brazil In: *Proceedings of the electronic conference.*

FAO. (1988) Mulberry cultivation. FAO Agricultural Services Bulletin No. 73/1, Rome. p.127.

FAO. (1990) Sericulture training manual. FAO Agriculture services Bulletin 80, Rome. p.117.

Gongfu, Zhong; Zengqi, Wang and Houshui, Wu. (1997) Land-water interactions of the dike-pond system. *Presses Universitaires de Namur,* Namur, Belgique, p.130.

Huo, Yongkang. (2000) *Mulberry cultivation in China. Proceedings of the electronic conference.*

Jolly, M.S.(1986) Economics of sericulture under irrigated conditions. Central Sericultural Research and Training Institute (Central Silk Board-Ministry of Textile -Govt.of India), Mysore-570008.p.19.

Katsumata, (1972) *Mulberry species in West Java and their peculiarities. Journal Sericultural Science Japan,* 42(3): 213-223.

Katsumata, (1973) Mulberry species in South Vietnam. *Journal Sericultural Science Japan,* 42(1): 81-88.

Manuel D. Sanchez (2004) World Distribution and Utilization of Mulberry: Potential for Animal Feeding. Animal Production and Health Division, FAO, Rome, Italy.

Ottman, Yvonne (1987) Rediscovering the realm of fruiting mulberry varieties. *Fruit varieties* Journal, 41: 4-7.

Patnaik, S. N. and Chand, P. K., (2000) Protoplast techniques for plant genetic engineering. *Plant Science Research,* 18:77-92

Rao, L. S. P.; Rao, T. P. and Narayanan, E. S.(1963) Response of mulberry seeds to gibberelic acid treatment. *Current Science,* 32:348-49.

Ravindran, S.; Ananda Rao, A.; Girish Naik, V.; Tikander, A.; Mukherjee, P. and Thangavelu, K. (1997) Distribution and variation in mulberry germplasm. *Indian Journal Plant Genetic Resources,* 10(2): 233-242.

Saraswat, R. P.; Chandra Sekhar, M.; Jayappa, T. and K. Thangavelu (2004) Management of mulberry field gene bank, its significance and cost analysis. *Indian farming*, 54 (6):24-26.

Sastry, C.R. (1984) Mulberry varieties, exploitation and pathology. *Sericologia*, 24(3):333-359.

Sharma, D. D. (1999) Root knot disease of mulberry and its management. *Indian farming*, 49(5) : 20-24.

Singh, H.B. and Arora , R. K. (1978) Wild edible plants of India. ICAR, New Delhi.

Singh, J. (1952) Preparation of jam from mulberry.

Srivastava, R. P. (1984) *Nutrition of fruit tree.* Pub.ICAR, New Delhi.p.276-277.

Sturrock, D. (1940) Tropical fruit for southern Florida and Cuba and their uses. Arnold Arboretum of Harvard Univ. Jaimaica Plain mas. Preparation of wine in Europe.

Tipton, J. (1994) Relative drought resistance among selected south-western landscape plants. *Journal of Arboriculture*, 20(3):151-155.

Yongkang, H. (2000) Mulberry cultivation in China. *Proceedings of electronic conference.*

—x—

13

PHALSA

Phalsa (*Grewia subinaequalis* DC.) is basically a small statured small fruited fruit crop but it is also used as a folklore medicine. Ripe fruits are consumed fresh, in desserts, or processed into refreshing fruit and soft drinks enjoyed during hot summer months in India (Salunkhe and Desai, 1984). But post harvest shelf life of phalsa fruit is very short hence; these are suitable mainly for local marketing. Its fruits are rich in carbohydrates (14.7g), calcium (12.9 mg/100g), phosphorus (39 mg/100g), iron (3.1 mg/100g), carotene (419 mg/100 g) and fibre (1.2%). Analyses made in the Philippines established that the phalsa fruits contain 725 calories/kg edible fruit; along with moisture, 81.13%; protein, 1.58%; fat, 1.82%; crude fibre, 1.77%; and sugar, 10.27% (Morton, 1987). At Fort Valley, Georgia (USA) every 100 g edible portion of phalsa fruit has been found to contain moisture 76.3 %, calories 90.5 Kcal, fat 0.1g, protein 0.57 g carbohydrates 21.1g, dietary fibre 5.53 g, ash 1.1g, calcium 136 mg, phosphorus 24.2 mg, iron 1.08 mg, potassium 372 mg, sodium 17.3 mg, vitamin- A 16.11μg, Thiamin 0.02 mg, Riboflavin 0.264 mg, Niacin 0.825 mg and Ascorbic acid 4.385 mg.

Phalsa fruits are cooling, aphrodisiac and tonic and alley thirst and burning sensation, remove biliousness, cure inflammation, heart and blood disorders and fever. Fresh phalsa leaves are used as animal fodder. The phalsa fruit are astringent and stomachic. Unripe phalsa fruits alleviate inflammation and are administered in respiratory, cardiac and blood disorders and; in fever management (Morton, 1987). An infusion of the bark is given as a demulcent, febrifuge and in treatment of diarrhoea. Barks of root are utilized in treatment of rheumatism. The leaves are believed to have antibiotic properties hence, applied on skin eruptions. The bark is also used as a soap substitute in Myanmar. A mucilaginous extract of the bark is useful in clarifying sugar. Fibre extracted from the bark is made into rope. The wood is yellowish-white, fine-grained, strong, flexible and is used for preparing archers' bows, spear handles, shingles and poles for carrying loads on the shoulders. Basal portion of pruned out branches are utilized in orchards for staking of fruit plants while upper portion of branches with little more than pencil thickness are used for basket making. The phalsa flowers contain grewinol, a long chain keto-alcohol (Lakshmi and Chauhan, 1976). The phalsa seeds yield approximately 5% bright yellow oil which contains 8% palmitic acid, 11% stearic acid, 13.5% oleic acid, and 64.5% linoleic acid with 3% un-saponifiable portion (Morton, 1987).

Origin and Taxonomy

The phalsa plant is native to the Indian subcontinent and Southeast Asia (Hays, 1953; Chundawat and Singh, 1980) but is cultivated on commercial scale mainly in the northern and the western states of India (Hays, 1953; Sastri, 1956). Smith (2002) opines that it to be native to India and Nepal, also tropical Asia: Cambodia, Laos and Thailand. Phalsa is a member of family Tiliaceae which belongs to the order *Malvales*. Other important families of the order Malvales are *Bombacaceae, Sterculiaceae* and *Elaeocarpaceae*. Tiliaceae family consists of about 400 species under 41 genera spread across tropical regions of the world. Other than phalsa {(*Grewia subinaequalis* DC Syn. *Grewia asiatica* L.(Masters)}, *Chorchorus capsularis* and *Chorchorus olitorius* (Jute) are major species of economic importance in this family. Member of this family are mainly trees and shrubs, rarely herbs having usually mucilaginous bark with strong fibres. *Grewia* is a large genus with about 100 species, of which 33 are found in India. In this genus, *Grewia asiatica* L. (*G. subinaequalis* DC. syn. *G. asiatica* Mast.) and *G. hainesiana* Hole are the only members that yield edible fruits. According to Sastri (1956), phalsa is the most commonly used vernacular name in trade for these fruits in India but there are several other names in customary usage like *dhamin, parusha,* and *shukri* in Hindi, *dhaman* in Punjabi, *man-bijal, bijol goah and phuhura* in Assamese, *phalsa* and *shukri* in Bangala, *mirgi chara* and *pharasakoli* in Oriya, *phalsa* in Gujarati, *phalsi* in Maharashtra, *jana, nallajana, phutiki* in Telagu, *palisa, tadachi* in Tamil, *buttiyudippe* and *tadasala* in Kannada. It is known by various names in other countries of the world like *phalsa and dhaman* in Pakistan; phalsa and siyal phusre in Nepal; pophlie in Combodia, nhap in Laos, bariuan-gulod in Philippines, malai, ya-khithut and pa tao hai in Thailand; and co-ke-a in Vietnam.

The phalsa plant is bushy in nature but can grow to a height of four to five meter as a small tree, if left un-pruned (Sastri, 1956). The stem is rough with numerous long, slender, drooping branches. Bark greyish-white or greyish-brown, sapwood whitish, heartwood small, irregular shaped, dark brown. The young branch lets are densely covered with a coating of hairs. Branch lets and underside of leaves varying from glabrous to densely and softly tomentose. Stipules are linear to foliaceous and broadly falcate. The leaves are alternate, coarsely toothed, deciduous, widely spaced, thick, broadly heart-shaped or ovate, pointed at the apex, oblique at the base, measure up to 20 cm in length and 15 cm in width with whitish blush on the dorsal surface. Flowers are 10 to 19 mm across, bright orange to yellow in colour and; are borne in dense cymes during late spring as well as during rainy season in the leaf axils (Plate 13.1). Flowers have 4-5 sepals, 4-5 petals, 70-80 stamens and well developed gynoecium. Peduncle axillary, in fascicles of 2-10, varying in length from 1.2cm -5.0cm, each bearing 3-5 flowers (Plate 13.2). Sepals 6-12 mm, petals 3-6mm, yellow or red and yellow, blade as long as or longer than claw. Ovary four loculed and stigma irregularly lobed. Fruits globose and indistinctly lobed (Brandis, 1907; Robertson, 1989). The fruits are small, almost round in shape and a drupe like blueberry.

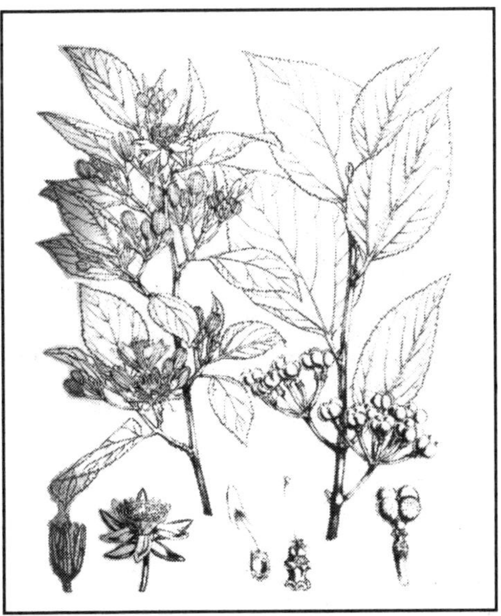

Plate. 13.1. Phalsa inflorescence and twig

Plate. 13.2. Flowers of phalsa

The green coloured fruits are borne on a 2- to 3-cm-long peduncle in great numbers in open, branched clusters and turn purple, crimson or cherry red at ripening (Plate 13.3). Individual fruits measure from 1.0 to 2.0 cm in diameter, 0.8 to 1.6 cm in vertical height and 0.5 to 2.2 g in weight. Fruits ripen gradually on bushes during the summer months. During ripening, the fruit skin turns from light green to cherry red or purplish red and finally becomes dark purple or nearly black.

Plate. 13.3. Fruiting in phalsa bushes (i) Immature green fruits on the plant (ii) Cherry red ripe fruits of phalsa

The ripe fruits are soft and tender which are covered with a very thin, whitish blush. The flesh is delicate and light greenish to nearly white in colour near seed which turns purplish to red near the skin. Later on over ripe fruit flesh becomes suffused with purple colour and shrivels on account of moisture loss. The flavour of phalsa fruits is pleasantly astringent but delicious and tasty due to very appropriate sugar-acid blend in juice. Generally big fruits contain two hemispherical, hard, buff-coloured seeds measuring up to 5 mm in diameter while small fruits are normally single-seeded.

Seeds germinate in 15-20 days. The juvenile phase lasts for 15-18 months. Flowers develop only on new shoots of the current growing season. The period between flowering and fruit maturity ranges from 45-55 days. Under sub-tropical conditions there is a brief period of dormancy lasting 4-6 weeks when the plant sheds its leaves during extreme winter.

Area and Distribution

It is widely cultivated in tropical and subtropical parts of India, Pakistan, Bangladesh, Thailand and Philippines. Phalsa is a drought hardy fruit and is grown successfully in the hot and dry plains having distinct summer and winter and is, thus, the most suited short duration fruit crop for arid parts of the country for the production of quality fruits even under the extreme summer conditions. Around the beginning of the 20th century, it was introduced into the East Indies including the Philippines where, it is naturalized at low elevations in dry zones of the island of Luzon. It is grown in northern and central Thailand. It reached western countries much later in the 20th century. In the United States also, few phalsa plants have been established as specimens for experimentation at Homestead and Miami (Florida) and at the former Federal Experiment Station, Mayaguez, and Puerto Rico. Lately, it is becoming widespread in northern Queensland (Australia). In Australia, it is found in drier woodlands and on most soils as well as drier vine thickets and coastal regions (Smith, 2002). Since it is under utilized fruit crop and its regular and commercial orchards are very rare and are scattered, reliable data on area and production is not available.

Soil and Climate

The phalsa plant grows vigorously and produces satisfactorly yield under variable soil types including fine sand, clay or even limestone, when soil fertility is not very poor. Phalsa is often grown in marginal lands close to city markets to facilitate prompt marketing of fruits. But loam soils are ideal for phalsa cultivation. It can be grown on moderately alkaline soils. It thrives in medium fertile soils but will not sustain salinity. The soil pH range for growth is reported to be 5.5-8.0 with the optimum between 6.0-7.5. Fertilizer application, although casual, improves plant health and production. The plant is drought-tolerant, but occasional irrigation during the fruiting season and in dry periods, is profitable for growers.

The phalsa is a warm climate fruit plant. In India, this plant grows satisfactorily and produces well up to an elevation of 1,000 m. It can tolerate light frost although at the cost of defoliation. The plant is deciduous and normally loses its leaves slowly in those areas with mild winter season. The plant grows both in tropical and subtropical climates. However, is does best in regions having distinct summer and winter seasons. The phalsa plant flourishes well under variable climatic conditions but it requires protection from the freezing cold temperatures. Adequate sunlight and warm or hot temperatures are required for fruit ripening, development of appropriate fruit colour, and good eating quality. In the absence of a prominent change of seasons the plant does not shed its leaves, flowers erratically throughout the year and fruits poorly. The temperature range for growth is reported to be 10-44°C with the optimum between 23-34°C. Photoperiodically it is a short-day plant. Observations in the Middle Georgia (USA) indicated that growing phalsa is feasible in temperate areas also if cold protection is assured. Indian phalsa was more fruitful and had tastier fruits than the MIA-12489 line.

Plant Genetic Resource Management and Varieties

The tall-growing wild phalsa plants produce fruits which are of marginal quality and are not relished by most of the consumers. The low-growing dwarf and/or bushy type of phalsa plants which develop a good blend of sugar and acid in the fruit flesh are preferred for cultivation (Hays, 1953). There are no well-known cultivars available but there are local favorites for different growing regions (Sastri, 1956; Yadav, 1998). However, some local selections such as Local, Sharbati, Tall, Dwarf, etc. are known. Dwarf type phalsa is high yielder having 34.8 per cent juice, 12.1° Brix TSS and 3.6 per cent acidity in its fruits. The Tall type gives comparatively lower yield but has higher TSS (19.0° Brix) and lower acidity (1.1%).

Observations at Fort Valley Fruit Research Station, Georgia (USA) revealed that the Indian phalsa plants are shaggy shrub or small tree, bushy in appearance with numerous branches, leaves-large, thick, dark green and ovate, pointed apex, coarse teeth with prominent leaf veins, flowers-small, prominent orange petals, light yellow sepals with 10.5mm diameter. Fruit size was 10.45 ± 0.25 cm with 0.98 ± 0.43 g weight and; the seed weight was 0.16 ± 0.15g. Contrary to this Miami 12489 types were tall shrub or small tree with fewer and stronger branches, leaves large with thick leaf blades, broadly ovate, base oblique, coarse toothed with white blush under side; flowers- small, inconspicuous, orange petals, bright yellow sepals of 18.4mm diameter. The fruit size was 14.26 ± 0.38cm, fruit weight 1.15 ± 0.61g and seed weight 0.23 ± 0.02g. Indian phalsa bushes had greater flowering and fruiting intensity than the MIA-12489.

The astringent fruits of the Indian type phalsa had more pleasing flavour but the fruit of Miami phalsa (MIA-12489) was sweeter. The majority of MIA-12489 phalsa fruits had double seeds while only a few large fruits of the Indian phalsa had double seeds. The Indian phalsa had smaller seeds than MIA-12489, which is preferred. Fruits of Indian phalsa remained fresh for 7 days or longer under refrigeration but spoiled in a few days at room temperatures.

Plant Propagation and Nursery Management

Seeds spread when ingested and expelled by birds and animals. Also spread as ornamental plantings and by flood waters (Smith, 2002). Conventional propagation of phalsa plant is by seed. The seed loses viability after 90-100 days at ambient storage conditions. Therefore, these should be sown immediately after extraction from fruit during May-June. Under cold storage, seed viability can be maintained up to 6 months. The phalsa plant is readily propagated by rooting of hardwood cuttings as well as layering (Samson, 1986). Seedlings produce the first crop of well developed fruits after 12 to 15 months from planting. Seeds stay viable for years and germinate in less than three weeks (Sastri, 1956). Wood type and planting date influence rooting of phalsa (Singh *et al.*, 1961). Treatment with auxins (IAA, IBA, NAA) improve rooting of difficult-to-root hardwood cuttings of phalsa (Yadava and Rajput, 1969), ground layers and air layers (Mohammed and Chauhan, 1970).

Seeds take 15-20 days to germinate and 3-4 months old seedlings are suitable for planting. Phalsa can also be propagated by vegetative means. Stem cuttings are difficult to root. Only 20% of semi-hardwood cuttings from spring flush, treated with 1,000 ppm NAA and planted in July under Indian conditions rooted and grew normally. Cuttings taken from 5-6 months old shoots during July-August give better rooting if treated by quick dip in 1000 ppm IBA. Stool layering during July gives profuse rooting after treatment either with 1000 ppm IBA alone or with combination of 5000 ppm IBA and 5000 ppm catechol. For stooling, bushes are headed back during February-March. The newly emerging shoots are ringed leaving one unringed nurse shoot, growth regulator paste is applied on the upper end of the ring and these treated rings are covered with moist soil by mounding around the bush. The rooted shoots are separated after about 2 months and can be planted in the field after proper hardening in the nursery beds. Phalsa can also be propagated by air layering. Ground-layers treated with growth regulators have given 50% success where as air-layers success 85%.

Planting

Phalsa is a bushy plant and can be planted very close, often in hedge rows. The common spacing is 2x2 m. The best time of planting is during July-August and February. After preparation of pit of 40x40x40 cm size and filling them with top soil mixed with 5-7 kg FYM in each pit, the planting is done and plants are irrigated for establishment.

Training and Pruning

Phalsa bears on current season's growth in axillary cymes which can be induced by pruning or defoliation. The presence of mature leaves on the unpruned shoots inhibits axillary bud sprouting (Reddy, 1983). Apical dominance is weak in phalsa hence; the number of new sprouts is more if more number of nodes is left on a pruned shoot. (Anon., 1987). Flowering and fruiting is confined to 15-20 nodes from the base depending upon vigour of new shoots. The number of new shoots formed after pruning actually determines the yield of fruit per bush. Generally, there are 2-9 peduncles of 2.5 cm length per node and 18-47 per branch. Each peduncle bears 3-6 pedicellate flowers (Swamy *et al.*, 1986).

In South India, phalsa sheds leaves rather slowly but in the North India it begins to shed leaves by middle of the winter. Accordingly, pruning under North Indian conditions should be started from the late winter and may continue up to early spring (Singh and Sharma, 1961). The fruit yield decreases due to delayed pruning on account of decreased number of days available for sprouting, flowering and picking. Late December or early January has been found to be the best time of pruning under North Indian conditions because early pruning may result in killing of new growth of shoots due to winter (Hays, 1957). Maximum yield of 0.98 kg/ bush from a two year old plantation has been recorded in Maharashtra if pruning is done during last week of January due to production of large number of shoots while yield decreased if pruning was done earlier (Shinde *et al.*, 1976). Pruning on First January significantly increased the number of new shoots, fruiting area and yield over six dates of pruning from November to January at fortnightly interval under Hyderabad conditions (Rao and Reddy, 1989) while under Bangalore conditions, two crops could be taken per year buy pruning at 100-150 cm height during January and again during July - August (Anon., 1986).

Generally, the severity of pruning should be reduced with delay in the pruning time. The number of vigourous shoots increased in phalsa when pruned to 75-120 cm bush height as compared to 45-60 cm pruning height (Singh, 1967). At Parbhani (Maharashtra), 20-30, 18.7 and 18.5 number of new shoots per pruned bush was observed when pruning was done at 75 cm, 100 cm and 125 cm above the ground level (Shinde *et al.*,1976). While under Bangalore conditions, increased number of new shoots per pruned shoot with decreased severity of pruning and the number of node up to which fruiting occurred was highest in severely pruned bushes with 6-8 fruits per node. This operation caused faster growth with increased length but reduced number of shoots (Anon.,1985). Generally, fruit yield increases with light pruning at 120-135 cm from ground level and minimum fruit yield has been obtained due to pruning at 15-22.5 cm bush height (Singh and Sharma, 1961). Pruning the bushes to 75-120 cm height has been found best for fruit yield under New Delhi conditions (Singh, 1967). At Parbahni (Maharashtra) also, pruning the phalsa bushes to 75 cm above the ground level (426g fruits/pruned stem) has been found best while further decrease in pruning severity reduce yield (Shinde *et al.*, 1976). Singh (1979) observed good yield in phalsa by pruning at 100cm plant height and further decrease or increase in pruning severity reduced the fruit yield. Under Bangalore conditions, maximum fruit yield (850g/bush) was obtained at 1.5 m pruning height (Anon., 1982, Anon., 1985) and later on pruning to 100 cm bush height has been found to be ideal (Anon., 1988). Fruit size has been observed to increase but fruit yield decreases with increasing severity of pruning while T.S.S. and acidity increases but juice content and TSS: Acidity ratio decreases with decreasing severity of pruning (Singh and Sharma, 1961). Fruit weight, juice content and Brix: acid ratio increased by increasing severity of pruning whereas, TSS, acidity, reducing sugars and ascorbic acid content decreased (Naidu, 1987). At Bangalore, no significant differences in the TSS and acidity of phalsa fruits were observed due to pruning severity.

Foliar spray of either thiourea or potassium nitrate @ 2 to 4 per cent before pruning of phalsa bushes has been observed to influence shoot growth and fruit yield under Bangalore conditions. Spray application of potassium nitrate to phalsa bushes pruned to 75 cm height induced more number of sprouts per node (1.74 against 1.16 in unsprayed) and resulted with more number of nodes with two sprouts, highest fruit yield (1.73kg/bush) and higher TSS. Increased concentration of thiourea spray to 4% induced further increase in number of sprouts per node (Anon., 1987a). Therefore, annual pruning is essential to induce good vegetative growth and fruit yield. Under north Indian conditions, the plants remain dormant during winter therefore; pruning is done towards the end of this period (January) so that new growth starts during spring (February-March). Under south Indian conditions growth is continuous and, therefore, pruning is done twice, i.e., during December-January and June-July. The bushes are pruned at a height of 75-100 cm from ground level to induce more number of lateral shoots on which vegetative growth, flowering and fruiting takes place.

Orchard Management

One-year-old seedlings are usually spaced 2 to 4.5 m apart. Bushes flower progressively during the spring months. Since phalsa bears fruit on current season's growth there is a need for regular but severe annual pruning before the on-set of spring. Annual pruning to a height of about 1 m encourages new shoots and higher yield of marketable fruit than does more drastic trimming (Singh and Sharma 1961). Gibberellic acid has been reported to improve fruit set and increase fruit size (Randhawa *et al.,* 1959). The phalsa plant shows good response to nitrogen applications. High levels of phosphorus supply increases sugar content in the fruit while higher potassium suppresses sugar and promotes acidity. Phalsa is considered stress-tolerant and is commonly grown under neglected (Hays, 1953) conditions.

Nutrient Management

Vegetative growth and fruit production in phalsa is greatly improved by the application of manures and fertilizers. Pareek and Vishal Nath (1996) suggested that the nutrients in phalsa plant should be applied as per diagnosis of leaf tissues. 4th leaf from the tip of shoot, 30 days after bud brust considered to be the best representative tissue for nutrient diagnosis. Under semi-arid conditions of Udaipur, application of 40 g N and 10g K_2O per bush is considered optimum whereas, in the arid part of Haryana 15 kg FYM per plant should be applied soon after pruning followed by 250g calcium ammonium nitrate (CAN) before sprouting and after fruit set. Two foliar sprays with 60 ppm GA_3 + 1.0 per cent urea, first at flower bud opening and second after 15 days of the first spray increases growth, yield and quality of fruits. In sandy soils, basal application of 50g Zn SO_4 and 0.4% borax where boron deficiency occurs should be done during April and May.

Water Management

Phalsa is a drought tolerant species and it can withstand high temperatures and water stress. However, the bushes must be irrigated starting just after pruning and application of manure and fertilizer during January-February. Thereafter, during the fruit development period, two irrigations are required at 15 days interval. Irrigation or heavy rainfall at the time of fruit ripening reduces sweetness and quality of fruit.

Orchard Floor Management

Green manuring with *moong, cowpea, urd, dhaincha* (*Sesbania* sp.) can be done during rainy season by turning them into the soil at early growth stage. Owing to its bushy mature and close planting coupled with heavy annual pruning, no intercrop is taken in phalsa orchard. One or two shallow ploughing after pruning is desirable to control weeds and to mix the FYM into the soil. One ploughing is also done after fertilizer application.

Disease Management

In some areas, leaf spots are caused by *Cercospora grewiae*. These can be controlled by two sprays of 0.3 per cent Diathane M-45 at 15 days interval after the symptoms are seen.

Pest Management

Leaf-cutting caterpillars attack the foliage at night. A blackish caterpillar causes galls on the growing shoots. Termites often damage the roots. Fruits are occasionally damaged by black ants during the ripening periods. Bark eating caterpiller (*Indarbela* sp.) infects phalsa bushes. Plugging the holes made by the caterpillar with cotton swab wetted in petrol or kerosene and sealed with wet mud should be done to kill the caterpillar. The affected branches should be pruned and destroyed.

Flowering, Fruiting and Yield

In phalsa, summer is main fruiting season. On current season growth, flowers emerge during February-March from the axils of leaves. The inflorescences with 3-7 peduncles are borne in each axil. Each peduncle contains 3-6 flowers. The flowers are yellow or pink in colour. It is a cross-pollinated crop where honeybees and flies play major role in pollination. In north India, fruits ripen after 45-70 days of flowering during May-June while in south India the fruits ripen twice, during November-December and June-July. Gibberellic acid has been reported to improve fruit set and increase fruit size (Randhawa *et al.*, 1959). All the fruits on a bush do not ripen at one time. Therefore picking of individual fruits is done several times when they turn purple in colour. The fruits keep poorly and must be marketed within 24 hours. In north India a well established *phalsa* bush produces 6-7 kg fruits per year. Average yield per plant in USA is 20 to 25 lbs (9-11 kg) in a season. The peak fruiting period in Fort Valley Fruit Research Station, Georgia (USA) ranged from July 15 to August 5. The average fruit production from the Indian type was 2.7 kg fruits/ bush compared to only a few dozen fruits (approx 25 g) on the MIA-12489. Phalsa can produce air dried fuel wood of 4.8-8.0t/ha from two prunings in a year depending upon pruning severity (Reddy, 1988). Application of plant growth regulators has been found to increase the fruit set and yield in phalsa. Spray of 10ppm GA_3 and 5ppm 2,4,5-T increased fruit setting and yield (Randhawa *et al.*,1959). Spray of cycocel twice at an interval of 50 per cent fruit set increased, the fruit size of phalsa and delayed the fruit maturity by 3-5 days. Application of 1,000 ppm ethrel resulted in maximum (71.1%) ripening of fruits in five days after application (Singh *et al.*, 1979). Fruit yield and quality improved and plant height decreased due to spray application of SADH 100 ppm at early stage of plant growth (Chundawat and Gupta, 1974). Fruit size and quality improved due to spray of GA_3 (50 ppm), 2,4,5-T (5 ppm) and 2,4,D @ 2.5 ppm (Singh and Singh, 1981). Spraying of GA_3 at 60 ppm once at the beginning of flowering, another after 15 days and ethrel at 1000 ppm when ripening of berries had just started, increases fruit retention inform and early ripening and yield. It also reduces the harvesting span and increases Total Soluble Solids content of fruits (Singh *et al.*, 1986).

Harvesting and Post Harvest Management

Gradual but steady fruit ripening during the summer of only a few fruits in a cluster necessitates frequent harvesting (Salunkhe and Desai, 1984) in phalsa. The fruits which resemble blueberry have a very short shelf life, and must be marketed within a day or two. Phalsa berries are highly perishable and should be consumed within 24 hours after picking. The fruits are small, ripen gradually which makes the harvesting of phalsa very tedious and expensive. The crimson red to dark purple berries of phalsa have a pleasing taste with very good sugar acid blend and are enjoyed as a dessert fruit. The red colour of the fruit is due to anthocynin pigment. The fruits are packed in baskets using phalsa leaf as lining material. However, phalsa fruit has a short shelf life and is considered suitable only for local marketing (Anand, 1960).

Processing and Value Addition

The phalsa fruits can be processed to prepare juice, nectar, squash, syrup. For preparation of beverage from phalsa fruits juice is prepared as per the procedure mentioned in Fig. 13.1 by Shrivastava and Kumar (2002).

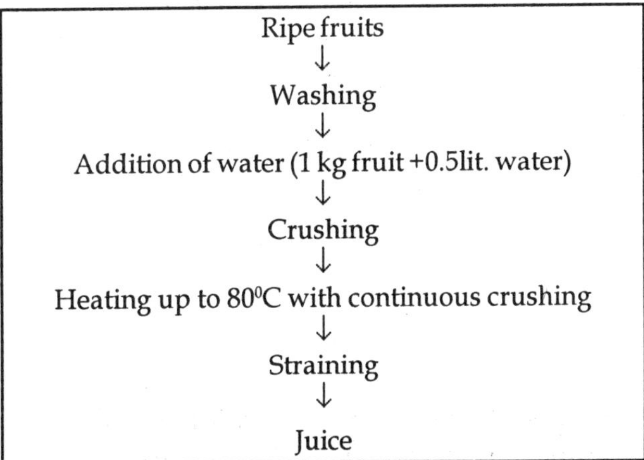

Fig. 13.1. Flow chart for extraction of phalsa juice

RTS

This is a type of fruit beverage which contains at least 10 per cent fruit juice and 10 per cent total soluble solids besides about 0.3% acidity. The calculated amount of sugar and citric acid is dessolved in water and the juice is then mixed thoroughly. The sodium benzoate 600ppm is mixed in the RTS and then the beverage is filled in 250 ml capacity bottles. The bottles are pasteurized at 80°C boiling water for 20 minutes after working.

Squash

For preparation of squash from 1 liter phalsa juice, 1.8 kg sugar, 5g citric acid is desolved in 1.0 liter water (Shrivastava & Kumar, 2002). After mixing the phalsa juice with sugar syrup, 4.0 g Sodium Benzoate is uniformly mixed (Fig. 13.2). Phalsa juice ferments so readily, therefore sodium benzoate must be added as a preservative (Anand, 1960).

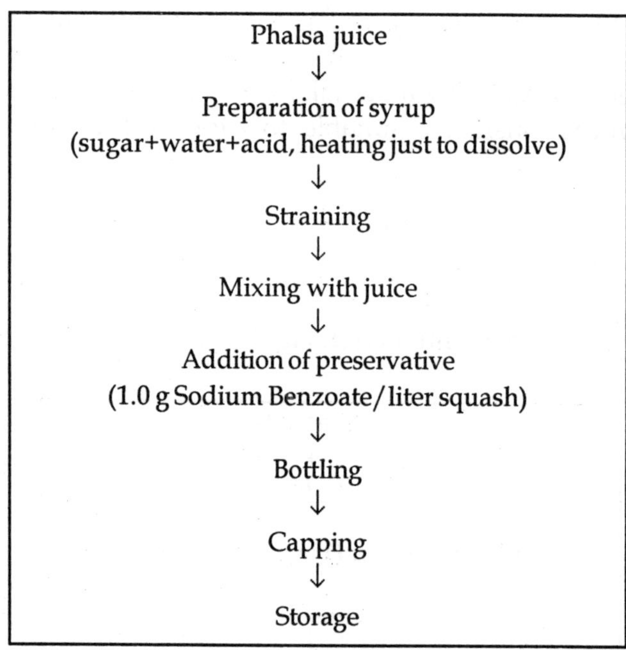

Fig. 13.2. Flow chart for phalsa squash preparation

Syrup

Phalsa syrup should contain 25% fruit juice, 65 % TSS and 1.5 % acidity. At commercial level, 50 kg sugar is dissolved in 13 liter water and 25 liter juice and 250 g citric acid is mixed with sugar syrup to prepare 100 liter phalsa syrup. At home scale however, 1 liter juice should be mixed with sugar syrup prepared by mixing 2 kg sugar and 10 g citric acid in 0.5 liter of water. Prepared syrup should be bottled and stored.

Future Research Needs

Phalsa is a dwarf stature fruit tree which can very well fit to mixed plantations in arid and semi arid areas. Agro techniques of this crop have been standardized, however suitable varieties with bigger fruit size and harvesting and post harvest management techniques for this crop is very much essential.

REFERENCES

Anand, J.C. (1960) Efficacy of sodium benzoate to control yeast fermentation in phalsa (*Grewia asiatica* L.) juice. *Indian Journal Horticulture,* 17:138–141.

Anonymous. (1982) Annual Report, IIHR, Bangalore, pp. 24.

Anonymous. (1985) Annual Report, IIHR, Bangalore, pp. 35.

Anonymous (1986) Annual Report, IIHR, Bangalore, pp. 24-25.

Anonymous. (1987) Annual Report, IIHR, Bangalore, pp. 24-25.

Anonymous. (1988) Annual Report, IIHR, Bangalore, pp. 26-27.

Chundawat, B.S. and Gupta, O.P. (1974) *Haryana Journal Horticulture Science,* 3:113-15.

Chundawat, B.S. and Singh R. (1980) Effect of growth regulators on phalsa (*Grewia asiatica* L.). I. Growth and fruiting. *Indian Journal Horticulture,* 37: 124–131.

Csurhes, S. and Edwards R. (1998) Potential environmental weeds in Australia: Candidate species for preventative control. Queensland Department of Natural Resources. pp. 109.

Hays, W.B. (1953) Fruit growing in India. 2nd Revised edition. Kitabistan, Allahabad, India.

Hays, W. B. (1957) Fruit Growing in India. Kitabistan, Allahabad, pp.411-13.

Lakshmi, V. and Chauhan J.S. (1976) Grewinol, a keto-alcohol from the flowers of *Grewia asiatica. Lloydia* 39:372–374.

Mohammed, S. and Chauhan K.S. (1970) Vegetative propagation of phalsa (*Grewia asiatica* L.). *Indian Journal Agriculture Science,* 40:581–586.

Morton, J.F. (1987) Phalsa. In: Fruits of warm climates. Julia Morton, Miami, FL. pp. 276–277.

Naidu, L.N. (1987) Studies on the effect of time and severity of pruning on phalsa (*Grewia asiatica* L.) under Hyderabad conditions. M. Sc.(Agri.) thesis, APAU, Hyderabad.

Pareek, O. P. and Vishal Nath (1996) Phalsa. In : Coordinated fruit research in Indian Arid Zone- Two decades profile. NRCAH, Bikaner: 1-105.

Randhawa, G.S., Malik R.S., and Singh J. P. (1959) Note on clonal propagation in the phalsa (*Grewia asiatica* L.). *Indian Journal Horticulture*. 16:119–120.

Randhawa, G.S., Singh J.P., and Khanna S.S. (1959) Effect of gibberellic acid and some other plant growth regulators on fruit set, size, total yield and quality of phalsa (*Grewia asiatica* L.). *Indian Journal Horticulture,* 16:202–205.

Rao, L.J. and Reddy, M.G.R. (1989) Effect of time and severity of pruning on yield of phalsa (*Grewia asiatica* L). *South Indian Horticulture,* 37:115-17.

Reddy, Y. N. (1983) *Proc. Second National Workshop, Arid Zone Fruit Res. Cell III AICFIP, ICAR,* pp. 235-38, 242-43, AND 253-57.

Reddy, Y. N. (1988) Potential of dryland fruit crops to make social forestry programme more productive. National Seminar, Dryland Horticulture, Hyderabad, Abstr., pp. 37.

Robertson, S. A. (1989) Flowering plants of Seychelles. Royal Botanic Gardens, Kew. p.25.

Salunkhe, D.K. and Desai B.B. (1984) Phalsa. In: Post harvest biotechnology of fruits. Vol. 2. Salunkhe and Desai (eds.), CRC Press, Boca Raton, FL. pp. 129.

Samson, J.A. (1986) The minor tropical fruits. In: Tropical fruits. Longman Inc., New York. pp. 316.

Sastri, B.N. (1956) *Grewia* Linn. Tiliaceae. In: The wealth of India: Raw Materials 4. Council of Scientific and Industrial Research, New Delhi, India. pp. 260–266.

Singh, J.P. and Sharma H.C. (1961) Effect of time and severity of pruning on growth, yield and fruit quality of phalsa (*Grewia asiatica* L.). *Indian Journal Horticulture,* 18 (1):20–28.

Singh, J.P., P.S. Godara, and R. P. Singh. (1961) Effect of type of wood and planting dates on the rooting of phalsa (*Grewia asiatica* L.). *Indian Journal Horticulture,* 18 (1): 46–50.

Singh, H. and Singh, R. (1981) *Indian Journal Horticulture,* 38: 157-61.

Singh, I.S.; Singh, H.K. and Chauhan, K.S.(1979) *Haryana Journal Horticulture Science,* 8:196-97.

Singh, M.; Gautam, R.K.S.; Prasad, Y. and Singh, A.K. (1986) *Haryana Journal Horticulture Science,*15:196-9.

;h, D. (1967) Cultivation of some less known fruits in India. *Indian Horticulture*, 11 : 52.

;h, R. (1979) Fruits. National Book Trust, New Delhi, pp. 142-43.

de, N. N.; Santakke, M.B. and Warke, D. C. (1976) Effect of severity of pruning at different periods on growth and yield of phalsa under Parbhani, (Maharashtra) condition. *Journal Maharashtra Agriculture University*, 1:157-58.

vastava, R. P. and Kumar, S. (2002) Fruit and vegetable preservation-Principles and practices. 3rd Edition. IBDC, Lucknow.

h, N. M. (2002) Weeds of the wet-dry tropics of Australia - A field guide. Environmental Centre NT, Inc. pp. 19.

my, P. N.; Mokashi, A.N.; Mohan Kumar, G.N.; Nalwadi, U.G.; Prabhakar, N. and Devar, K.V. (1986) Floral Biology of Phalsa (*Grewia asiatica* L.). *South Indian Horticulture*, 34:115-16.

v, A.K. (1998) Phalsa. Commodity sheet FVSU-004 of the Fort Valley State Univ., Ag. Res. Station.

v, A.K. (1999) Phalsa: A Potential New Small Fruit for Georgia. p. 348–352. In: J. Janick (ed.), Perspectives on new crops and new uses. ASHS Press, Alexandria, VA.

v, U.L. and Rajput C.B.S. (1969) Anatomical studies of rooting in stem cuttings of phalsa (*Grewia asiatica* L.). *Horticulture Science*, 1:19–22.

—x—

14

LASORA

Lasora (*Cordia myxa* Roxb. *Syn. Cordia dichotoma*) is a drought hardy fruit tree spread in arid and semi arid regions of North India. It is characterized by sunken stomata and other drought tolerance characters which enable the plants to perform better in climatically challenged areas. Plants are deciduous in nature and new flush comes in spring (March) when plant inters in floriferous phase. Medium size tree having dense foliage distributed throughout the country especially in deciduous forests and range lands of arid and semi arid ecosystem. Bunches of light yellow coloured hermaphrodite fragrant flowers are born auxiliary on current season growth. Immature green fruits are available in April-May which is used for pickle and vegetables preparation. Ripe fruits are available during June-July which are very nutritious, mucilaginous and medicinally important.

Cordia plants are sensitive to frost but can tolerant the extremely high temperatures. Unripe fruits are cooked or processed whereas the ripe fruits are eaten fresh. Green fruits are used for preparing vegetable and delicious mixed pickles. The fruits are rich in sugar and minerals. It has anti-helmenthic, diuretic, demulcent and expectorant properties (Anon., 1950). Other plant parts are also used for different purposes. Leaves are used to cure ulcer and headache. The bark is used for the treatment of fevers. Lasora is used in disease of the chest and urinary passes. The kernels are used in external application for ringworm. A decoction of the bark is used in dyspepsia and fevers. Glue is also prepared from the mucilaginous pulp of lasora fruit (Anon., 1950). Plant gives light timber used for various domestic purposes. However, wood is light in weight and are used for making boat and agricultural implements. Stem barks contains 2% tannin.

Fruits are highly nutritious and also contains fair amount of tannin and flamenoids. Its 100g ripe fruit contains 82.5 g moisture, 1.8g protein, 1.0g fat, 12.2 g carbohydrates, 40 mg calcium and 60 mg phosphorus besides mucilage. In spite of high nutritive and medicinal value in fruits, leaves and barks the lasora is commercially not been used except in arid and semi arid parts of the country.

Origin and Taxonomy

Lasora (*Cordia myxa* Roxb.) is native of India and occurs throughout the warmer regions of the world. It belongs to family Boraginaceae. It is also known as Indian cherry, sebestem, lehsua and gonda/goondi. Another related species of lasora or goonda, having small fruits is known as *Cordia gharaf* Forsk. (*gondi*) which is found distributed in arid and Semi arid zones of the country. Other economically important species of family Boraginaceae except that of *C. gharaf* are *C. rothii, C. macleodii, C. vestita* and *C. wallichii.*

Lasora is a medium size tree having dense foliage (Plate 14.1) with crooked trunk. The plant has drought tolerance characters like sunken stomata and leaf scars, deciduous nature during winters, deep and extensive root system. New flush comes in spring (March) when plant inters in floriferous phase. Gonda has cyme inflorescence (Plate 14.2). Bunch of light yellow coloured hermaphrodite fragrant flowers are

Plate. 14.1. Well grown up plant of Lasora

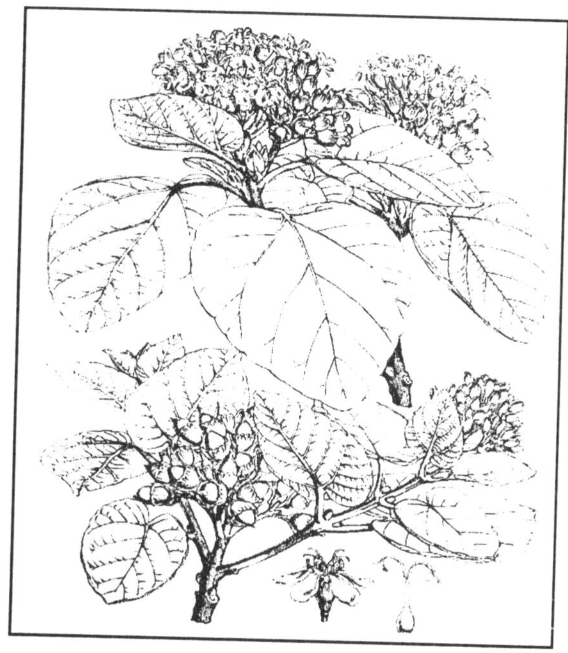

Plate. 14.2. Infloresence and fruiting twig of Lasora

borne axillary on current season's growth (Plate 14.3). Flowers are pedicillate, complete, perfect, actinomorphic, hypogymous and pentamerous. Fruit is drupe. The fruit is 2.0-3.0 cm long green at unripe and yellowish brown at maturity stage (Plate 14.4). Immature green fruits are available during April-June. Ripe fruits are available during June-July which is mucilaginous and medicinally important.

Plate. 14.3. Flowering is Lasora (i) Bunches of axillary cymes on crooked current season growth (ii) Full blooms of Lasora

Plate. 14.4. Fruiting branches of Lasora (i) Developing Lasora fruits (ii) Immature fruits ready for culinary purpose (iii) Ripe fruits of Lasora

Area and Distribution

Lasora is generally planted along with farm boundary and road side and in garden as shade tree. It is not been planted as sole orchard, however, its commercial plantation is available in Pali, Sirohi, Udaipur, Jodhpur and Jalore districts of Rajasthan. Being a drought hardy frui tree, it is suitable for arid and semi-arid region of the world. This species is widely distributed in India and Sri Lanka especially in warme regions.

Plant Genetic Resources Management and Varieties

Genetic variability exists in plantation of lasora. However, ther is no improved variety so far. In general, two types of plant viz. Larg Fruited and Small Fruited are found. Therefore, work on the selectio of better genotypes are essential to increase the productivity. Variabilit with respect to fruit size, skin and pulp colour (yellow, pink, dar pink), flesh content and pulp-stone ratio exists as a result of see propagation. Polyploidy (x=7, 8) has been reported in lasora whic also widens the genetic base. Some elite types have been identifie which have large fruits, small stone and less mucilage content. Som small fruited types have prolific bearing. Large fruited types are calle 'Paras Gonda' in Gujrat and 'Naruvili' in Tamil Nadu. Some plus tre have been identified by the CHES, Bhubaneswar (Plate 14.5) NBPG Jodhpur and CIAH, Bikaner. At Bhubaneswar, CHL-1 a promisir lasora genotype recorded the high content of TSS (17° Brix), pulp conte and phenols (Plate 14.5).

Plant Propagation and Nursery Management

Lasora plants are commercially raised from seeds which sho heterogeneity in the resultant population. Seeds are extracted fro eshly harvested ripe fruits and are sown immediately during April-ne in the nursery in polythene tubes of 25x10 cm size filled with a ixture of FYM + clay + sand in equal proportion. Seeds germination lasora is low. The seed germination percentage varies from 25-30%. goondi (*Cordia gharaf*) the germination is 70-80%. Directly sown eds take longer period (20-25 days) for germination. Pre-treatment seeds in cow dung for 24 hrs or soaking in water enhanced the per nt germination and seedlings growth (Singh *et al.*, 1996). Seeds tracted from freshly harvested ripen fruits can be used for sowing irpose in June-July.

Plate. 14.5. CHL-1 : A promising genotype of lasora
(i) Fruiting twig (ii) Ripen fruits ready for harvest.

To maintain true-to-type plants and uniform stand in the orchard, sora should be propagated by vegetative methods. Patch and I-idding during July to September on 8-10 months old seedling otstocks of *C. myxa* or *C. gharaf* are successful. T-budding has been und as the most successful with 96.7 per cent success during the onth of August. In addition, cleft grafting has also been found ccessful and the best period of grafting was observed during August llowed by July and September (Pundir, 1987). Twelve month old ock and 2 months old scion were found to be the most congenial for afting. *Cordia gharaf* was found superior root stock over *Cordia hotoma* in respect of the growth of buddling (Pundir, 1987). By getative propagation, even Small Fruited trees of inferior fruit quality ay be transformed into Large Fruited trees by top working. Cutting ken from one year old shoots during July-August and treated with 00 ppm IBA have produced roots.

Inferior seedling population of *lasora* can be converted in superior types by top working. For top working, the technique w standardized at Jobner, where 5 year old road side Small Fruited lasc trees were transformed into Large Fruited tree. The trees were head back in April and the new shoots emerging from these shoots we budded with the Large Fruited scion bud during the month of Ju The emerging shoots from the scion bud might often break due their own weight, being very vigorous and therefore pruning frc top was recommended to develop a stout framework (Sharma *et* 1983).

Planting

Lasora is planted at 5-6 m spacing in square system in w prepared 2x2x2 feet size pit filled with the mixture of surface soil, 20 kg FYM and 50 g Methyl parathion-5% dust. In comparatively hea soils, wider spacing (8 x 8 m) facilitates the intercultural operations bearing orchards. Application of organic manure in plants pit beneficial for establishment and moisture retention. Planting shou be done during rainy season i.e. July-September. It is advisable transplant about one year old plant for better establishment. *In-s* planting and patch budding at appropriate time is successful meth for establishment of lasora in arid region (Singh and Vishal Nath, 199

Training and Pruning

Lasora does not require regular pruning. During initial yea young plants should be trained to provide a strong frame by allowi 4-6 scaffold branches from a height of 1.0 m from ground level. Otl shoots emerging from the main stem in the initial 2-3 years should removed. Lasora do not require regular pruning however, suck emerging from main shoots/rootstock, undesired, week and diseas branches should be pruned. Defoliation by spray of potassium iodi (KI) has been found to induce early flowering in lasora plantation

Orchard Management

There is lack of systematic plantations of lasora, therefore information on orchard management is scanty in this crop.

Nutrient management

Systematic studies have not been conducted to standardiz nutritional requirement for lasora. However, to achieve optimu growth and fruiting in arid and semi-arid areas, an annual dose of 100:50:50 g NPK per plant should be applied. The same doses should be added in the previous year's dose. Doses for well grown up plant should be fixed after 5 years. To achieve vigorous plant growth and high fruits yield in arid regions, application of 100 g N, 125 g P_2O^5 and 50 g K_2O along with 10 kg FYM per tree/year has been recommended during monsoon period. Application of 20-30 kg rotten FYM and 1.0 kg DAP to 5 years old plant is found beneficial to increase fruit yield in semi-arid region (Singh *et al.*, 1996). Full dose of FYM should be applied with half dose of NPK during December-January at the time of commencement of new flush while remaining doses of fertilizer should be given during monsoon.

Water Management

In lasora, basins of 1 m diameter should be made around each plant in the orchard where watering and fertilization can be done. The basin size should be increased as the plant grows. Lasora can be grown rainfed with *in situ* water harvesting by providing 5 per cent slope to the land surface around the plant to collect run off near the main stem. In rainfed areas, *in situ* water harvesting by providing 5-10% slope in two directions of plants has been found beneficial to increase productivity of the plant (Singh *et al.*, 1996). Hoeing and removal of weeds from pits and provision of mulching is also useful to check the loss of water through soil surface. Mulching of entire basin should be done with locally available organic materials or black polythene to conserve soil moisture. Irrigation should be done where ever possible for extending flowering and fruiting period. Being salt tolerant, Lasora can be irrigated by saline water (3-6 dsm) in sandy soils, however, it is a drought hardy plant and does not require much water after establishment.

Orchard Floor Management

For additional income from orchard and proper utilization of land, intercropping is done till full bearing stage. Lasora trees are grown with annual crops under irrigated as well as rainfed crops in arid condition. In rainfed arid conditions, crops like *bajra, moong bean* and *cluster bean* can be grown during rainy season whereas *mustard* and *taramira* can be grown successfully during *rabi* season. Crops like wheat, raya and some vegetables and spices (cumin) can also be grown during *rabi* seasons if irrigation facilities are available. Lasora is a suitable fruit tree for Agri-horti., Horti-pastoral and Agro-forestry system in arid region (Singh *et al.*, 1996).

Disease and Pest Management

Lasora is comparatively free from insect pests and diseases. Tender leaves are damaged by insects which can be controlled by 2 sprays of 0.03 per cent monocrotophos or 0.1% Dimecron at 15 days interval during February-March. This is also effective against leaf gall midge causing pustules on the under side of leaves leading to severe defoliation. Two spray of Bavistin (2g/lit.) at 10 days interval is found effective to control minor diseases occurig on lasora plants. The problem of pustule formation on leaves and brown leaf spot (*Drechslera australiensis*) during establishment and early growth stage some times become serious. This can be controlled by removal and destruction of infected leaves. 2-3 altermate spray of copper fungicides (Blitox-50 or Fytolan @ 2g/lit.) at 10 days interval followed by insecticide Dimethioate 0.1% spray is an effective control measure. Lasora is a host plant for lac insect.

Flowering, Fruiting and Yield

The budded plant of lasora starts bearing in third year but the seedling plant gives few fruits in third-fourth year only under good care. In North India, flowering takes place during March - April and fruits ripen during May- July. In Southern tropical plains, flowering is earlier compared to that in subtropics. A 15 years old tree yields about 1.5-2 quintal fruits. Fruits are harvested at immature tender green stage for pickle making and culinary purposes. On full maturity, the fruits turn yellowish- brown and contain mucilaginous sweet pulp. For dessert and processing, fully ripe fruits are harvested when the skin becomes deep pink in colour.

Harvesting and Post harvest Management

Lasora fruits are harvested manually by picking the individual clusters. In mature fruits are generally harvested as economical yield. A very little information is available on post harvest handling of lasora fruits.

Processing and Value Addition

Green fruit can be stored for 2-3 days at ambient temperature. Immature, green tender fruits of lasora are generally processed as mixed pickle or used as vegetable (Chundawat, 1989). The fruits can not be stored for longer period at room temperature as these turn yellow thereby becoming unsuitable for utilization (Chandra *et al.*, 1994). Ripe fruits are used for preparing liquors. Lasora pulp is a rich source of pectin and could be utilized for making jams and jellies.

Future Research Needs

Lasora is an under utilized fruit crop in which least work has been done for standardization of varieties, agro-techniques, plant protection and value addition. Research efforts needs to be focused on following aspects to bring this drought hardy crop as common man's crop in the waste land and drought affected areas of the country.

1. Identification of cultivars with larger fruit size having small stone and higher pulp content.
2. Standardization of convenient and faster multiplication technique for commercial orcharding of lasora.
3. Standardization of complete production technology in view of the available natural resources in the potential production areas.
4. Work on post harvest technology needs be strengthen especially on keeping quality, processing and value addition.

REFERENCE

Anonymous (1950) Lasoda (*Cordia dichotoma*), Wealth of India, Vol.II, CSIR, New Delhi, p 346-347.

Chandra, Atul; Chandra Anju and Gupta, (1994) Arid Fruit Research, Scientific Pub., Jodhpur, p. 235-239.

Chundawat, B.S. (1989) Arid Fruit Culture, Oxford IBH Pub., New Delhi.

Pundhir, J.P.S. (1987) Studies on growth, flowering and fruiting behaviour of Lehsua (*Cordia dichotoma* Forst.). Ph.D. Thesis, Sukhadia University, Udaipur.

Singh, R.S. and Vishwanath (1991) *Shushka Kshetra mein labhkari lasoda lagayein. Phal-Phool*, April-June issue, p. 8-9.

Singh, R.S.; Vishal Nath and Tewari J.C. (1996). Lasora- a promising fruit tree for arid ecosystem. *DECO MIRROR*, 3(2): 5-11.

Vashishtha, B.B., Raja Ram and Ranvir Singh (1985). *Haryana Journal Horticulture Science*, 14: 156-159.

—x—

15

WOOD APPLE

Wood apple (*Limonia limonia* Syn. *Feronia limonia* (L.) Swingle) is a globose woody fruit produced on a medium sized tree and belongs to family Rutaceae. Wood apple is also called as elephant apple and monkey fruit because elephants are fond of this fruit. For an animal that wholly survives on plant parts, including the bark of several trees, it is not a surprise that elephants are especially attracted to this fruit. There is more to it than mere fondness of elephants for this fruit. The rural folk believe that the digestive tract of the animal has a peculiar capacity for digesting the inner contents of the fruit without, in any manner, affecting the fruit's woody rind. So, as the animal defecates, the fruit comes out looking as if it is whole. How far this is true remains verifiable but this belief is not wholly confined to tribal but has percolated even into the urban milieu is borne out by the fact that in the literature of local languages (at least Telugu) lends some credence to the belief.

Wood apple is cultivated mostly in Rajshahi and Kushtia areas of Eastern Uttar Pradesh, however, as homestead tree it is grown throughout the country. Wood apples of Gazipur and Rajshahi area are fairly large. Its pulp is used for preparation of sherbet, chutney and pickles. The gum obtained from the trunk and branches are used as a substitute of gum arabic.

The pulp represents 36% of the whole fruit. The pectin content of the pulp is 3 to 5% (16% on the basis of dry-weight). The seeds contain non-bitter oil high in unsaturated fatty acids. Fruits contain 7.3 per cent protein, 15.5 per cent carbohydrates, 170 mg/100g riboflavin, 2mg/100g vitamin C and minerals, especially calcium and phosphorus.

The fruits are highly acidic and contain 7.6 per cent acidity when unripe and 2.3 per cent when ripe. They also contain 3-5 per cent pectin making it suitable for preparation quality jelly. The food value of fruit pulp and seed has been given in table-15.1. Its fruits is also used in various *ayurvedic* preparations for curing diarrhoea, dysentery, gum and throat problems.

Table 15.1. Composition of wood apple (per 100 g edible pulp)

Constituents	Pulp (ripe)	Seeds
Moisture	74.0%	4.0%
Protein	8.00%	26.18%
Fat	1.45%	27%
Carbohydrates	7.45%	35.49%
Ash	5.0%	5.03%
Calcium	0.17%	1.58%
Phosphorus	0.08%	1.43%
Iron	0.07%	0.03%
Tannins	1.03%	0.08%

The wood apple has several medicinal properties. The leaves and bark have medicinal values. It is antiscorbutic. It is an antidote for poisons and also helps in curing sore throat. After the rains, the trunk and branches give off a gum called '*Feronia gum*', which counteracts diarrhoea, dysentery and diabetes. However, this statement should not be understood that the gum totally cures these diseases. It serves to mitigate the problem. The fruit is much used in India as a liver and cardiac tonic, and, when unripe, as an astringent means of halting diarrhoea and dysentery and effective treatment for hiccup, sore throat and diseases of the gums. The pulp is poulticed onto bites and stings of venomous insects, as is the powdered rind. Juice of young leaves is mixed with milk and sugar candy and given as a remedy for biliousness and intestinal troubles of children. The powdered gum, mixed with honey, is given to overcome dysentery and diarrhoea in children. Oil derived from the crushed leaves is applied on itch and the leaf decoction is given to children as an aid to digestion. Leaves, bark, roots and fruit pulp are used against snakebite. The spines are crushed with those of other trees and an infusion taken as a remedy for menorrhagia. The bark is crushed with that of *Barringtonia* and applied on venomous wounds. The unripe fruits contain 0.015% stigmasterol. Leaves contain stigmasterol (0.012%) and bergapten (0.01%). The bark contains 0.016% marmesin. Root bark contains aurapten, bergapten, isopimpinellin and other coumarins.

The rind must be cracked with a hammer. The scooped-out pulp, though sticky, is eaten raw with or without sugar, or is blended with coconut milk and palm-sugar syrup and taken as a beverage, or frozen as ice cream. It is also used in chutneys and for making jelly and jam. The jelly is purple and much similar to that made from black currants. Bottled nectar is made by diluting the pulp with water, passing through a pulper to remove seeds and fibre, further diluting, straining, and pasteurizing. A clear juice for blending with other fruit juices has been obtained by clarifying the nectar with Pectinol. Pulp sweetened with syrup of cane or palm sugar, has been canned and sterilized. The pulp can be freeze-dried for future use but it has not been satisfactorily dried by other methods.

The reddish pectin of wood apple has potential for multiple uses in India after purification. The fruit shell is fashioned into snuffboxes and other small containers. The trunk and branches exude a white, transparent gum especially after the rainy season which can be utilized as a substitute for adulterant of gum arabic and also for making watercolors, ink, dyes and varnish. It consists of 35.5% arabinose and xylose, 42.7% *d*-galactose, and traces of rhamnose and glucuronic acid (Rao, 2004).

The wood is yellow-gray or whitish, hard, heavy, durable, and valued for construction, pattern-making, agricultural implements, rollers for mills, carving, rulers, and other products. It also serves as fuel. The wood is also used in making agricultural implements and handles of different tools. The heartwood contains ursolic acid and a flavanone glycoside, 7-methylporiol-D-xylopyranosyl-D-glucopyranoside.

Origin and Taxonomy

The wood-apple is native and common in the wild state in dry plains of India and Ceylon where it is planted along roads and edges of fields and occasionally in orchards. It is also frequently grown throughout Southeast Asia, in northern Malaya and in Penang Island (Morton, 1987). In India, the fruit was traditionally a "poor man's food" until processing techniques were developed in the mid-1950's.

The wood-apple, *Feronia limonia* Swingle is the only species of its genus, in the family Rutaceae. The other species viz. *F. elephantum* Correa; *Limonia acidissima* L.; *Schinus limonia* L. are considered as synonymous to *Feronia limonia* Swingle now named as *Feronia feronia* or *Limonia limonia*. Besides wood-apple, it may be called as an elephant apple, monkey fruit, curd fruit, *kath bel* and other dialectal names in India. In Malaya it is known as *gelinggai* or *belinggai;* in Thailand as *ma-khwit;* in Cambodia as *kramsang;* in Laos as *ma-fi;* and in French, it is called as *pomme d' elephant, pomme de bois,* or *citron des mois.*

The slow-growing tree is erect, with a few upward-reaching branches bending outward near the summit where they are subdivided into slender branches drooping at the tips. The bark is ridged, fissured and scaly and there are sharp spines: 3/4 to 2 inches (2-5 cm) long on some of the zigzag twigs. The deciduous, alternate leaves, 3 to 5 inches (7.5-12.5 cm) long, dark-green, leathery, often minutely toothed, blunt or notched at the apex, are dotted with oil glands and slightly lemon-scented when crushed. Dull-red or greenish flowers are ½ inch (1.25 cm) wide and are borne in small, loose, terminal or lateral panicles. They are usually bisexual. The fruit is round to oval, 2 to 5 inches (5-12.5 cm) wide with a hard, woody, grayish-white, scurfy rind about 1/4 inch (6 mm) thick. The pulp is brown, mealy, odorous, resinous, astringent, acid or sweetish, with numerous small, white seeds scattered through it (Morton, 1987). It is diploid plant with chromosome number 2 n = 18.

The tree is moderate in size, deciduous with straight and strong axillary spines. It has a much-branched dense crown of dark foliage and dark grey bark which is longitudinally furrowed, rough and wrinkled. The leaves are compound with 3-7 paired leaflets arising from either side of the rachis ending with one unpaired leaflet. The main rachis as well as the stalks of leaflets are narrowly winged, i.e. narrow, green leaf-like structures from a border to the stalks (Morton, 1987). On the leaves, faint oil-gland dots are common. The pinnate leaves bear spine at the base. The flowers are reddish; bark thick, comparatively soft and brownish. The fruit is quite large, about 3 inches in diameter, globose to globular with diameter 5-16 cm (Plate 15.1) and single-chambered enclosing several seeds in the dark green pulp, traversed by fibres. The rind is woody, hence, signifies the name 'wood apple'. The ripe fruits have soft, fleshy, yellowish edible pulp with numerous seeds embedded within the pulp. The pulp of ripe fruit is less sweet and more acidic, therefore, it is consumed by mixing with sugar/honey. The pulp of unripe fruit is however, sour to taste and is made into chutney by mixing it with a paste of green chillies, salt and suitable spices.

Plate. 15.1. Fruit of Wood apple

Soil and Climate

Wood apple is a hardy crop which is almost in the natural form of its existence. The trees have been found growing at an elevation of 450 m in the western Himalayas and also in the tropical forests of peninsular India. Throughout its range of availability, there is a diversity of soil in which it performs and produces the fruits, but it is best adopted to light soils.

Although wood apple prefers dry conditions but in Malaysia and Indonesia, it is predominantly cultivated in coastal regions and therefore it is advised to grow wood apple in those places where monsoon climate prevails as usual (at the time of fruit development) with a distinct dry season (Pareek and Vishal Nath, 2006). It tolerates drought and saline conditions and can also be grown on marginal stony soils up to an elevation of 1500 m above mean sea level.

Area and Distribution

The wood apple tree is quite common through all the dry districts of India where they are naturally growing in the compounds of some institutions and residential bungalows or being planted at homestead level particularly in Chennai (Rao, 2004). Wood apple, as one of its aliases 'elephant apple' suggests that it is the favourite of elephants and therefore, the wood apple has been planted extensively or being allowed to grow naturally in the reserved forests and elephant sanctuaries. Its fruits are also used to pray the Hindu Elephant-headed God, Lord Vinayaka, and therefore it is extensively grown near the Lord Ganesh temple all over the country. Another related species, *Feronia limonia* popularly known as *kathbel* is a highly demanding fruit. Its trees are medium-sized, spiny, with small leaves and reddish flowers. The fruit is about the size of a tennis ball, greenish at the early stage but appears whitish when ripe. The little sourly pulp of the ripe fruit is very tasteful and widely used as chutney. Numerous seeds are embedded in the aromatic fleshy pulp. This species is extensively grown throughout Bangladesh with varying plant densities. However, *Kaitha*, the true wood apple is found throughout the tropical and subtropical plains of Sri Lanka. It also grows wild in drier regions of Myanmar and Indo-China (Vietnam).

Plant Genetic Resource Management and Varieties

Wood apple is a highly cross pollinated crop. This has resulted in considerable variability with respect to tree architecture, fruit size, shape, pulp colour, sweetness, acidity and fibre content in the resulting seedling population (Pareek and Vishal Nath, 2006). There are 2 forms/ ecotypes in wood apple i.e., one with large, sweetish fruits and the other with small, acidic fruits. From the existing genepool, some elite genotypes have been identified in the country. At Parbhani in Maharashtra, wood apple variety HB 19, having large fruits and high pulp content has been selected. Three high yielding and one thorn less type have been identified at Aruppukottai in Timil Nadu. At central horticultural experiment station (IIHR), Bhubaneswar, also five genotypes have been collected from the existing natural variability of wood apple in the coastal region of Puri district of Orissa. The genotypes have exihibited variation in fruit weight (29 g to 365 g) and fruit size with fruit length measuring from 40 mm to 100 mm and fruit width from 36 mm to 90 mm.

Plant Propagation and Nursery Management

The wood-apple is generally grown from seeds though seedlings will not bear fruit until at least 15 years old. Multiplication may also be by root cuttings, air-layers, or by budding onto self-seedlings to induce dwarfing and precocity. Fresh seeds are sown in nursery which takes about 20 days to germinate. Pre-soaking in water improves germination. Seeds are sown 5 cm apart at 2 cm depth in raised beds and when the seedlings attain 3-10 cm height, these are planted at 25 cm apart in another nursery. Seedling of 10-11 month age is used for planting. Plants can also be regenerated through cuttings, layering and budding. Air layering and budding are the most successful methods for propagation of wood apple in moderate climate of south India. Under rainfed conditions, however, *in situ* raised rootstocks are budded by patch or shield methods during July-August (Pareek and Vishal Nath, 2006). Wood apple as rootstock induces bloom in *citrus* scion.

Planting

Wood apple is planted at 8x8m spacing in square system (Chundawat, 1990). Monsoon is the best time for planting of one year old saplings in well prepared pits of 1x1x1m size. Into each pit 15 kg FYM along with 50 g insecticide chlorpyriphos check termite attack. For raising an orchard *in situ,* six month old seedlings are planted in the field for budding when these attain proper thickness where as for seedling plantations in waste lands generally one year old healthy plants are selected.

Training and Pruning

Wood apple trees are allowed to grow along a central leader with well spaced branches in all directions. The trees require no pruning except removal of criss-cross branches. At initial stage, pruning of plants to provide a desired shape is essential. When planted as windbreak and shelter belt, the trees are allowed to grow tall.

Orchard Management

Although no systematic work on orchard management of wood apple has been carried out however, plant seems to respond to various management practices. Based on the work under taken mainly in southern peninsula, some management practices have been suggested.

Nutrient Management

Precise recommendations on mineral nutrition have not been developed. However, application of 25kg FYM or compost per plant during monsoon helps in improving fruit size and quality. The trees sometimes exhibit micronutrient deficiencies, especially of zinc, boron and iron. Therefore, to obtain good productivity from bearing trees, 500gN, 250g P_2O_5 and 100 K_2O along with 50g Zn SO_4 per tree per year should be applied during June-July when the fruit growth takes place. In rainfed orchards, manures and fertilizers is applied in 30 cm wide trenches around the center of tree canopy.

Water Management

Under rainfed conditions, provision for *in situ* water harvesting by providing 5 per cent slope to the basin towards the tree trunk in all directions of the plant proves beneficial. In areas having comparatively higher rainfall, clean cultivation coupled with mulching of the exposed soil with locally available organic material is practiced. If water is available, irrigation can be done at 20-25 days interval during the plant establishment and during the fruit set and developmental period.

Orchard Floor Management

Under rainfed conditions, clean cultivation is preferred. Under irrigated conditions leguminous vegetables, pulses, cucurbits, etc. can be grown as intercrop during the pre bearing period. In grownup orchard however, some shade loving crops can be grown in the inter spaces.

Disease and Pest Management

No serious pest or disease has been observed on wood apple. Citrus butterfly (*Papilio demolius*) causes some damage to its foliage, especially to young plants. At young stage of plants individual larval of the insect can be collected and destroyed. Spray of insecticides like endosulphan or malathion @ 2 ml /liter is quite effective in managing this pest espcially when new vegetative growth takes place.

Flowering, Fruiting and Yield

Seedling trees grow slowly and will not produce fruits until the 15th year but budded plants start fruiting earlier. Small terminal or axillary panicles with dull red or greenish yellow flowers are borne on mature shoots. In north India, flowering takes place in March and the fruits ripen from November to March (Plate 15.2). It takes about 12 months from flowering to fruit maturity. Under south Indian conditions, fruits take comparatively less time to attain maturity, i.e. 210-240 days after flowering (Pareek and Vishal Nath, 2006). Here

Plate. 15.2. Plant of wood apple with mature fruit

the flowering occurs between February and May. In Malaya, the leaves are shed in January, flowering occurs in February and March, and the fruit matures in October and November. In India, the fruit ripens from early October through March (Rao, 2004). Average yield from a well managed tree is 200-250 fruits each of 400-500g weight. Higher yield of 600 fruit per year can also be obtained by better crop management practices.

Harvesting and Post Harvest Management

Shell hardening, detachment of fruit from plant, pulp separation from the shell and change in colour of rind with black patches are some indices of fruits maturity. Fruits are harvested with bamboo poles having knife and baskets at the top. The fruit is tested for maturity by dropping into a hard surface from a height of 1 ft (30 cm). Immature fruits bounce, while mature fruits do not. After harvest, the fruit is kept in the sun for 2 weeks to fully ripen. Since the fruits have hard covering, these can be stored for longer duration at ambient temperatures.

Processing and Value Addition

Sweet and aromatic pulp of the wood apple is used fresh as well as for preparation of various products. Wood apple is a good source of natural protein which can be processed into an excellent quality jelly having clear, bright purple appearance with firm consistency and nice flavour. Fruit pulp is used for garnishing and saucing the dishes. Soft drinks can also be prepared from ripe wood apple pulp. Soft drinks, jelly, Chutney, etc. can be prepared from the pulp of mature fruits. Wood apple pulp is little astringent and can be used for blending with guava pulp to prepare mixed fruit jelly.

Future Research Needs

Wood apple is an under utilized fruit, drought hardy fruit plant. In the natural system, it is found growing throughout the country, however, focused attention to this crop has not been given for which it deserves. Therefore, to promote wood apple following researches are need of the time.

1. Collection of the variability and its evaluation for promising horticultural traits.
2. Identification and popularization of high yielding varieties for different eco-regions.
3. Development of faster multiplication techniques to supply the true-to-type plants to the farmers.
4. Standardization of complete package of practices for commercial orcharding of wood apple.
5. Development of post harvest handling value addition methods and recipes for better utilization of the wood apple fruit.

REFERENCES

Chundawat B. S. (1990). Arid fruit culture. Oxford and IBH Public Company Pvt. Ltd. New Delhi

Morton, J. (1987). Wood-apple. *In:* Fruits of warm climates. J. F. Morton, Miami, FL.:p. 190–191.

Pareek, O. P. and Vishal Nath (2006). Arid and minor fruits. *In:* Text book of Horticulture (K. L. Chadha, Eds.). MPH, New Delhi. *In press.*

Rao, K. N. (2004). Fruit-bearing Trees - Wood apple

—x—

16

KER

The Ker or Karil (*Caparis decidua* (Forsk.) Edgew) is one of the prominent wild plant species found in the Indian desert. It is an ancient fruit of Indian subcontinent. At present, the medicinal and nutritive value of ker has been appreciated which is used to prepare various Ayurvedic medicines and nutritive dishes. In *Panchkuta,* the important dish of Marwad region, ker makes the major component. It is also used for pickling and dehydration. Ker is rich source of protein, phosphorus, potash, calcium and magnesium. The ker bush is drought hardy and can grow under most hazardous conditions of soil, water and climate. It can produce fruit under degraded and sandy soils in arid and semi arid tropics and sub tropics. Ker has special xerophytic characters like deep tap rot system, scanty foliage, mucilaginous sap and tough conical spines which impart considerable hardiness to it even under the most adverse conditions. Studies conducted at Sariska National Park in Western India by Sharma and Prasad (1992) revealed that *Capparis decidua* plants were not at all debarked by Indian crested porcupine (*Hystrix indica*) and therefore, it is well proved that ker is the only fruit plant which produces economical yield under the extremely adverse conditions of the Indian desert without any care. Among hardy fruit species, ker requires almost no care and input except to harvest the fruits at right stage. This plant thus gives assured income to the local habitat of the region and improves their socio-economic and health standard levels. Ker fruits are being produced rather gathered by the poorest people of the region i.e. tribals and numads and therefore, it provides nutritional security to poor people of the society. This in fact assumes greater significance in view of our determination to ameliorate the prevailing malnutrition among the weaker section of the society.

Ker is a nutritive fruit which is consumed at green immature stage (for pickle and vegetable) as well as after ripening. Chundawat (1990) reported that ripe fruits of ker are sweet and acrid and enjoyed by children in natives while immature fruits are used in pickling or as vegetable. Khera and Singh (1975) reported that fruit contains about 55 per cent pulp rich in protein and mineral matter. In Wealth of India, (Raw material, 3:211) it is mentioned that ker fruits are rich source of protein, carbohydrates and ascorbic acid besides other vitamins and mineral matters. Duhan *et al.*, (1992) reported that ker fruit contains appreciable amount of ß-carotene and vitamin C. The nutritive value of ker fruits varies at various stages of its development (Table 16.1) has been compiled from available literature.

Table 16.1 : Nutritive value of ker fruits

Composition (per 100 g)	Values	
	Immature fruits	Ripe fruit
Moisture (g)	28.0	56-70.1
Calories (Kcal)	-	100
Protein (g)	14.9	13.6-17.0
Fat (g)	7.4	1.2-5.0
Carbohydrate (g)	59.4	20.9-71
Fibre (g)	12.3	1.0
Calcium (mg)	90.0	55-210
Phosphorus (mg)	779.0	50.8-360
Iron (mg)	3.5	2.0-6.0
Vitamin A (IU)	8964	3486-3600
Ascorbic acid (mg)	120.7	119-133.3

Source : (1) Chauhan *et al.*, (1986), Duhan *et al.*, (1992), Gopalan *et al.*, (1987), Khera and Singh (1975)

Ker is a multipurpose plant species in which all part are used in one or the other way by the native people. It is a boon for desert. Fruits are highly nutritious and medicinally important. Fruits are processed into quality products. Stem wood of ker is very strong and durable. It is used to make the pivotal of stone mill. Thick wood of ker is used to make foundation around the well (Chandra *et al.*, 1994) where as smooth thin wood of light yellow colour is used for making small agricultural implements (Singh, 1993). Ker wood resists the termite attack and therefore, thinner branches are used for fencing the field and mulching. It is also used as fuelwood in rural areas. Ali and Ali (1987) mentioned that *Capparis dicidua* bases are important source of fuelwood in Sudan. The young twigs serve as a feed for camels and goats (Anon; 1992).

Immature flower buds and flowers of ker are used for vegetable and pickling purpose. Anon (1992) mentioned that the flower buds of ker are pickled and cooked as vegetable. Pareek *et al.*, (1998) have also described the use of immature flower buds and flowers of *Capparis decidua* as vegetable. Ker flowers have also been tested for cercaricidal properties (Nama *et al.*, 1990). Rai (1987) reported that ker flowers and buds contain 14% surface wax straight chain saturated hydrocarbons and ketones.

Whole plant of ker has medicinal importance. It is reported to be a good source of nutrients for desert animals. As mentioned in Wealth of India, Raw material, blanched fruits when fed to rats at 10% dietary fibre level, showed a significant hypocholesteraemic effect on the serum and liver-cholesterol and a corresponding increase of cholesterol and bile acid in faecal matter. Fruit husk and flowers contain l-stachydrine, n-pentacosane, n-triacontane, n-triacontanal, n-nonacosane, pelargonidin-3-galactoside, glucocappasalin, b-sitosterol alkaloid and phthalic acid. Presence of ß-carotene and traces of choline is also reported in fruit husk. The flowers yield a steam-volatile sulphur compound which is active against several micro organisms. The roots contain the indole bases capparin, capparilin and capparinim. The root bark contains l-stachydrine (Anon., 1992).

Ahamad *et al.*, (1989) have isolated isocodonocarpine, a new spermidine alkaloid from root bark of *Capparis decidua*. Two other new spermidine alkaloids viz. 14-N-acetylisocodonocarpine and 15-N-acetylcapparisine have also been isolated from root bark of *Capparis decidua* (Ahmad *et al.*, 1992). Garg *et al.*, (1993) isolated two hydrocarbon compounds viz. n-nonacosane and n-hentriacontane from C. *Decidua* flowers. They also isolated and characterised some glycosides from *Capparis decidua*. Bark of ker plant possess analgesic, diaphoretic, akheteric, laxative, anthelmintic properties (Chandra *et al.*, 1994) and cure various diseases and disorders in body (Table 16.2). Tanira *et al.*, (1989) reported that C. *decidua* caused an increase in urine volume in rats tested for diuretic activity whereas Agrawal and

Chauhan (1989) mentioned that teent (*C. decidua*) is a good source of dietry fibre.

Table 16.2 : Medicinal importance of various plant parts of *Capparis decidua*

Plant parts	Effective in Disease/disorders	References
Roots		
Root bark	intermittent fever and rheumatism	Chundawat, 1990
Stem and branches		
Tender shoots	plastering boils	Chundawat, 1990
Tender twigs	Toothache	Chundawat, 1990; Chandra *et al.*, 1994; Singh, 1993
Stem bark	cough, asthama, inflammation	Chundawat, 1990; Chandra *et al.*, 1994; Ahamad *et al.*, 1989
Stem bark	ulcer, vomiting, piles	Chandra *et al.*, 1994
Leaves		
Tender leaves	plastering boils	Chundawat, 1990; Singh, 1993
Tender leaves	affliction of pain	Chandra *et al.*, 1994
Flowers and fruits		
Fruits	cardiac troubles and biliousness	Chundawat, 1990; Chandra *et al.*, 1994; Singh, 1993
Fruits	foul breath, discharge of urinary purulent	Chandra *et al.*, 1994
Fruits	Gastric	Meghwal and Vashishtha (1998); Vashishtha, 1987; Singh, 1993
Seeds and Seed Oil		
Seed oil	skin diseases	Chandra *et al.*, 1994

Scanty vegetation in arid part of the country is possing the threat of ecological imbalance. Under arid ecosystem, *Capparis* is an important species in almost all habitats (Mann and Saxena, 1980). Pareek (1978), however, mentioned the natural habitat of ker in pediment plains all over the dry region with an average of 22 plants/ha.

Apart from saline patches of land and brackish under ground water found extensively in arid region, extreme high temperature (45-48°C) and diurnal variation of temperature (14-16°C) along with low and erratic rainfall (100-400 mm punctuated by long dry spells) creates problem in the processes of taming sandy waste lands for agricultural use and promotes wind erosion (Gupta *et al.*, 1987). Under such hostile conditions, Gupta *et al.*, (1987) recommended that xerophytic plants like *Capparis decidua* is most suitable because of its wider adaptability and soil binding capacity apart from other uses. Sharma and Gupta (1989) reported that *Capparis* bushes are much effective in stabilizing the sand dunes in Bikaner region of Thar Desert whereas Pandey and Rokad (1992) considered that *Capparis decidua* is the best species for shelter belt to check the movement of sand. Ker bushes have also been found to improve soil health. Sharma and Gupta (1989) reported that soil pH has decreased from 8.58 (in open soil) to 7.55 (under *Capparis* cover) in arid part of Western Rajasthan. Gupta *et al.*, (1987) mentioned the extent of soil improvement with ker plantations (Table 16.3). Singh *et al.*, (1992) described that *Capparis decidua* is one of the important fodder tree species for rehabilitation of degraded arid ecosystem of Western Rajasthan.

Table 16.3. Chemical properties of dune sand as affected by ker plantation

Soil properties	**Barren dune sand**	**Dune sand under ker**
pH	8.6	7.6
Ece (d Sm^{-1})	0.3	0.4
Organic carbon (%)	0.03	0.3
Total nitrogen (%)	0.007	0.360
Available nitrogen (kg/ha)	87.5	168.0
Available P_2O_5 (kg/ha)	14.9	24.1
Available K_2O (mg/100 g)	0.4	0.8

Source : Gupta *et al.*, (1987)

Area and Distribution

Although ker plants are available in different parts of the country its systematic cultivation is limited. Some natural groves of ker are available in arid part of Rajasthan, Haryana and Gujarat which are producing fruits at limited scale. To extend the area under ker orchards greater awareness is required among growers with regard to economic, nutritional and ecological importance.

Caper plant is a hardy fruit plant which usually grows in drier and exposed habitat of India, North East Africa, Arabia, Egypt, Ethiopia, Iraq, Persia, Baluchistan and Pakistan (Kritkar and Basu, 1918; Bhandari, 1995). Its considerable population has been noticed in degraded and waste lands and on foot hills. It is found growing on sand dunes either alone or in association with other drought hardy species like *Prosopis cineraria* (*Khejri*), *Calotropis procera* (*Ak*), *Leptadenia pyrotechnica* (*Kheep*), *Calligonum polygonides* (*Phog*), etc. Chundawat (1990) described that ker is grown in wild state in arid and semi arid region of the country where hardly few species can survive. Chandra *et al.*, (1994) mentioned that ker plant thrives well in most adverse climatic conditions of Rajasthan, Punjab, Haryana, Gujarat and Uttar Pradesh. According to Pareek (1978), natural habitat of ker is pediment plains all over the dry region and its density ranges from 8 to 44 bushes per hectare. Singh (1993) reported 20-30 ker bushes/ha in pasture and community lands of semi arid and low rainfall areas of Rajasthan, Gujarat, Haryana, Punjab, Madhya Pradesh and Uttar Pradesh. Gupta *et al.*, (1987) reported that flat aggraded plains with saline-alkali soils in Jodhpur are dominated by *Solvadora oleoides* - *Capparis decidua* community whereas the frings of natural saline depression in western Rajasthan are widely occupied with the *Capparis decidua*. Singh and Jha (1993) reported wild bushes of ker in forests of Mathura-Manipuri-Agra region of Western Uttar Pradesh. Under sandy undulating hummocky plain habitat of Jaisalmer in Western Rajasthan, ker was a major woody plant species (Kumar, 1992). Kumar and Bhandari (1993) reported that perenial ker bushes are the major sand dune plant community in arid and semi arid parts of Rajasthan. Presently, plant population of ker varies at different locations due to soil and climatic conditions and human interference. Table-16.4 describes the density of ker bush at different location in arid parts of the country.

Origin and Taxonomy

The original abode of *Capparis decidua* (Forsk.) Edgue is India (Pareek *et al.*, 1998). The natural diversity of the species is available in Thar Desert of Rajasthan (Bhandari, 1995). Variability of *Capparis decidua* is also found in parts of Haryana, Punjab, Gujarat, arid and semi arid parts of Uttar Pradesh, Madhaya Pradesh and Southern peninsula.

Table 16.4. Population density of *Capparis decidua* at different locations in India

Location	District	State	Nature of soil	Plant population (ha)
Bikaner	Bikaner	Rajasthan	Sandy pediments	10-15
Napasar	Bikaner	Rajasthan	Sand dunes	15-17
Kanasar	Bikaner	Rajasthan	Sand dunes	15-20
Fatehpur	Sikar	Rajasthan	Pediment plains	20-25
Mandava	Sikar	Rajasthan	Pediment plains	20-25
Abusar	Jhunjhunu	Rajasthan	Pediment plains	25-30
Sardar shahar	Churu	Rajasthan	Sandy dune	15-22
Nagaur	Nagaur	Rajasthan	Transinal plain	10-12
Khivsar	Nagaur	Rajasthan	Transinal plain	20-25
Chandi	Nagaur	Rajasthan	Sand dunes	12-17
Phalodi	Jodhpur	Rajasthan	Sand dunes	10-15
Shergarh	Jodhpur	Rajasthan	Sand dunes	20-25
Bap	Jodhpur	Rajasthan	Pediment plain	10-12
Pokharan	Jaisalmer	Rajasthan	Rocky land	4-5
Ram Devra	Jaisalmer	Rajasthan	Pediment plain	10-12
Nachana	Jaisalmer	Rajasthan	Sand dunes	8-10
Shiv	Barmer	Rajasthan	Pediment plain	20-25
Shindri	Barmer	Rajasthan	Rocky land	12-15
Sambhar	Jaipur	Rajasthan	Saline soil	2-5
Jardon	Pali	Rajasthan	Pediment plain	25-30
Hisar	Hisar	Haryana	Sandy loam soil	20-22
Narnaul	Mahendra garh	Haryana	Sandy soil	10-15
Palampur	Banaskantha	Gujarat	Sandy soil	12-15

Source : Anon., (1995)

In India, ker is known for its medicinal and religious uses from sim antiquity. In different Indian languages it is known as Karer, Kureel (Hindi), Kerada (Gujarati), Chippur, Karira, Kariuppi-gidda, Niovatge-gidda (Kanadda), Ker (Marwadi), Karira (Sanskrit), Sengam (Tamil) and Kariramu (Telegu). In various Indian states it is known by several vernacular names such as Dela in Delhi; Kair in Punjab; Kair, Kareel, Kerro, Taint in Rajasthan and Karil in Uttar Pradesh. The detailed use of Ker has been described in the ancient Indian literature. In ancient

Sanskrit text on the science of medicine, ker is known as KARIRA, KRAHARA, APATRA and GRATHTHILA. Our ancient medicines literature is full with praise of ker plant. The SUSHRUTA SAMHITA, mainly dealing with surgery has enumerated the properties of ker flowers and fruits. In CHARAK SAMHITA, also, properties of ker plants have been mentioned in detail. Bhava-Prakash-Nighantu, Kaiyavadeva-Nighantu and Gada-Nighraham also describe the antiseptic, anticough, antigastic and antidote properties of ker plant and its different part.

In vedic literature, this plant figures prominantly in the Samhitas and Brahmanas of Yajurveda. Karira plants have been main cause for rains and ecological balances, leading to production of food grains in vedic period. According to Kathak Samhita (11.10), the very existance of Karira plants amount to the sacrificial oblations of Soma, causing rains from heaven. Ker seems to have some special properties which enable it to play the sacred role in the vedic yagnas and meditation process. Ker does have cultural and social importance in Indian history. RASKHAN, a medeaval poet was so much enamoured of it that in one of his poem he says "I can squander crores of planets like the moon if I can get an opportunity to graze Lord Krishna's cows in the KARILA grove of Briz".

The ker or capper berry (*Capparis* species) belongs to family Capparidaceae of order Capparideae. *Capparis* is a large genus of climbing shurbs or small trees comprising more than 300 species (George Usher, 1974; Caind *et al.*, 1976) distributed throughout the warm regions of the earth. 31 species have been described by Hooker (1987) whereas occurrence of 40 species in India have been reported by Chundawat (1990) and Shekhawat (1994) and 26 species by Anon (1992). *Capparis* is a polyploid species in which basic chromosome number is X = 7 or 8.

This is a densely branched spiny shrubs or tree of 6-10 m height with a clear bole upto 2.4 m found chiefly in dry and arid regions from Punjab in North upto Dakshina Kannada in South (Anon., 1992). Generally, branches of ker remain tender with waxy bloom however in nature twigs are grey, rough and corky bark which has straight or curved, 3-7 mm long paired thorns. 1-2 cm long, linear stiff, pale mucro-like prickle and caduceus leaves having short apex are found only on young branches. Inflorescence is cymes having red or pink or yellow flowers arranged in lateral corymbs. Flower; compact, Sepals four;

free, imbricate in two series or 2 outer sub valvate; petals four, sessile, imbricate; stamens indefinite, inserted on the tours at the base of the long gynophores; ovary-stalked, 1-4 celled; stigma-sessile; ovules-many on 2-6 parietal placentas. Fruit are fleshy berry with globose or avoid shape having 1-2 cm diameter and dull red colour. Seed many in number having globose shape and 2-5 mm diameter and imbedded in pulp. Testa remains crustaceous or curvaceous with convolute cotyledons inside (Hooker, 1987; Anon., 1992). Some important species of genus *Capparis* have been described.

Capparis decidua (Forsk.) Edgew (Syn. C. aphylla Roth)

This is the most important species of genus *Capparis* native to India. It is known by several vernacular names in various languages of the country. (Gujarati-kerada; Hindi-karer, kurrel; Kannada- chippur, karira, kariuppi-gidda, niovatge-gidda; Marwadi- ker; Sanskrit-karira; Tamil-sengam; Telgu- kariramu) in India. Dela in Delhi, Kair in Punjab, Kair, Kareel, Kerro, Taint in Rajasthan, Teent in Haryana and Karil in Uttar Pradesh are some common name of this species. It is the most useful and popular species which is economically very important.

Capparis zeylanica Linn. *(Syn. C. Horrida* Linn. *f. C. brevispina* DC)

This is a thorny, climbing shrub growing in most parts of India, Burma and Ceylon. It is known as Karambha in Sanskrit, Ardanda in Hindi, Kalokera in Bengali, Govindi in Marwadi, Adonda in Telegu, Adondai in Tamil and Tottulla in Kannada. The fruits of *C. zeylanica* are pickled while leaves are used as a counter-irritant and as a cataplasm in boils, swellings and piles. The root bark is bitter, sedative, stomachic, cholagogue, anti-hidrotic (Kirtikar and Basu, 1918) and is used in cholera. It contains an alkaloid, a phytosterol, a muciloginous substance and water soluble acids. The wood is yellowish white, moderately hard and heavy which is used as fuel of high caloric value.

Capparis spinosa Linn.

It is a small, prostrate shrubs found generally on rocky and hilly localities in the Deccan peninsula, Rajputana and North-Western India. It is known as kabra (Hindi), kokilakshamu (Telegu), mullukattari (Kannada), kalvari (Sind) in various parts of the country. Its flower bud is pickled which has acrid burning taste and useful to cure scurvy. It is known as European capers in western countries. In India, flower buds and fruits are used for picking. Flower bud of *C. spinosa* contains several glycosides and organic acids (Anon, 1998). Caper seed yields 34-36% oil whereas root bark contains rutic acid and volatile substances. The bark is bitter, aperients, diuretic, expectorant, emmenagogue and tonic. It is used in rheumatism, paralysis, tooth ache, affections of liver and spleen and tubercular glands. The bruised leaves are used as poultice in goat (Kirtikar and Basu, 1918). The leaves and ripe fruits are relished by goats and sheep. The leaf contains 69.6% moisture, 13.8% protein, 1.5% ether extract, 7.9% crude fiber, 53.7% N-free extract, 5.5% insoluble ash, 17.7% soluble ash.

Capparis grandis Linn.

This is an armed, crooked shrub or tree species with 3.6 m height. It is known as bhuti or orghuti in Gujarati, revapi or torate in Kannada, woghunty in Malyalam, kauntel in Marwadi, himona in Sanskrit, nakkulinjan in Tamil and oridonda in Telegu. It also recognised by the name of Avaguda or gulli in East Godavari (AP) and kantore in Mt. Abu (Rajasthan). This species is found from Rajasthan in North West to Karnataka in South. The plant has elliptic or ovate leaves and bears white flowers in terminal corymbs or racemes and 2-6 seeded purple fruits. Owing to presence of several glucosinolate, quercetin, rhamnetin and phenolic acids in different parts of the plant, it is used as one of the ingredient in Ayurvedic drugs.

Capparis micracantha DC var. *micracantha*

This is a large stout thorny shrub or small tree species found extensively in Burma and Andaman Island. The plant has elliptical leaves and bears white flowers and bright red or yellow berries with many seeds embeded in juicy sweet fruit pulp (Anon., 1992). The ripe fruit has a sweet aromatic flavour and is eaten by local people whereas unripe fruits are not safe to eat. The roots, leaves and seeds have medicinal properties and are used for the cure of various diseases like asthama, bronchitis, cardic pains. The leaves and fruits are used for poulticing swelling and the roots are diuretic.

Capparis moonii Wight.

This is a large, armed climber or shrub of 2-3 m height which is known as rudanti in Hindi, luthikai or mullukatri in Kannada and poorwi or wagati in Marwadi. The species is commonly occuring in Deccan Peninsula particularly along the Western ghats in South Maharashtra. The plant has elliptic-oblong leaves and produce showy white flowers and globose reddish brown fruits (Anon., 1992). Two strains, viz. Khandala and Konkan of this species have been distinguished. The leaves are used in curries while fruits are used to cure weakness, cough and many other diseases by the tribals of Maharashtra and North-West Mysore. It is also used in various drugs.

Capparis sepiaria Linn.

This is a much branched, armed shrub or woody climber species occuring rarely in the dry regions of Northern India but commonly found throughout the Deccan Peninsula, Andaman Islands, Burma and Sri Lanka. It is known as kaliakara in Bengali, kantharil in Gujarati, heens or kanthari in Hindi, basingi or kadukatri in Kannada, kanthar or mostondi or waghati in Mrawadi, himsra or kakadani in Sanskrit, karunjurai or shenkathari in Tamil, hingerna or hins in Punjabi and chhail or kather in Rajasthani. The plant has oblong, elliptic or obovate leaves and white flowers. Fruits are globose, deep pruple, fleshy berry which are edible. Plants of this species are used for compact hedge and feed for goats while durable wood is used for various domestic purposes. The plant has antipyretic and antiseptic property and therefore used to cure various diseases particularly skin diseases.

Soil and Climate

Ker is a very hardy plant species which can thrive well in sandy wastelands having poor organic matter and fertility level. Chundawat (1990) reported that it grows well in light sandy soils where most of the crop would fail to grow. Vashishtha (1987) mentioned that ker bushes grow well in lower pediment under saline-alkaline soil at certain level. Gupta *et al.*, (1987) reported that in saline depressions of Jodhpur, Nagaur and Shekhawati region of Rajasthan having 50-100 dSm^{-1} Ece, ker is the main vegetation. They also observed that ker bushes improved the fertility status of nearby soil. Singh (1993) reported that the ker develops deep and extensive root system within a short period and therefore, the unproductive, stony, marginal lands which are otherwise unfit for growing other fruits or even crops can be efficiently used for production of ker fruits. Singh and Jha (1993) reported that under Mathura conditions of Western Uttar Pradesh, ker bushes grow well on fine mixed hyperthermia strongly Metric Camborthrids having soil pH range of 7.7-8.2. Studies of Singh (1994) at 20 sites of Aligarh, Mathura, Agra and Manpuri districts of Western Uttar Pradesh reveal that *Capparis decidua* bushes grows well on soils with lower sodicity (pH < 9, ESP < 35, Ece < 4 dSm^{-1}) devoid of clay kankar pan throughout the root zone (20-140 cm depth).

Ker is extremely drought hardy shrub which grows well in the water scarce region of the country. Megwal and Vashishtha (1998) reported that the ker is most drought tolerant fruit crop and grows well under harsh climatic conditions of arid region. The growth pattern of the ker bush coincides with the climate of the region. Ker plant is so adopted to arid environment that it put forth maximum growth during most dry period of the year i.e. between March to June as flowers, fruits and produce economical yield during April-June. Ker also does well in wastelands and foot hills of Aravali with *Calatropis, Prosopis, Ziziphus* and *Lepteclinia* (Mann and Saxena, 1980; Anon, 1992). Ker can flourish well under extremes of temperature and solar radiation but susceptible to frost during winter. Chundawat (1990) reported that being a xerophytic species ker has ability to grow under hot and dry climate of tropics and subtropics spread over Deccan plateau, Rajasthan and Punjab but arid and semi arid regions are highly suitable for ker production. Singh (1993) also mentioned that hot arid dry climate is most suitable for fruit growth and development in ker. Chandra *et al.*, (1994) described that in Thar Desert of Rajasthan where ker grows widely, the mean maximum temperature goes around 50°C with a monthly average of 45-48°C during May coupled with low and erratic rainfall (100-400 mm from 19 rainy days between July-September) which shows the temperature and moisture stress tolerance limit of ker plant. Thus, it can be inferred from the available literature that ker requires hot and dry climate for optimum growth and productivity.

Plant Genetic Resource Management and Varieties

Ker is a ployploid plant species (X = 7, 8) which exists in wild and semi wild form through out the arid and semi arid part of the country. Chundawat (1998) mentioned that there is no any recognize cultivar of ker but great variation with respect to fruit size and pulp content exists in the biodiversity regions (Plate 16.1). Variability with respect to plant growth, fruiting and fruit characters has been identified by CAZRI, Jodhpur and NBPGR, RS, Jodhpur. 20 promising genotypes at CAZRI, Jodhpur, 5 at CIAH, Bikaner and 22 at NBPGR, RS, Jodhpur have been identified (Plate 16.2) and collected on which evaluation process is under progress (Pareek and Vishal Nath, 1996). Chandra *et al.*, (1994) mentioned red and green fruiting types in ker. Chundawat (1990) laid thrust on identification of suitable cultivars by screening

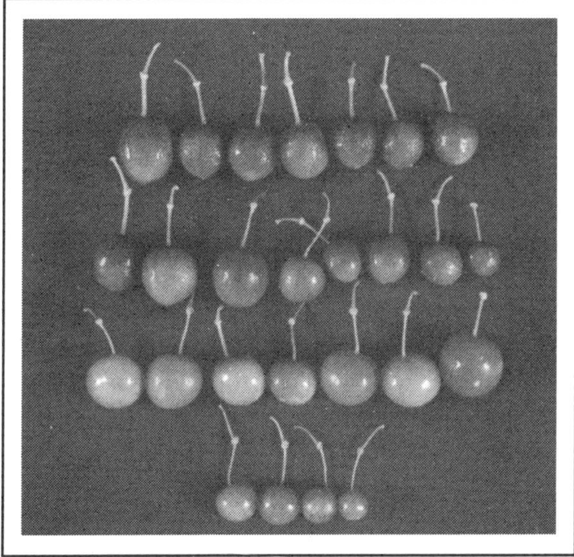

Plate. 16.1. Variability in Ker frutis

Plate. 16.2. Promising genotype of Ker

the local material with a view to select large fruit size with small and soft seeds, low acidity and high protein content. Some plus bushes of ker have been identified in Shekhawati and Marwar region of Rajasthan (*Singh R.S., 2002 Personal Communication*).

Plant Propagation and Nursery Management

Natural regeneration of ker bush occur through seeds and root suckers commonly known as clump (Pareek, 1978; Chandra *et al.*, 1994; Vashishtha, 1987, Chundawat, 1990). However, the species can artificially be propagated through sexual and asexual means. Meghawal and Vashishtha (1998) mentioned that since propagation through seeds does not produce true to type plant due to variable seed lot obtained from cross pollinated fruits, emphasis should be laid to develop vegetative propagation technique for ker. Hartmann and Kester (1983) also mentioned propagation through seed leads to long gestation period and variable population in fruit plants. Shekhawat (1994) reported that seedling plant of ker comes in bearing after 7-8 years of planting.

Although, the information on propagation of ker is limited but in recent past attempt has been made to regenerate ker plants through vegetative techniques including micro propagation. The important methods of ker propagation have been described.

Propagation of ker through seed is an age old practice. Seeds are collected manually from ripe fruits during June and washed properly in clean running water. Natural dispersal of seed also takes place through birds, bats, animals and human being. Seeds are shown in open field or in nursery beds during monsoon period. Singh (1993) reported that the best time for sowing ker seed is July-August under arid and semi arid condition. Ker seed has very hard seed coat and takes 15-30 days for germination. The germination per cent is also very low. To ensure quick and higher seed germination, soaking of seeds in water for 24 hrs before sowing is recommended. Putting seeds in raw cow dung or fresh cow urine for 24 hrs also improves the seed germination. Seeds should be sown in 300 gauge polythene bags/tubes of 25 x 10 cm filled with sand, clay and FYM in equal proportion. This technique allows better root growth of young plants and thus, improves the field survival.

Different methods of vegetative propagation have been attempted for true to type regeneration of ker plants. The successful among are root suckers, cuttings and micro propagation techniques.

Propagation through Root Suckers/Clump is a traditional method of ker regeneration. Root suckers emerging from the sides of well established old bushes are removed carefully during rainy season and planted in the field. It has been observed that heading back of plants at ground level coppices well and produces root suckers freely (Anon, 1992). The main disadvantage of this method is the limited availability of rooted suckers and its poor field survival probably due to absence of functional / absorbing roots. Pareek (1978) suggested that only rooted clumps should be planted in field. Inspite of absorbing roots, the success of clump depends upon its weight, method, time of planting and after planting care. Shekhawat (1994) reported that clumps of 1-3 kg weight planted during July after onset of monsoon gave 63 per cent survival in comparison to August planting (43 per cent survival). Further, it was observed that clumps weighing more than 3 kg gave 60 per cent survival when planted during second week of July and 35 per cent during second week of August (Table-16.5). The planting of clumps should be done in well prepared and properly filled pits of 60 x 60 x 60 cm size (Singh, 1993). After planting, proper hoeing, irrigation and protection against termite is essential to ensure higher success.

Table 16.5 : Effect of planting time and weight of clump on survival of ker under Jodhpur condition.

Planting time	Survival (%)	
	Clump wt. (1-3 kg)	Clump wt. (> 3 kg)
Mid July	63	60
Mid August	43	35

Source : Shekhawat, 1994

Ker can be propagated through semi hard wood and hard wood cuttings with application of rooting hormones. Soni (1970) reported that IBA is better for treatment of cuttings because of its ability to produce fibrous roots, chemical stability and low mortality of plants. Vashishtha (1987) obtained 65 per cent initial sprouting in semi hard wood cuttings with quick dip (2 seconds) in 1000 ppm IBA solution where as the sprouting was only 52 per cent in case of hard wood cutting dipped for 2 minutes in the solution. Shekhawat (1994) reported that quick dip in 5000 ppm IBA solution is the best which gives 30 and 53 per cent sprouting, respectively in semi hard wood and hard wood cuttings. Anon., (1994) reported 25% sprouting (2500 ppm IBA) and 16% rooting (5000 ppm IBA) in hard wood cutting of ker at Jodhpur. They reported that initially thin roots are coming from basal portion of cutting which helps them to survive upto 7-8 months after that it is dried in May. Meghwal and Vashishtha (1998) also observed the effect of IBA on rooting and sprouting of ker cuttings. Dipping the base of cuttings for 24 hrs in distilled water and treating them with 5000 ppm IBA gives better rooting in hard wood cutting under Bikaner condition (Bhargava *et al.*, 1999).

Time for taking cutting in ker is directly related with the sprouting and rooting of cuttage. Vashishtha, (1987) suggested that August-September is the best time for taking cutting under Jodhpur condition. Shekhawat (1994) and Meghwal and Vashishtha (1998) reported that July- August is the best season for cutting. In a year round study conducted at NRC for Arid Horticulture, Bikaner to identify the suitable month for taking ker cutting (Anon., 1998) reveal that October is the best month for taking cuttings. Ker plant is believed to have high level of inhabitor in its stem, branches, twigs, etc. which inhibits the root initiation and growth in cutting. Hartmann and Kester (1983) suspected that difficult to root plants may have high level of internal growth inhibitor and continuous scleranchyma ring which inhibits the root initiation and development. Aqueous extract of ker stem has been reported to have high level of inhabitor (Gautam and Bishnoi, 1991; Bhargava *et al.*, 1999). Gautam and Bishnoi (1991) reported that aqueous extract from *Capparis decidua* root and shoot checks 100 per cent seed germination in *Calligonum polygonoides* L. (*Phog*) and *Lasiuras scindicus* Herr. (*Sewan*). Further it was observed that degree of inhibition was directly proportional to the concentration of extract. Bhargava *et al.* (1999) reported that aquaceous extract from ker stem cutting contains Abscisic Acid (ABA). In an experiment they applied the aqueous extract of ker stems (dipped for 24 hours in distilled water and then stained) to *Nerium odorum* cuttings, an easy to root plant and pre germinated seeds of clusterbean (*Cymopsis tetragonoloba*) to study its effect on root initiation and development. It was observed that treated cuttings were having on an average 2 roots of 5.5 mm length where as the untreated cutting were having 16 roots with 26.5 mm roots. Similarly, the pre germinated treated seeds were having 2.37 ± 1 cm root length while untreated controls were having 4.71 ± 1.5 cm root length. Thus, it was concluded that, its aqueous extract has high level of inhibitor, which may be responsible for poor rooting and root growth in ker cutting. Therefore, the practice of the dipping the cuttings base for 24 hours in distilled water followed by IBA treatment and planting in nursery tubes should be attempted to improves sprouting and rooting because the level of inhibitor is washed out due to water. They also suggested that for better success in cutting it should be taken after monsoon (September-October) because the level of inhibitor in plant is reduced during rains or diluted due to humid weather during rainy season.

Rooting medium has important role in propagation of ker through cutting. Nutritionally rich, porous and germfree medium having good water holding capacity have been considered better for ker. Polythene tubes (25 x 10 cm both end open) or bags (25 x 10 cm one end closed) filled with a mixture of sand, clay and FYM in equal proportion is generally used for planting of ker cutting. Meghwal and Vashishtha (1998) reported that polythene tubes filled with a mixture of sand and powdered goat manure in 2:1 ratio gives better result. Medium containing sand and FYM in equal proportion was found best by Shekhawat (1994).

Visualising the difficulties in traditional methods of ker propagation, efforts are now switching over to newer techniques for rapid and true to type plant production in bulk. Tissue culture technique has been successfully used for regeneration of ker in India. Shekhawat (1994) reported that in ker multiple shoots (3-5 shoots/ explants) from the single nodal shoot segment are differentiated within three weeks in MS medium supplimented with IAA 0.1 mgl^{-1} + BAP 2.5 mgl^{-1} at 28 ± 2^0C temperature and 2000-3000 mX photon intensity for 12 hrs photo period. For sub culturing, the medium having 0.1 mgl^{-1} IAA + 1.5 mgl^{-1} BAP + additives was better where as for root initiation half strength MS basal salt medium with 2.0 mgl^{-1} IBA medium was considered optimum.

Tyagi *et al.*, (1997) reported *in vitro* clonal propagation in *Capparis decidua* using nodal explant from mature tree. Explant cultured on MS medium supplemented with benzyladenine (BA 1, 3 and 5 mgl^{-1}) showed differentiation of multiple shoots and shoot buds in 4-5 weeks in the primary culture. The medium with 5.0 mgl^{-1} IBA was the best for shoot bud proliferation from the nodal explants as well as from seedling explants. Shoot multiplication was best with cotyledonary nodes. With nodal explants, shoot multiplication was best on medium supplemented with 5 mgl^{-1} IBA and after the second sub culturing, further multiplication of shoot buds was highest on the medium containing 3 mgl^{-1} IBA. Best rooting was obtained in MS medium using 1.0 mgl^{-1} IBA. Rooted plantlets were transferred to earthen pots with garden soil and a peat moss mixture.

Planting

Ker plant grows as bush or small tree which can be planted at closer space. Chundawat (1990) and Chandra (1994) mentioned that although there is no systematic work on planting technique, time and method in ker but considering its importance, these need to be standardized for systematic orcharding. Further, Chundawat (1990) suggested that, looking to its size and bushy nature, regular plantation of ker in field as orchard or at boundary as compact hedge or for sand dune stabilization, can be done at 3 x 3 m in the begining of showers. Singh (1993) reported that ker should be planted during July-August in pre prepared pits of 60 x 60 x 60 cm size at 3-4 meter apart. In pediment plains, Singh (1993) reported that pits should be dug before on set of monsoon during June and kept open for 15 days. After that it should be filled with top soil, FYM (10-15 kg) and termiticide (50g) per pit. At present, ker has diversified importance as hedge, sand dune stabilization, and boundary plantation and as natural habitat in arid and semi arid region. Gupta *et al.*, (1987) mentioned three natural associations viz. (i) ker - *Ziziphus nummularia* (ii) ker - *Tephrosia purea,* (iii) *Prosopis cinerasia* - *Acacia senagal* - ker in Thar Desert.

Training and Pruning

The ker plant is self stature and naturally pruned. It does not require regular pruning, however, training of plants to achieve a desired shape is essential. To develop a tree form, trunk should be kept branchless up to 60-90 cm at initial age which also facilitates the inter culture operations. In bush shape plants however multiple shoots should be allowed from ground. Once a plant achieves desired shape, only broken, diseased and insect affected branches are removed after fruit harvest regularly. Ker bears on old branches and therefore, it does not require regular pruning.

Orchard Management

Ker is a drought hardy plant and therefore requires less care after plant establishment. To ensure high plant establishment, regular monitoring of plant health and application of inputs and inter culture operation is essential in the initial 2-3 years.

Ker plantation is generally found associated with agro-forestry systems or natural habitats of the arid and semi arid region. Therefore, at present no specific orchard management practices are adopted by the growers except the minor pruning and training for the benefit of associated crop. To ensure higher production, general orchard management practices do require in ker plantation.

Nutrient Management

Ker plants are seldom manured. To improve productivity of a bush it should be applied with 15-20 kg well rotten FYM or sheep manure along with inorganic fertilizers. 50 g nitrogen and 25-30 g phosphorus have been considered optimum for ker plant in sandy soils of Rajasthan and Haryana. Manures should be applied during monsoon whereas fertilizers should be given after rainy season during October by mixing them below and around the bush in a ring form.

Water Management

Ker is a hardy plant which often does not require irrigation. For initial establishment of plant and after application of manures, particularly in drought years light irrigation is essential. At the time of fruit development in spring crop during April light irrigation improves the immature fruit yield. Irrigation or rains at the time of flower bud initiation hampers the yield of ker bush. Mulching of plant basins with available organic materials (dry weeds, bui, sania, khneep in arid region) improves plant growth and soil health.

Orchard Floor Management

In arid and semi arid region, mixed cropping is a common practice to provide sustained income and life support. In case of failure of one crop other may yield better. Therefore, ker plantations are also mixed with a number of annual crops. During kharif season, crops like moth, guar, til, mateera, kachari and bajra are sown in ker plantations. These crops apart from generating additional income improves soil properties and minimize soil erosion. Leguminous crops are generally preferred as intercrop in systematic ker plantations.

Disease and Pest Management

Intense solar radiation and environmental stresses in arid region allowed very few or no pest and disease to attack on ker plants. Except some biotic agents, ker plants, generally do not suffer from any serious melody (Chundawat, 1990). Chandra (1985) reported that the plant is subjected to attack by the caterpillars of *Belenois aurota* Fab. In Rajasthan (Anon, 1992), stink is another harmful insect for ker in which the affected parts of plant gradually shriveled and dried. Sticky secretions of insect (*Stenozygium pseudospeciosum* Ghauri) deteriorates the market value of fruits. Besides, young plants of ker are attacked by termite particularly during drought. Soil drenching with Chlorpyriphos with irrigation water (5 ml/10 lit.) checks the problem effectively (Singh, 1993). Die back problem is also observed in some area which leads the drying of shoots (Singh, 1993). Chandra *et al.*, (1994) emphasized that work on disease and insects needs to be under taken in order to ascertain resistance/tolerance against prevalent insect-fauna of the region. Ripe fruits are damaged by birds and bats.

Flowering, Fruiting and Yield

Although ker plant flowers in three seasons i.e. February-March (*Ambe bahar*), July-August (*Mrig bahar*) and October-November (*Hasth bahar*) but the main flowering season in arid region is *Ambe bahar* (Plate 16.3). Flowers appear on young as well as old shoots but one year old twigs bear maximum flowers which retain fruits. Shekhawat (1994) reported that older shoots produced more flowers than the new shoots. In *Ambe bahar* crop, flower bud initiation starts by second fortnight of February and continues till the first week of

Plate. 16.3. Ker tree in full bloom

March under arid conditions of Bikaner and Jodhpur. Since ker has cyme inflorescence, in which all flowers do not open at one time, therefore flower buds of different stage can be seen in single bunch. Opening of flower takes place after about 2 weeks of flower bud initiation while the peak of flowering occurs after a month of first flower bud initiation. Shekhawat (1994) reported that the maximum flowering in ker bushes takes place by 20-25 March.

Peak growth in flower buds take place within 8 days of its initiation. During the development of a flower bud, the colour of bud changes from greenish white with hairy base to dark brownish red with greenish white tinge. Anthesis of flower commence at 2 PM and continues upto 7 PM with maximum opening at around 3 PM. Before anthesis, the mature buds can be distinguished from immature bud by their size and colour. Mature bud become plump before opening. At the time of opening, the top of the bud cracks in the petal which subsequently widened downwards and the petals get separated.

Ker flowers are generally 2.5 cm in diameter with brick red, pink or yellow colour. Flowers are borne in the bunch of 2-8 flowers arranged in terminal fascicles or corymbose racemes (Plate 16.4). Pedicles are slender, 1-1.5 cm long, densely white or fulvous. Ker has complete flower in which calyx, corolla, androecium and gynoecium are arranged on floral disc. Sepals four; petaloid, scarlet, ciliolate and

Plate. 16.4. Ker-flower

pubescent on outer surface. The anterior outer sepal measures 10-12 x 5-7 mm in size, deeply sacate, enclosing the remaining floral parts where as the upper sepal are smaller measuring 10-12 x 6-8 mm in size, ovate, oblong and concave. The lateral sepals are 8-10 x 2 mm in size, elliptic, acute with floccose ciliate margin and posterior sepals deeply concave with 8 x 3 mm size. Petals four separated in two parts. Two posterior petals are measuring 12-13 x 3-4 mm size, narrow, oblong, bright scarlet, obtuse, pubescence while two anterior petals are almost included in the hooded sepal. These are obliquely ovoid having 10-12 x 5-6 mm in size, with one bifid pocket like nectary at the base of each which contains a drop of sweet nectar for attracting the insect pollinators. Stamens 10-18; filaments red, glabrous, unequal; anthers basifixed, introrse; disc glandular which remain present on the posterior side between petals and stamen. 4 mm long, glabrous, superior, ovary with four placenta remain attached on 12-15 mm long gynophores; style 1 mm long.

Dehiscence of pollen in ker initiates at 10 AM with the appearance of longitudinal slit in the anther lobes. The pollen completely dehisces within 3-4 hrs i.e. between 10 AM and 2.30 PM. The round pollen grains are transported by insects and winds from one flower to another. Receptivity of stigma in ker is very poor; therefore, the flowers pollinated within 4-6 hrs of anthesis only set fruits (Shekhawat, 1994). Fruit setting in ker is about 18-20 per cent under open pollination where as only 8 per cent self pollinated flowers set fruits. Under self pollination fruit drop is to the tune of 75 per cent where as under open pollination condition 100 per cent fruit retention has been noticed. Peak of fruit setting takes place by the end of March. Shekhawat (1994) reported two rapid fruit development period in ker i.e. between 1-5 days and 16-20 days after fruit set (Plate 16.5). The ker fruit attain maximum size within 34-36 days of fruit set. It is therefore, it appears that ker has double sigmoid curve for fruit development (Fig.16.1).

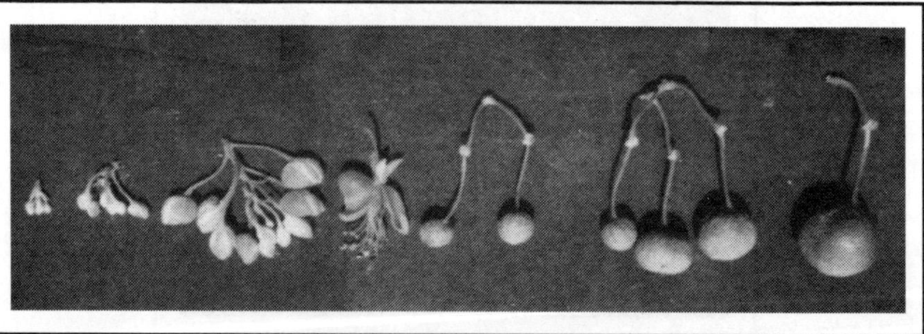

Plate. 16.5. Ker friut development stages

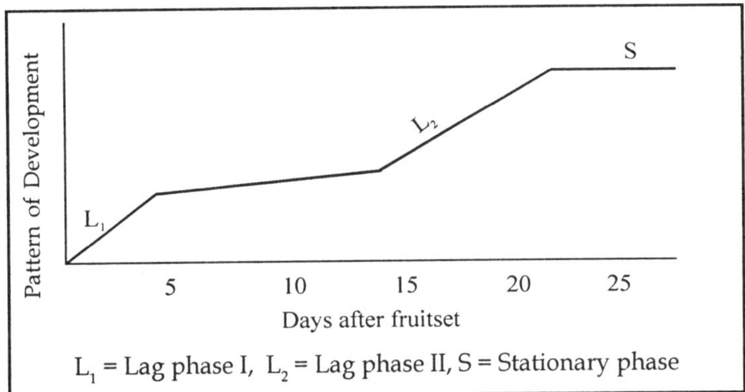

Fig 16.1. Fruit growth and development pattern in *Capparis decidua*

Botanically, the ker fruit is a berry. Fruits are ovoid to sub globose measuring 0.7-1.5 cm in diameter. Glabrous, beaked fruit of scarlet red colour (ripe) contains many seed. Seed size in ker is 2.5 x 2 - 4.0 x 2.0 mm. The immature fruits remain green with soft seeds.

Ker fruits do not attain economical maturity at one time because of long duration of flowering and fruit set. For pickling, dehydration and vegetable purposes, fruits attain required maturity in 30-45 days after fruit set. Because of ununiform maturity and selectivity for quality product, immature green fruits are picked manually and graded according to their size and tenderness. Ker seeds become hardy within a few days therefore, to maintain quality, harvesting should be very quick. Due to thorny nature of plant, harvesting is a difficult process in ker. Chandra *et al.*, (1994) estimated that one man can harvest approximately 10 kg fruits per day.

Ripe fruits are harvested after attaining cherry red colour during June-July. Ker plant starts bearing after about 5 years of age. Yield of a ker plant is a function of plant age, management practices and climatic factors at a particular location. Chundawat (1990) reported that a five year old bush can yield about 5 kg immature green fruits whereas Pareek (1978) estimated 10-20 kg fruits from well grown up ker bush. Singh (1993) reported that a well managed ker bush can produce 10-15 kg fruit in one season.

Harvesting and Post harvest Management

Ker fruits are acrid in taste; therefore, freshly harvested fruits are not consumed as such. The fruits are kept in brine solution for few days and send to market afterwards which gives better market price and also minimizes the physiological loss in weight during transportation and storage. Fruits can be stored in brine solution for months together. To store the fruits for longer period, these are drained from brine solution and sun dried. Chandra *et al.*, (1994) reported 20 per cent recovery of dried fruits. Dried as well as brined fruits can be use for preparation of various value added products.

Ker fruits are sold in local as well as in distant market. In market, ker is sold at a very high price. Because of lack in systematic production technology and high picking cost (manual picking), the price of ker fruits remains much higher and it is considered as delicacy of Marwari diet all over the country. Processed products from ker (dehydrated, pickled, etc.) are sold at a very high price. In local market of arid parts of Rajasthan, brined fruits and flower buds of ker are sold at a price ranging from 40 to 60 rupees a kilogram whereas the dried fruits are sold @ Rs.120-160 per kg. In distant market, only dehydrated and processed form of ker is sold. Dried fruits of ker are sold in markets of Delhi, Calcutta, Bombay, Bangalore at the rate of 150-200 rupees per kg. Pickle prepared from ker is an international items. Gupta *et al.*, (1987) estimated that cultivation of ker is very remunerative with a net income of Rs.10, 000 per hectare, if the fruits are properly collected, processed, dried and marketed.

Processing and Value addition

Immature tender green as well as pink or red ripe fruits of ker are used in various ways (Anon; 1992; Chundawat, 1990; Chandra *et al.*, 1994; Vashishtha, 1987; Meghwal and Vashishtha, 1998; Pareek *et al.*, 1998; Singh, 1993). Anon (1992) mentioned that unripe ker fruits are pickled and or cooked as vegetable which forms an integral part of the diet of people in desert and semi-desert areas of the country where as the ripe fleshy fruits are eaten by birds. Chundawat (1990) however, described that ker fruits are mainly used in pickling or as vegetable when immature while sweet and acrid fruits are enjoyed by children in natives after ripening. Chandra *et al.*, (1994) described the use of ker fruits at various stages. They mentioned that unripe green fruits after processing are eaten as vegetable and also used for making pickle, where as, mature unripe fruit contain high level of glucosinolates and can only be used at limited scale after proper processing. Singh (1993) and Tiwari (1996) mentioned that ker fruits are used as an important ingredient in the famous delicious mixed vegetable (*Panchkutta*) of the "Marwar" region of India. Pareek *et al.*,

(1998) described that green, immature, acrid fruits of ker are rich in protein content and are used for pickle.

Seeds of ker are rich source of fat, protein and sugar. Rai (1987) reported that ker seed contains 20% saturated hydrocarbons and ketones with C_{28} to C_{32} chain lengths, and 0.7% sugar alongwith 8.6% protein.

Apart from cooking as vegetable of various tastes, flower buds, flowers and immature fruits of ker can be processed into various other value added products. Ker makes important component of mixed pickle of sour taste. Besides, brining and canning and brining and dehydration are other important value addition techniques to the ker fruits in Rajasthan. After harvesting, flower buds and fruits are properly washed and dipped in 2% salt solution for 24 hours. Whole flower buds are cured in 4% salt solution for next 2-3 days whereas fruits are dipped in 4, 6 and 8 per cent salt solution alternatively for next 72 hours. Finally, the flower buds are kept in 6 per cent salt solution where as the fruits are preserved in 10 per cent salt solution in wide mouth containers. Before use, these need to be washed in running tap water thoroughly. To improve the marketability of brined product, sun drying is a local practice in ker growing region of the country, however, they can also be mechanically dried in dehydrators. Chundawat (1990) reported that immature fruit of ker are dried in Rajasthan for subsequent use as vegetable in off season. Singh (1993) and Pareek *et al.*, (1998) mentioned that ker fruits can dried and kept for longer period.

Future Research Needs

In spite of very high economic value, the ker a gift of nature in the desert has not been paid due attention in past. Very limited research work has been done on various cultural aspects of ker, therefore, intensified research is needed to develop improved package of practices for scientific cultivation and future line of work has been suggested. Apart from its capability to stabilize sand dunes and combat drought with tolerance to biotic and abiotic pressure, no systematic attempts have been made for its scientific cultivation so far. In ker, the following research thrust should be addressed in future.

1. Identification of cultivar with large fruit, less acridity and soft seed.
2. Development of economical and easy method of vegetative propagation to provide better planting material to growers.

3. Standardization of agro-techniques particularly with respect to planting techniques, nutrition, water management and plant protection.
4. Development of marketing set up for better disposal of produce with suitable return to growers.
5. Development of cottage as well as big industries for processing and value addition in ker fruits at commercial scale.

REFERENCES

Agrawal, V. and Chauhan B.M. (1989) Effect of feeding some plant foods as source of dietry fibre on biological utilisation of diet in rats. *Plant Foods for Human Nutrition,* 39 (2): 161-167.

Ahmad, V. U.; Ismail N. and Amber A. (1989) Isocodonocarpine from *Capparis decidua. Phytochemistry,* 28 (9): 2493-2495.

Ahmad, V. U.; Ismail N; Arif S. and Amber A. (1992) Two new N-acteylated spermidine alkaloids from *Capparis decidua. Journal Natural Products,* 55(10) : 1509-1512.

Ali, A.E. and Ali A.E.S. (1987) Topography mapping of vegetational features from land-set imagery : an experiment from Sudan. *Sudan-Silva,* 6 : 26.

Anonymous (1992) The Wealth of India - Raw materials (Vol. II). Council of Scientific and Industrial Research, New Delhi, India: 210-215.

Anonymous (1994) Central Arid Zone Research Institute, Jodhpur, India: Annual Report, 1993-94:83

Anonymous (1998) NRCAH Annual Report, 1997-98. National Research Centre for Arid Horticulture, Bikaner, India.

Anonymous (1995) Tour Report. Submitted to Indian Council of Agricultural Research, New Delhi.

Bhargava, R; Vishal Nath and Pareek O.P. (1999) Role of endogenous plant hormones on sprouting of ker (*Capparis decidua*) cuttings. *Current Agriculture,* 23 (1-2): 59-60.

Bhandari, M.M. (1995) *Flora of Thar Desert.*

Chandra, A. (1985) *Indian Farming,* 111: 568.

Chandra, A; Chandra A. and Gupta I.C. (1994) Ker. In: Arid Fruit Research. Scientific Publication, Jodhpur, India: 298-302.

Chauhan, B.M.; Duhan A. and Bhat C.M. (1986) Nutritive value of ker (*Capparis decidua*) fruit. *Journal Food Science Technology,* 23 (2): 106-108.

Chopra, R.N.; Nayar S.L. and Chopra I.C. (1956) Glossary of Indian medicinal plants. CSIR, New Delhi.

Chundawat, B.S. (1990) Kair. In: Arid Fruit Culture, Oxford and IBH Pub. Com. Ltd., New Delhi, India: 187-189.

Duhan, A; Chauhan, B.M. and Punia, D. (1992) Nutritional value of some non-conventional plant foods in India. *Plant Food Human Nutrition,* 42 (3): 193-200.

Garg, S.P. and Rai S. (1993) Chemical examination of medicinal desert plant - *Capparis decidua.* Part 1. Isolation and mass spectral studies of n-nonacosane and n-hentriacontane from its flowers. *Current Agriculture,* 17 (1-2): 97-99.

Garg, S.P. and Rai S. (1993) Chemical examination of medicinal desert plant - *Capparis decidua.* Part 2. Isolation and characterization of glycosides. *Current Agriculture,* 17 (1-2): 101-105.

Gautam, D.D. and Bishnoi S. (1991) Influence of root and shoot extracts of some desert plants on seed germination of *Calligonum polygonoides* Linn. and *Lasiurus scindicus* Henr. *Journal Phytological Research,* 4 (2): 177-181.

George, U. (1974) A dictionary of plants used by man. Constable and Company Ltd. London: 121.

Gopalan, C; Rama Shastri B.V. and Balasubramaniam S.C. (1987) Nutritive value of Indian Foods. National Institute of Nutrition. ICMR, Hyderabad.

Gupta, I.C.; Harsh L.N., Shankarnarayana K.A. and Sharma B.D. (1987) Note on *Capparis decidua* (Kair) from waste lands. *Current Agriculture,* 11: 87-89.

Hartman, H.T. and Kester D.E. (1993) Plant propagation: Principle and practices. Printis Hall, London.

Hooker, J.D. (1987) Flora of British India. Reeve and Co. Henrietta Street, Convent Garden, London.

Kirtikar, K.R. and Basu B.D. (1918) *Indian medicinal plant,* 1: 256.

Khera, A.P. and Singh J.P. (1975) *Indian Horticulture,* 20 (2): 10.

Kumar, M. and Bhandari M.M. (1993) Impact of human activities on the pattern and process of sand dune vegetation in the Rajasthan desert. Desertification Control Bulletin No. 22: 45-54.

Kumar, S. (1992) Comparing classification and different coordinations of vegetation in the thar desert of India with implications in resource management. *Tropical Ecology.* 33 (1): 110-131.

Mann, M.S. and Saxena S.K. (1980) Khejri (*Prosopis cineraria*) in the Indian desert. CAZRI Monograph No. 11, IV: 25.

Meghwal, P.R. and Vashishtha B.B. (1998) Effect of time of planting and auxins on rooting of cutting of ker (*Capparis decidua* Forsk.). *Annals of Arid Zone,* 37(4): 401-404.

Nama, N.S. and Bhatnagar B. (1990) Laboratory evaluation of cercaricidal properties of certain plant extracts. *Indian Journal Parasitology,* 14 (1):72-82.

Pandey, A.N. and Rokad M.V. (1992) Sand dune stabilisation : an investigation in the thar desert of India. *Journal Arid Environment,* 22 (3): 287-292.

Pareek, O.P. (1978) Development of Horticulture in arid region. *Indian Horticulture,* 23(2): 23-30.

Pareek, O.P. and Vishal Nath. (1996) Coordinated Fruit Research in Indian Arid Zone - A two decades profile (1976-95). National Research Centre for Arid Horticulture, Bikaner, India: 1-104.

Pareek, O.P.; Suneel Sharma and Arora R.K. (1998) Under utilized edible fruits and nuts : An inventory of Genetic Resources in their region of diversity. IPGRI Office for South Asia, New Delhi, India: 83.

Rai, S. (1987) Oils and fats in arid plants with particular reference to *Capparis decidua. Transaction of the Indian Society of Desert Technology,* 12 (2): 99-105.

Singh, K. (1994) Site suitability and tolerance limits of trees, shrubs and grasses on sodic soils of Ganga-Yamuna Doab. *Indian Forestor,* 120 (3): 225-235.

Singh, K. and Jha M.N. (1993) Trees, shrubs and grasses on saline soils of Indo-Gangetic plains. *Indian Forestor,* 119 (8): 630-647.

Singh, R.S. (1993) Maru kshetra ka upyogi phal-Kair. *Uddyan Lok,* 5 (4): 1-2.

Sharma, B.D. and Gupta I.C. (1989) Effect of trees cover on soil fertility in western Rajasthan. *Indian Forestor,* 115(5): 348-354.

Sharma, B.D. and Gupta I.C. (1989) Stabilising sand dunes with tree cover. *Indian Farming,* 38 (12): 19 & 21-22.

Sharma, D. and Prasad S.N. (1992) Tree debarking and habitat use by porcupine (*Hystrix indica* kerr) in Sariska National Park in western India. *Mammalia,* 56 (3): 351-361.

Shekhawat, J.S. (1994) Studies on growth, fruit development and quality of ker [*Capparis decidua* (Forsk) Edgew]. Ph.D. Thesis submitted to Rajasthan Agricultural University, Bikaner, India.

Soni, S.L. (1970) Use of growth regulators in horticulture. *Farmer and Parliament,* V (II): 7.

Tanira, M.O.M.; Ageel A.M. and Al-Said M.S. (1989) A study on some Saudi medicinal plants used as diutretics in traditional medicine. *Fitoterapia,* 60(5): 443-447.

Tyagi, P.; Kothari S.L. and Tyagi, P. (1997) Micro propagation of *Capparis decidua* through *in vitro* shoot prolification on model explants of mature tree and seedling explants. *Journal of Plant Biochemistry and Biotechnology,* 6(1):19-33.

Vashishtha, B.B. (1987) Vegetative propagation of *Capparis decidua. Annals of Arid Zone,* 26 (182): 123-124.

—x—

17

PILU

Pilu (*Salvadora oleoides* Decne.) is an under-exploited multipurpose tree species. It is an evergreen, small, twisted, bushy, drought and salinity tolerant tree and grows successfully in ravines and degraded lands. Ripe pilu fruits are eaten fresh; however it can also be used as dehydrated fruits. It is believed to produce tingling fruits which are useful to cure small mouth ulcers when eaten as fresh. Seeds contain 40-45 per cent non-edible greenish yellow fat (oil) which is a potential industrial substitute for coconut oil (Singh and Tiwari, 1995). The fat is rich in lauric and myristic acids and is used in medicines. Seed also contains 18.9% albuminoides, 23.5 % carbohydrates, 5.8 % fibre and 3.5 % ash (Pareek, *et. al.*, 1997). Ripe pilu fruits contains 6.0 g protein, 2.0 g fat, 76.0 g carbohydrate, 2.0 g fibre, 630 mg calcium, 167 mg phosphorus, 8.0 mg iron and 2.0 mg ascorbic acid in 100 g pulp and the seeds contain 45-46% fat, 18.9% albuminoides, 28.5% carbohydrates, 5.8% fibre and 3.5% ash. Besides the edible fruits and non-edible but medicinally rich oil, pilu provides fuel wood from branches and leaves fodder for camels (Pareek, *et. al.* 1997). It is also planted as a windbreak and fence.

Salvadorin, a new dimeric dihydroisocoumarin was isolated from the chloroform fraction of *Salvadora oleoides*. Tariq *et al.*, (2005) have established its chemical structure as B-benzyl-6-[6-(6-ethyl-7-methyl-5,8-dihydro-Z-naphthalenyl)-1-oxo-3,4-dihydro-1H-isochromen-8yl]-3, 4 dihydro-1H-isochromen-1-one through spectroscopic technique and chemical.

Origin and Taxonomy

Salvadora is native of warmer parts of the Middle East particularly near to Bay of Aden. It belongs to the family Salvadoraceae of which two species, viz. *S. oleoides* (*meetha jal*) and *S. persica* l. (*khara jal*) are common in India (Bhandari, 1995). The former is a small to medium tree with twisted trunk and drooping branches while the latter produces evergreen and are much branched shrub or tree (Plate 17.1). In general, the plant can be described as a shrub or tree, branches stiff,

Plate 17.1. Plant of Pilu

rough, whitish. Leaves opposite, coriaceous, petiole 0.2-1.2 cm long, lamila 1.5-7.5 cm long, 0.4-1.5 cm broad, elliptic- lanceolate, mostly acute rarely obtuse or mucronate glabrous with obscure lateral veins. Inflorescence axillary panicles or branched spikes, 2.5-4.0 cm long (Plate 17.2). Flower comes in bunches which are perfect. Flowers are greenish white, 2.0-3.0 mm across, pedicel 0.1 mm long or abscent.

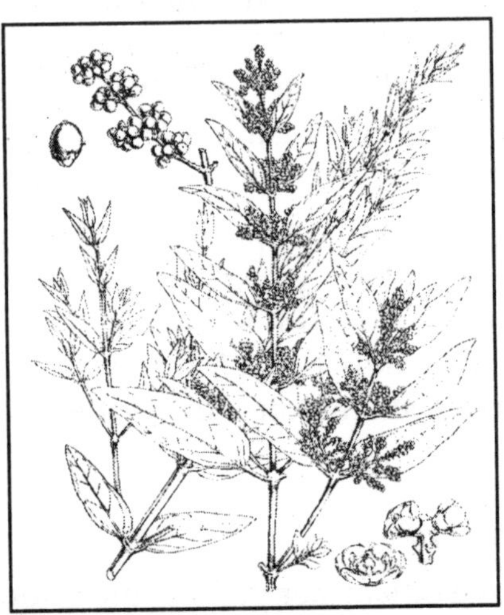

Plate 17.2. Sketch of Pilu Infloresence

Calyx 1.5-2.0 mm long with round lobes and wavy margin, glabrous (Plate 17.3). Corolla 2.5 mm long, obovate or oblong, lobes sub-acute and recurved. Stamens 4, incerted at the base of the corolla tube. Style abscent, stigma peltate. Fruit is a drupe, 5.0 mm in diameter, globose, yellow on maturation with high oil content in seeds. Fruit ripens with in 4 months of set. All the fruits do not mature at the one time. It is commonly found growing in saline patches in semi-arid and arid region of India.

Plate 17.3. Flowering paniel of Pilu

Area and Distribution

No systematic plantation of pilu has been reported, however, natural grooves of this under exploited fruit species are available in its original abodes in arid and semi arid parts of the world. The species is distributed in Pakistan, Aden, United Arab Emirates and India. In India, pilu is commonly found in Rajasthan, Gujarat, Haryana, Delhi and parts of Madhaya Pradesh.

Soil and Climate

Pilu is a hardy plant species which grows in ravines and degraded form of land under natural condition. It prefers medium to high saline soil condition of alluvial or canker soils prevalent in arid and semi arid ecosystem. In heavy soils, however, pilu plant produces luxuriant growth and fruiting.

Jhal prefers drier climate with high aridity index. It is quite clear from its growth habit. The plant inters in floriferous condition during most dry period. It does not require high humidity during flowering and fruit ripening. Light cyclonic showers which are quite occasional in arid and semi arid region help increased fruit size during fruit development. Development of high quality in jhal related to high temperatures during fruit ripening. Warmer days and cool nights help sugar development in pilu fruits.

Plant Genetic Resource Management and Varieties

Pilu grows wild in degraded habitats under arid ecosystem. It is a naturalized species where natural regeneration takes place through dispersal of seeds through birds and nomads. This creates high genetic variability. A high degree of variability has been observed in natural population of pilu in arid and semi arid tract of Rajasthan and Gujarat (Pareek, *et. al.* 1997). Efforts are being made to identify high yielding, sweet and high pulp containing genotypes from the available genetic diversity. Variability with respect to fruit size, quality and oil content in seed has been collected (Pareek, *et. al.* 1997). So far there is no recognized cultivar of pilu.

Plant Propagation and Nursery Management

Pilu is commonly propagated by seeds and root suckers (Pareek and Vishal Nath, 2006). Seedlings of pilu can be raised by sowing seeds in polythene tubes/bags of 25x10 cm size during July-August. Tubes should be filled with a mixture of FYM and sand in equal proportion. Seeds are sown during monsoon which germinates after 7-10 days. Because of slow growth, seedlings take 10-11 months to attain the proper size for planting. Root suckers emerging around the tree bases under the trees canopy are also used as planting material. Plants are also regenerated by layering at limited scale with comparatively low success. Advance techniques of plant propagation particularly the use of growth hormones has potential in regeneration of this species to have genetically uniform planting material.

Planting

Salvadora is planted at 5-6 m distance in square system as sole crop. Pits of 60x60x60 cm size are filled with soil and FYM mixture in 1:1 ratio before the onset of monsoon so that planting can be done during the monsoon period. One to two years old healthy saplings raised through seeds or suckers with well developed root system are preferred for planting.

Training and Pruning

Pilu has not yet attained the status of cultivated fruit plant, therefore information on training and pruning is lacking. However, initial frame working by allowing scaffold branches and promoting secondaries, tertiaries and fruiting terminals can develop better frame of the tree and enhanced production. Plants can also be trained as bush to prevent soil erosion through winds in desert areas. Regular pruning in pilu is not required because the flowers are borne on more than one year old shoots, however, removal of damaged and broken branches is considered essential. Pruning of few branches up to 50% once in three years help enhanced new shoots production.

Orchard Management

Since no systematic orchards of pilu is common, regular orchard management practices has not been standardized. Presently, the species is only planted in agro-forestry system as mixed planting where plants do require proper nourishment and care. Informations what so ever available are only on per plant basis and are only indicative for further intensification of work on these lines.

Nutrient Management

The pilu tree has extensive deep root system and can draw nutrients and water from the deep profiles of soil. Application of 10-15 kg FYM/plant along with 100 g nitrogen and 50 g phosphorus every year during monsoon is useful to improve the vigour of the plant and fruit yield. For optimum utilization of applied nutrients, it should be placed near the feeder roots by making 15-25 cm trench followed by covering with top soil.

Water Management

For successful establishment, plants should be watered during the initial period after planting. Once established, provisions of mulching and water harvesting (5% slope towards the plant trunk) would help plant to increase vigour and productivity. Preferably, the mulching with local dry weeds or farm waste should be done near the tree trunk covering the fertilizer ring. On sloppy lands half moon terracing against the gradient helps more accumulation of runoff in the tree basins.

Orchard Floor Management

Generally, pilu is planted as mixed crop with other species in agro-forestry system where different component crops are taking care of floor area. However, inter spaces between the tree rows can be used for raising intercrops like mung bean and clusterbean, sorghum and pearl millet. In arid parts of Rajasthan, grasses like *Ccenchrus ciliaris* (*sewan*) and *C. setigerus* (*dhaman*), cover crops such as moth bean and mung beans and intercrops like cluster bean are grown in between the pilu plant during monsoon season.

Disease and Pest Management

Pilu is a hardy tree and is generally free from diseases and pests. To check the attack of termites in young plantation, 50 g chloripyriphos dust (10%) should be applied at the time of pit filling and 5-10 ml chloropyriphos (36% EC) should be applied by mixing in water to each tree up to the age of 2 years. Beetles damage the developing leaves. These can be controlled by 2-3 sprays of 0.01% monocrotophos at 20 days interval after the appearance of the insect. Gall insect (*Thomaimiana salvadorae*) makes galls on the branches and checks extension growth of the plant. Phyllody or witches broom is a common disease in *S. persica* in Rajasthan.

Flowering, Fruiting and Yield

Fruiting in *Salvadora* starts after attaining an age of 6-7 years. Flowering takes place during January-March on more than one year old healthy twigs. Light yellow coloured flowers emerge in clusters (Plate 17.4). Pollination takes place naturally or through the visiting

Plate 17.4. Flowering twig of Pilu

insects at certain extent. The fruits ripen during May-June in Rajasthan. On ripening, the small, globose, aromatic berries become yellow to red. Pilu fruits are very soft and have delicate skin. Therefore, harvesting should be done manually and carefully. To reduce the harvesting cost, whole clusters are generally harvested. About 5-10 kg ripe fruits can be obtained every year from a tree.

Harvesting and Post Harvest Management

Pilu is a non-climacteric fruit which is harvested after full ripening on trees. It is an small fruiting species in which fruits are borne in bunches. The whole bunch of pilu is generally harvested when more than 80% fruits are pinkish in colour. The fruits are consumed fresh. The fruits are delicate and highly perishable and cannot be stored for long duration.

Processing and Value addition

Apart from fresh consumption, pilu fruits can be processed into juices and beverages. Ripe fruits can be converted into RTS, nectar and squash at limited scale however, various other value added products can also be attempted from the ripe pilu fruits.

Future Research Needs

Pilu is an under exploited fruit where no systematic research has been conduced so far. Visualizing the prospects of the species particularly in the degraded and waste land in arid and semi arid region, information on following aspects needs to be urgently generated.

1. Selection of high yielding cultivars with quality fruits.
2. Development of vegetative propagation techniques which can be commercially used for production of true-to-type saplings.
3. Development of complete agro-techniques for sustainable production of pilu in different areas.
4. Identification of pest and diseases and their management practices for successful crop production.
5. Development of suitable post harvest management and value addition techniques.

REFERENCES

Bhandari, M.M. (1995) Flora of Thar Desert.

Pareek, O. P. and Vishal Nath (2006) Arid and minor fruits. In: Text Book of Horticulture, MPH, New Delhi (In Press)

Pareek, O. P., Vashishtha B. B., Vishal Nath, Samadia D.K. and Singh R.S. (1997) Horticultural resources in Thar Desert. Proceedings: National Seminar on management of natural resources in Rajasthan. RAU, Bikaner: 4-9.

Singh, R. S. and Tiwari, J. C. (1995) *Salvadora* : A multipurpose tree of Arid tract. *Decoo Mirror*, 2 (2) : 19-22.

Tariq, E., Ahmad, E. and Malik, A. (2005) Structure determination of salvadorin, a novel dimeric dihydroisocoumarin from *S. aleoides* by NMR spectroscopy.

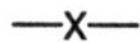

18

KHEJRI

Among the tree species, khejri (*Prosopis cineraria* L.) is scriptures in Hindu mythology. It is known as *Sami* in Sanskrit literature and trees are treated sacred as far as its religious value in Vedic era is concern. Its wood was used to kindle sacred fire for *Yagna* during that period. Mention about *Sami* has been made in ancient literature such as Rig Veda Samhita, Atharv Veda, Satpath Brahman, Mahabharata, Ramayana, etc. In the Vedas, it is mentioned that khejri (*Sami*) was used for generating fire or as Samidha at various occasions which clearly indicates that the plant has hard wood which can be used as fuel as well as for different agricultural purposes. Similarly the plants of khejri was used to hide the Bow of Arjuna 'the Gandiv' during the agyantvas of Pandava which indicates that the plant of khejri use to be very big and gigantic in early days. Thus it can be said that there were thick stand of khejri in past, during the ancient period which were put to various uses by ethinic races of that period. The khejri plants are highly valued for timber, fodder and fuel yielding species in the agro-foresting system of western India.

Apart from fuel and timber, khejri leaves are important source of fodder in desertic areas. It is popularly known as 'loong' in Rajasthan. Green nutritive leaves of khejri contains 11.9-18.0, 13-22, 43.5, 6-8, 2.9, 2.1 and 0.4 per cent crode protein, crude fiber, nitrogen free extract, ash, ether extract, calcium and phosphorus, respectively with a high calorific value (Bhimya *et al.*, 1964, Mathur, 1976 and Bhandari *et al.*, 1979) which are considered to be good fodder for animals. Twigs and branch woods of lopped trees of khejri are a major source of fuel for the rural population (Vishal Nath *et al.*, 2000). Unlopped trees of khejri produce pods (*sangria*). When immature, these pods are used for vegetable and preparation of pickles as fresh as well as after dehydration (Vishal Nath *et al.*, 2000, Pareek and Purohit, 2002, Vishal Nath, 2004). The ripe sweet and edible pods (*Khokha*) are used to prepare flour and many value added products Vishal Nath, (2004). Duhan *et al.*, (1992) recorded 18, 56, 26, 2, 0.4, 0.4 and 0.2 per cent crude protein, carbohydrates, crude fibre, fat, phosphorus, calcium and iron, respectively in immature pods while ripe pods contain 9-14 per cent crude protein and 6-16 per cent sugar (Arya *et al.*, 1991). Flowers, barks and other parts of the trees are also used for mechanical purposes.

Khejri tree holds an important place in the economy of Indian desert and other arid and semi arid parts of the world. This is only leguminous tree which grows well against all odds of fragile arid agro-ecosystems.

Origin and Taxonomy

Khejri (*Prosopis cineraria* (L.) Mac bride) is appears to be the native of western and southern part of India. Pallen strategraphic studies by Singh, (1970) revealed abudance of khejri in western Rajasthan for more that 10,000 years where it has a higher ecological status (Vishnu Mittre, 1975). It is an important member of family leguminoceae sub family mimosoideae. The oldest name of this species is therefore, *Mimosa cineraria* Linn. (Species Plantarum : 517 : 1753). Later Linnaeus himself treated this taxon under genus *Prosopis* and called it *Prosopis spicigera* Linn. (Mant : 68; 1767 and Flora of British India 2:288; 1878). Under the Botanical Nomenclature Code, Macbride (1919) validated the above combination under genus *Prosopis* and species *cineraria* and thus the correct name of the species is *Prosopis cineraria* (L.) Macbride. *Mimosa suma, Acacia suma* in earlier literature and *Prosopis cineraria* (L.). Druce in the recent literature are seems to be the synonymous of *Prosopis cineraria* (L) Macbride.

Khejri is known by several vernacular names in different parts of the Country. It is known as Khejri (Rajasthan), Janti, Chouka (Delhi), Jhind, Jhand, Jand (Punjab and Haryana), Banni (Karnataka), Sumri (Gujarat) and Kandi (Sindh). It is a large, much branched medium size tree with dark green foliage. Trunk dark grey, bark splitted with deep fissures and rough but attached to the trunk. Branches slender, glabrous with small compressed, straight or slightly curved scattered thorn of 3-6 cm length, pinnae mostly 2 pairs with a gland between each pair, opposite, 2.5-8.0 cm long, leaflets 8-12 pairs, oblong, obliquely rounded mucronate at apex, unequal sides, bare rounded and very oblique 12 x 4 mm size leaf let. Flowers-axillary, yellow in slander spike, 7-11 cm long mostly solitary or in terminal panicle (Plate 18.1). Calyx 5 x 1.5 mm long, cup shaped, faintly 5 toothed, corolla yellow, 3 mm long, stamen recurved, filament 3 mm long, pod straight or slightly curved at distal end, cylendric, glabrous, forulose, 10-20 cm long, thick with brown pulp. Seeds 10-15 embedded in the pulp, 3-8 mm long dull brown and oblong (Bhandari, 1978).

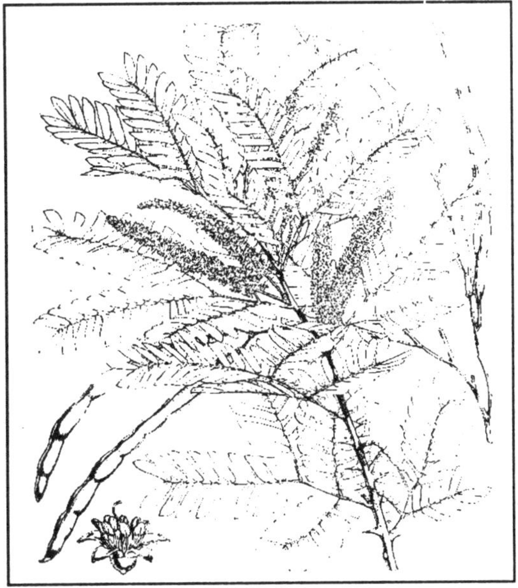

Plate 18.1. Khejri infloresence and twig

Area and Distribution

Khejri plants are found naturally growing in the arid and semi arid regions of the country. No systematic plantation of the species has been made for fruit or vegetable production points of view. What so ever plantation has been made, that it as agro-forestry or for wasteland utilization purpose. Therefore, the précised area as horticultural crop of this species is not available. In recent years, owing to development of different production technologies and fast growing high yielding varieties/genotypes and vegetaively propagation techniques, more area is expected under this unutilized tree species. The khejri is distributed in warmer parts of the Asia continent. Naturally occurring khejri has been reported from Afghanistan, Iran, Pakistan and India (Puri *et al.*, 1964).

In India, distribution of Khejri is discontinuous, however it is found to be concerntrated mainly on alluvial and sandy undulating plains of western Rajasthan (Satyanarayan, 1964) where as its scattered population can also be seen in other habitats in the state of Haryana, Delhi, Gujarat, Madhya Pradesh, Karnataka and Maharashtra. In Western Rajasthan, it is distributed from transect of Aravalis in the east to sandunes of Jaisalmer in the west with decrease in population pattern. Mann and Shankar (1980) reported the distribution of Khejri in Western Rajasthan (Table 18.1).

Table 18.1. Density distribution of Khejri in different habitat of Rajasthan

Region	Habitat	Land use	Soil depth (m)	Density range (Plant/ha)
Pali-Siroli-Jalore	Old alluvial flat-Lower pediment	Grazing land	0.6-1.0	10-250
Barmer-Jaisalmer	Rocky pediment-Dune	Grazing land	0.5-1.0	10-90
Nagaur-Bikaner	Old alluvial flat-Sandy undulating plain	Graging-cultivated land	0.8-4.0	10-170
Bikaner-Jaisalmer-Jodhpur	Sandy undulating-Pedimant Plain	Grazing-old fallow land	0.1-1.0	5-42
Ganganagar-Bikaner-Churu	Sandy undulating-Alluvial Plain	Old fallow-Arable land	0.4-1.1	10-250

Soil and Climate

Khejri is a common plant of arid region which is naturally available on most of the land forms except the hills and saline depressions. The laxurient growth of khejri has been seen in sandy soils of Barmer, Sikar and Jhunjhunu district of Rajasthan where the soil depth is comparatively high. Therefore, the best suitable soil for khejri would be sandy or sandy loam soil with more than 2 m depth and without injurious salts.

Khejri trees grows well under arid and semi arid waste lands and even sand dunes. The plants are very hardy and have set their morphology and growth to withstand highest climatic odds. The new foliar growth, flowering and fruiting in Khejri takes place during March-June which is the most dry period in arid region. The plant sheds their leaves during winter. Being well adopted to prevailing climatic conditions of arid region and its wider ecological amplitude, the khejri tree may be grown from 150 mm to 500 mm rainfall zone, however for optimum density 350-400 mm rainfall is considered sufficient (Saxena, 1980). The water logging for even few hours near khejri tree is harmful which cause severe leaf fall during August-September month. Therefore, high rainfall areas are not suitable for khejri cultivation.

Plant Genetic Resource Management and Varieties

Owing to continuous sexual regeneration, a lot of variability has been naturally created in khejri. Variability with respect to seedling vigour and growth have been reported from CAZRI Jodhpur (Muthana, 1980), Pareek and Vishal Nath (1997) reported variability in fruit and pulp characteristics of ripe pod and Vishal Nath *et al.*, (2000) reported the variability in vegetable traits (Plate 18.2) and nutrient content. Ripe pod (*Khokha*) quality in khejri genotypes vary with respect to taste (flat, bitter, acrid, sweet) where as vegetable quality vary with tenderness (tender, semi-hard and hard), seed number and size and content of protein and mineral.

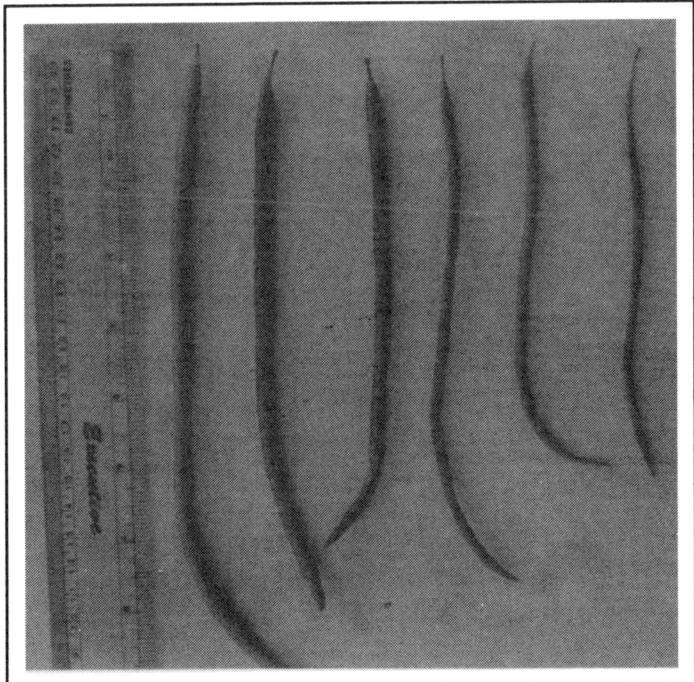

Plate 18.2. Variability in immature khejri pods

Therefore, different types of khejri genotypes are available in the variability zone. As per colour of pods, two type one with red or marron colour and the other with green or light green colour pod has been identified at CIAH, Bikaner (Anon., 1998). Muthana (1980) however, reported two types of khejri as far as growth habit and morphology of plant is concern. The one type has spines all over the branches with slightly up right growth, hard twigs and less foliage locally known as 'Khejra' in Shekhawati region. The other type has thronless twigs in upper and terminal branches with dense foliage, developing branches, locally called as 'Khejri'.

Survey in the variability region has resulted into collection of some promising genotypes of khejri. Vishal Nath *et al.*, (1993) reported the variation in ripe pod characterstics of sweet genotypes of khejri from Rajasthan (Table-18.2). *In situ* evaluation of 30 genotypes of Khejri for culinary purpose reveal that genotype 1, 2, 3, 13, 14, 20, 21, 26 and 29 have better characteristics (Vishal Nath *et al.*, 2000) (Table-18.3) and can be vegetatively propagated for commercial cultivation. Pareek (1997) mentioned that 200 plus trees of khejri has been identified and marked in the variability region by NRC for Arid Horticulture, Bikaner and emphasized that it is important to select location specific varieties of khejri for various purposes. The leads obtained from the survey and *in situ* evaluation in the variability in the region needs to be exploited for development of high yielding, quality podded variety of khejri.

Table 18.2. Variation in ripe pod characteristics of sweet podded genotypes of *P. cineraria* in various zones of Rajasthan

Agroclimatic zones	Pod length (cm)	Thick-ness (cm)	No. of seed per pod	Seed weight (g)
Arid Western Plain	9.0-39.0	2.5-3.4	3-35	0.5-1.2
Irrigated North Western Plain	10.0-31.0	1.7-2.3	6-25	0.7-1.5
Transitional Plain of Inland Drainage	12.0-25.0	1.3-3.0	10-20	0.4-1.2
Transitional Plain of Luni Basin	19.0-25.0	1.6-2.4	9-22	0.6-1.5
Semi-arid Eastern Plain	11.0-36.0	1.8-2.5	11-30	0.9-1.8

Plant Propagation and Nursery Management

Khejri is mainly propagated through seeds. However, vegetative techniques have also been attempted. Seed propagated plants have variation in their population and very slow in growth but due to ease of technique it is commercially used.

Table 18.3 : Variation in characters of immature pods (*Sangri*) of khejri (pooled data-1995-97)

Tree	Pod Length (cm)	Thickness (cm)	No. of seed Pod^{-1}	Weight of seed (mg)	Appearance	Tenderness	Taste
1.	20.1	0.34	19.8	30	Gr, Gl, R	Te, Fl	Flat
2.	22.8	0.39	17.1	30	Gr, Gl, F	Te, Fl	Sweet
3.	23.2	0.37	19.3	40	Lg, Gl, R	Te, Fl	Sweet
4.	23.1	0.40	20.0	40	Gr, Gl, F	St, Lg	Less Sweet
5.	13.5	0.20	16.0	30	Gr, Gl, F	Te, Fl	Slightly sweet
6.	17.4	0.34	19.9	20	Lg, S	Te, Fi	Slightly sweet
7.	17.0	0.36	15.9	40	Gr, Gl, F	Te, Fi	Slightly sweet
8.	13.5	0.35	12.6	30	Gr, C, Lo	H, Mf	Less sweet
9.	13.4	0.39	12.5	30	Lg, S, R	St, Lg	Less sweet
10.	23.3	0.35	18.5	30	Gr, Gl, F	Te, Mf	Flat
11.	16.1	0.45	13.7	40	Lg, Gr, R	Te, Kg	Less sweet
12.	14.8	0.35	19.0	30	Gr, Tr, Ro	Te, Fi	Acrid
13.	21.9	0.42	18.6	50	Lg, Gl, R	Te, Lg	Sweet
14.	7.7	0.44	5.3	50	Gr, C, Lo	Te, Fl	Sweet
15.	15.1	0.34	11.5	30	Lg, Gl, F	Te, Fl	Slightly sweet
16.	16.9	0.36	12.5	20	Gr, Gl, F	St, Lg	Slightly sweet
17.	18.2	0.40	19.0	30	Lg, Gl, F	Te, Fl	Acrid

Contd...

Table 18.3. Contd...

18.	15.6	0.39	12.2	40	Lg, Tr, Ro	Te, Lg	Slightly sweet
19.	15.6	0.26	15.0	20	Gr, Tr, Ro	Te, Fl	Flat
20.	20.3	0.38	14.5	30	Gr, Gl, F	Te, Fl	Slightly sweet
21.	22.3	0.45	22.8	50	Gr, Gl, R	Te, Lf	Less sweet
22.	18.2	0.42	18.2	50	Lg, Gl, R	Te, F1	Flat
23.	18.3	0.47	15.1	50	Lg, Gl, F	Te, Lg	Slightly sweet
24.	14.2	0.44	17.1	30	Lg, Gl, F	St, Fi	Slightly sweet
25.	13.2	0.40	16.8	30	Gr, C, Lo	Te, Lf	Acrid
26.	14.8	0.35	20.1	30	Gr, C, Lo	Te, Fl	Less sweet
27.	16.5	0.38	15.1	20	Gr, Tr; Ro	Te, Fl	Slightly sweet
28.	19.7	0.41	19.9	40	Lg, Gl, F	St, Fl	Flat
29.	18.9	0.51	14.7	50	Lg, Gl, R	Te, Fl	Sweet
30.	15.8	0.42	16.3	30	Gr, Tr, Ro	Te, Lf	Slightly sweet
Mean	17.4	0.38	16.3	35	-	-	-
SD	3.7	0.06	3.6	10	-	-	-
SEm±	0.7	0.01	0.7	2	-	-	-
CV(%)	21.3	15.63	21.7	28	-	-	-

Appearance : Gr.-Green; Gl-Glabrous; R-Round; F-Flat; Lg-Light green; S-Smooth; C-Curled; Lo-Lobed; Ro-Rough.
Tenderness : Te-Tender; St-Semitender; Tr-Tringed; Fi-Fibrous; Lf-Less fibrous; H-Hard; Mf-Medium fibrous, St-Semitender.
Source : Vishal Nath et al., (2000).

Seeds are soaked for 24 hours before sowing in polythene bag or in nursery beds. Saxena (1980) reported about 80% seed germination with maximum germinability after 3 weeks of sowing. Muthana (1980) reported that direct sowing of seed in field gave poor germination and survival whereas seedlings raised in tubes gave 69% field establishment with better growth and performance subsequently.

The khejri plant can be asexually propagated by cutting, layering and budding at limited scale. Arya *et al.*, (1993) reported very good success in soft wood cuttings of *Prosopis cineraria* under mist however, Pareek *et al.*, (1996) reported only sprouting in hard wood cuttings of khejri. Application of 6000 ppm BA resulted in 32% sprouting success and very poor survival due to development of thin and fragile roots. The age of shoot taken for cutting has great significance in rooting and ultimate survival of khejri plant. Arya *et al.*, (1994) reported that cutting from 6 month old shoots gave higher success than 8-10 month old shoots. Rooting medium and environment also play significant role in rooting of khejri cuttings. Arrangement of mist (Arya *et al.*, 1994) and hormonal treatment with highly porus nutritive rooting medium (Morsink and Smith, 1974), have been found to keep the cuttings alive for longer period and induce rooting in khejri, respectively.

Khejri can also be propagated through patch budding during March with more than 75% success on khejri (*P. cineraria*) rootstock. Pareek and Purohit (2002) described the detailed techniques for patch budding in khejri. They reported that *P. cineraria* is the best rootstock with 76% success. The days taken for sprouting on *P. cineraria* rootstock is 25.8 days (Pareek and Purohit, 2002). The complete technique involves, 1) Pruning of senile mother tree with proven records of producing good quality edible pods (elite plant) during December to induce new growth-bearing juvenile buds. 2) Raising of rootstocks of *P. cineraria* by sowing the seed in nursery bag or polytubes. The pencile thick rootstock (1.0-1.1 cm diameter) of 1-2 years age should be used for budding. 3) Budding should be performed through patch method during March. Pareek and Purohit (2002) mentioned that patch budding techniques can also be utilized for top working of the seedling plants.

Rawat *et al.*, (1983) got 70% success in air layering where as Solanki *et al.*, (1986) got 100% rooting success in air layering. The technique seems to be all right but Pareek and Purohit (2002) reported less than

10% survival in the air layered plants. Air layered plants from juvenile shoots of pruned plant produce better rooting (Solanki *et al.*, 1986). Goyal and Arya (1984) and Kacker *et al.*, (1991 and 1992) reported that khejri can also be propagated through micro-propagation. However, Pareek and Purohit (2002) emphasized that root system of cutting, air layering or micro-propagated plants lack storing tap root system and may not survive or perform well under extreme arid agroclimate. Therefore, asexual propagation techniques with strong tap roots i.e. budding or grafting should be adopted for khejri propagation.

Planting

Khejri is a tree species which produces many branches at primary, secondary and tertiarty level. As a result, the plant some times become huge and gigantic. The growth of the plant is dependent upon the soil condition and rainfall of a particular location. In heavy soils, the plant grow much whereas in sand dunes and hard soils, the plant do not spread much. Hence, the planting system and distance in khejri is location specific and purpose oriented. In agro-forestry system, it can be planted at 3 x 3 m, 5 x 5 m as sole crop in arid region. As mixed crop 8-10 m spacing in considered optimum. As such no systematic work on spacing and system of planting of this crop has been conducted and standardized. For fruit production from vegetatively propagated improved varieties 5-6 m spacing appears to be appropriate.

Training and Pruning

Presently, the khejri is mainly grow wild or if at all planted, it is under agro-forestry system. What so ever *sangria* or *khoka* is harvested, it is not from the systematic plantation of improved types. Now, for growing khejri for fruit production purpose by using the improved genotypes and vegetative propagation techniques, systematic training and pruning is of utmost importance. Seedling plants are very slow growing which needs frame working for 2-3 years. The plant bears fruit in the outer peripheri where abundance of solar radiation is available, therefore the training system should aim to provide a shape to the plant to harvest more light for flowering and fruiting. The apical dominance in the plant should be broken at 150-200 cm from the ground level and side branches between 150-175 cm should be promoted as main scaffolds. Well spaced secondary and tertiary branches should be decided and allowed to form the strong frame of the tree.

Khejri plant responds to pruning. At the same time flowering and fruiting in khejri takes place on one year old branches. No systematic study on pruning of khejri has been undertaken so far for fruit production. For fodder production, however, trees are lopped following a systematic lopping schedule in Western Rajasthan (Saxena, 1980) which starts from middle of November each year and continue upto middle of January. While lopping, older branches are removed and 1-2 years old branches are retained for making canopy. The tallest branch of 1-2 year old age is left with some foliage as 'Flag' in case of complete lopping in Shekhawati region of Rajasthan. For fruit production alternate tree or side of the tree should be pruned each year to have sufficient number of mature shoots for flowering.

Orchard Management

Although no systematic plantation of khejri orchard is common any where in the world however, the informations pertaing to water and nutrient management and floor management below the tree and plant protection put together to develop a comperhensive information for better management of the khejri plants in the arid region.

Nutrient Management

Khejri plant is a leguminous tree which fixes the atmospheric nitrogen. In arid region, generally khejri plants are not fertilized however, application of organic manure to new plantation improves growth. For systematic orcharding, nutrient doses and time of application needs to be standardized for this under exploited crop.

Water Management

Being a hardy species, Lahiri (1964) suggested that an established plant of *Prosopis cineraria* may not be affected with moisture stress despite the low moisture status in the upper layer of the soil. He further mentioned that 70-80% relative turgidity is maintained by leaves during March-May when the plant remain under floriferous condition. This indicates that, the khejri plant do not require irrigation after plant establishment. Lahiri (1980) recorded 'noon day drop' because of mid-day closure of stomata due to water stress or increase in incident radiation or temperature mediated rise in intercellular space carbondioxide which indicates the adaptive mechanism of the khejri under stress environment. Owing to spatial difference in root zone of khejri and other ground cover crops and capacity of the plant to maintain high moisture regime beneath the canopy even after light shower (Lahiri, 1977), the khejri tree do not suffer from water stress despite low soil moisture in the upper soil layers. Mulching with local weeds help increased soil moisture in the rhizosphere of khejri tree. The ability of plant organs to absorb moisture from light showers or vapour or dew (Lahiri, 1980) acquires importance in arid and semi arid regions as it often helps the plant to by pass the ragours of soil moisture stress.

Orchard Floor Management

Khejri is a leguminous tree with prinnatly compound leaves which enrich the soil and allows a lot of vegetation and crops under its shade (Lahiri, 1980, Singh and Lal, 1969). Although no systematic plantations and orchard floor management work has been taken in khejri, the farmers grow variety of crops below the khejri in Rajasthan. Mustard, wheat, gram are the major crop of rabi and ground nut, bajara, sesamum, guar, mooth, etc. are the major kharif crop which are grown with naturally growing khejri tree. Vishal Nath *et al.*, (1993) reported that khejri is well versed tree species of arid agro-ecosystem where number of crops can be grown along with khejri plants (Table 18.4).

Table 18.4 : Agro climatic zone; rainfall and annual crops with khejri in Rajasthan

Zone (districts covered)	Annual rainfall (mm)	Crops with *P. cineraria* in the agroforestry system
IA-Arid Western Plain (Jaisalmer Bikaner, Barmer, part of Jodhpur, Churu)	100-300	Pearl millet, kharif pulses, guar
IB-Irrigated North-Western Plain (Sri Ganganagar, Hanumangarh)	100-350	Wheat, mustard, gram, cotton, guar, pearl millet
IIA-Transitional Plain of Inland Drainage (Nagaur, Sikar, Jhunjhunu, part of Churu)	300-500	Pearl millet, guar, kharif, pulses, gram
IIB-Transitional Plain of Luni Basin (Jalore, Pali, part of Sirohi, Jodhpur)	300-500	Pearl millet, guar, mustard, kharif pulses, wheat, rape, til
IIIA-Semi-arid Eastern Plain (Ajmer, Jaipur, Dausa, Tonk)	500-600	Pearl millet, jowar, barley, wheat, mustard, gram, til
IIIB-Flood Prone Eastern Plain (Alwar, Bhratpur, Dhaulpur, part of Sawimadhopur)	500-650	Pearl millet, jowar, wheat, gram, mustard
IVA-Sub-humid Southern Plain and Aravalli Hills (Bhilwara, Rajsmand, part of Udaipur, Chittorgarh, Sirohi)	500-700	Maize, jowar, til, barley, groundnut, wheat, gram, mustard
IVB-Humid Southern Plain (Banswara, Dungarpur, part of Chittorgarh, Udaipur)	500-1050	Maize, paddy, kharif, pulses, wheat gram
V-Humid South-Eastern Plain (Bundi, Kota, Bara, Jhalawar, Part of Sawimadhopur)	650-1000	Jowar, maize, til, wheat, gram, mustard

Saxena (1980) reported that for growing wheat, trees are lopped during end of October to first week of November. Being a deep rooted drought hardy species, khejri tree is not much affected by weeds. *Prosopis cineraria* has been proved to be the best suitable species for mix plantation in different habitats of arid and semi arid ecosystem. Saxena (1978) reported 14 plant communities in association with *Prosopis cineraria,* the important among are *Ziziphus mauritiana* var. *rotundifolia, Capparis decidua, Salvadora oleoides, Tecomela undulata, Balanitis aegyptiaca, Caligonum polygonoides, Leptademia pyrotechnica,* etc.

Disease and Pest Management

Khejri plant is generally free from diseases and pests, however, some pests of minor importance have been noticed particularly on the unmanaged plants. The pod borers have been found to attack during flowering and the developing pods at immature green stage. The larve of the insect punctures the pod (Plate 18.3) and feeds on seed which makes the pod unusable. The pest can be controlled by sanitation in the vicinity of plant and removal of dense branches for proper light and air circulation. Foliar spray of monocrotophos 0.01% twice at 7 day interval just after the fruit set have been found effective to minimize the pest attack.

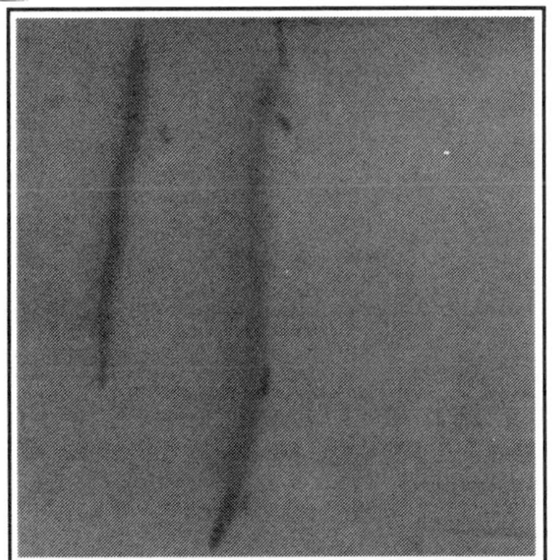
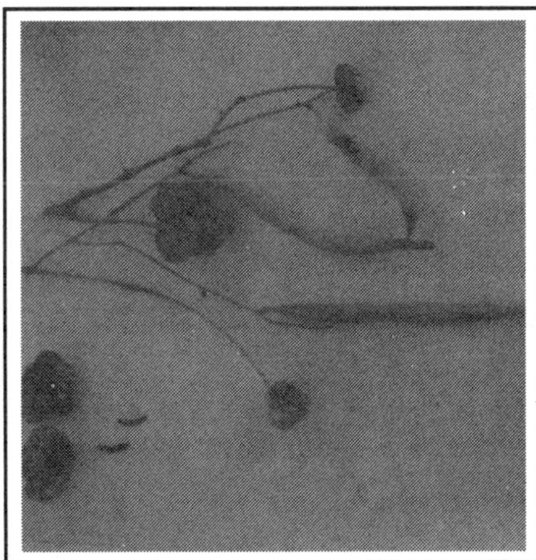

Plate 18.3. Infestation of pod borer on immature pods of khejri
(i) Pod with caterpillar (ii) Affected pods with galls

Shoot gall is another common problem in khejri. The gall is formed on the tender shoots and the flowering panicles due to the insect attack. The developing insect hide inside the gall and restricts the growth of the shoot and or fruits. Parihar (1980) reported this as gall fly caused by *Lasiopteraria* sp. which is a serious pest of khejri. The fly causes galls along the branches and rachis of leaves. The gall measures 1-2 cm length and 5-10 mm width and develops round or oval shape. The galls are green and tender at the time of formation which gradually turn brown and harden. When it is cut 1-3 maggots of the fly can be seen in each gall (Plate 18.4). Adult fly escapes from the gall by making minute pin holes. Field sanitation, removal of affected portion well in time and spray of systemic insecticide should be done for the management of pest. Parihar (1980) reported that gall fly is predated by ants (*Doryllus labiatus*) which reduces the pest population. 2-3 sprays of endosulfan (0.08%) at 7 days interval has been found effective. Nocturnal Chaffer bettles (*Holotrachia sp.*) cause foliar damage during pre monsoon period. High wind velocity and rain at the time of emergence delay the emergence of bettles (Annon, 1970). Pal (1977) reported that *Schizonycha, Aserica* and *Serica* species of chaffer bettle are also causing damage besides *Holotrachia*. Carbaryl 0.15% or Fenitrothion 0.05% have been found effective against the pest. Provision of light traps or petromax have been found effective to prevent the bettle. Desert locus (*Schistocera gregaria* Forsk.) and termite are also infesting khejri trees (Purthi, 1969 and Parihar, 1978). These can be controlled by systematic management.

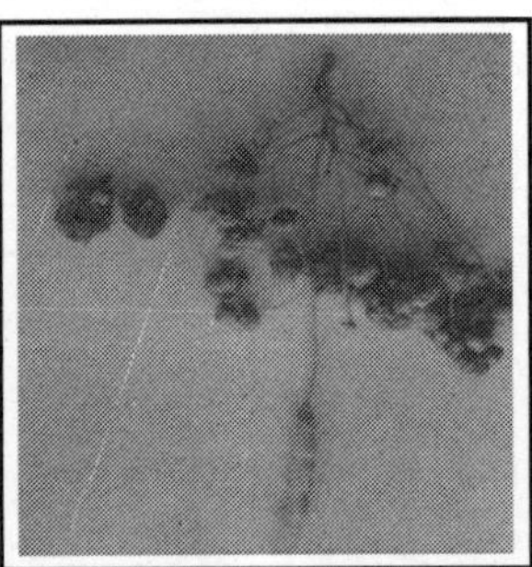
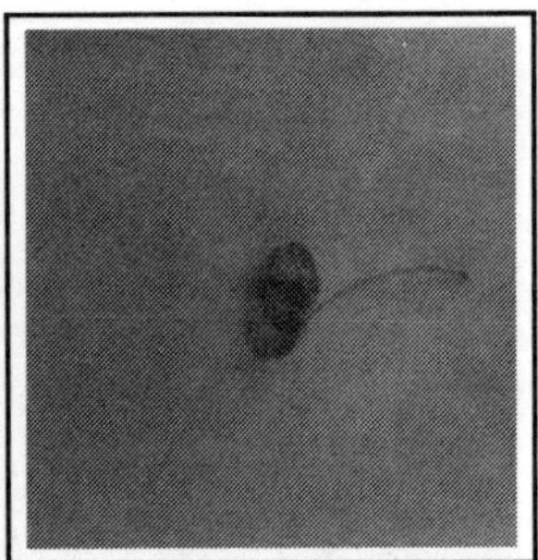

Plate 18.4. Khejri plant affected with shoot galls (i) view of the gall affected plant (ii) whole infrorescence converted into gall (iii) T.S. of the shoot gall

Flowering, Fruiting and Yield

Khejri flowers are borne on sufficiently mature almost one year old branches. The newly emerging branches after lopping in the same year do not bear the flowers and fruits. It is therefore essential to maintain optimum number of mature fruiting branches on a tree. Plant comes under floriferous condition from February to June. The flowering takes place during February-March and fruit (pods) ripes by June.

The physiological maturity in pods for culinary purpose takes place during April-May depending upon the cultivar (Plate 18.5). Green pods suitable for vegetable purpose are locally called as "Sangri' or 'Hangri' (Saxena, 1980). The ripening of pods takes place about 140-150 days after fruit set. The ripe pods are locally called as 'Khokha' which is eaten by the children and numads. Pre-monsoon wind and thunder storm breaks the pod and thereby dispersal of seeds takes place. Squirel, Gerbil and few birds which attack these pods also act as means of seed dispersal.

Plate 18.5. Plant of Khejri in bearing stage.

Almost all parts of the khejri tree is used and therefore, the yield of plant can be in various forms. For fodder, the dried leaf yield has significance while for fuel the lopped wood is important. For vegetable, the yield of pod has major significance. Kaul and Jain (1967) suggested an equation based on DBH of the tree to predict pod and fuel wood yield in khejri. Pareek *et al.*, (1998) reported that a full grown tree of khejri, if lopped, yields 50 kg dry wood as fuel and 30 kg dry leaves as cattle food and if not lopped, yields 5-7 kg dried pods for vegetable purpose. Similarly 20-25 kg mature dried pods of khejri can be harvested from a grownup plant of improved type which can be used for making *Khokha* flour.

Harvesting and Post Harvest Management

Khejri harvesting is done for culinary purpose (Immature, green, tender pods) and for *khokha* (ripe pod) purpose. Vishal Nath (1999) reported that 15-20 days old pods which have not developed bold seed, should be harvested for vegetable purpose as well as for dehydration. For *khokha* purpose full ripe pods having brownish-yellow colour should be harvested. Ripening of pods takes 110-120 days after fruit set. After harvest, the green pods are sorted as per their length, shape, etc. Vishal Nath (1999) reported that 15-20 cm long, soft skinned, less acrid and seeded pods have better market value as dried *sangri* (Plate 18.6). The ripe pods are also graded as per their sweetness and pulp content to make *khokha* flour. The dehydrated *sangria* can be utilized for various value added products as well as can be stored for longer period. The method of dehydration of *sangria* have been standardized by Vishal Nath (1999, 2004). Vishal Nath (1999) reported that blanching of immature pods in 2% salt solution produces high quality *sangri*. The detailed flow sheet for dehydration of *sangri* has been given in Fig. 18.1

Collection and cleaning of tender pods

↓

Blanching of pods in 2% salt solution for 5 minutes

↓

Removal of blanches pod and spreading on
clean cloth / paper / dehydration floor

↓

Covering of pods with paper/thin cloth to avoid dust

↓

Packing of dried Sangri in polythene packets and storage

Fig. 18.1. Flow sheet for *sangria* dehydration

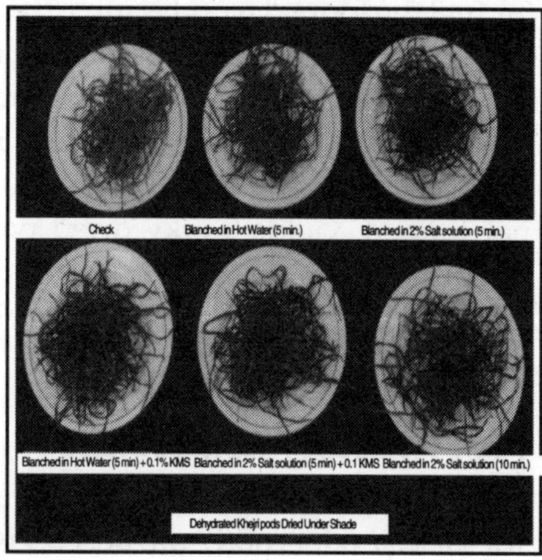

Plate 18.6. Dehydrated immature pods of khejri (*Sangria*)

Processing and Value Addition

Various delicious dishes can be prepared from dehydrated or fresh *sangri*. Tewari (1993) enumerated the recipes and methods to develop different value added products from dried and green *sangri*.

(i) Product from dried sangria

Sangri with onion

To prepare onion based product, take 200 g fine chopped onion and fry it in 70-80 ml oil in a sauce pan till the development of golden brown colour. Add 10 g each of turmeric, coriander, chilli and dried mango powder along with salt (to taste). Add 200 g rehydrated *sangria* into it and cook it in the pressure cook for 4-5 minutes.

Lal-ki-sangri

To prepare lal-ki-sangri, soak 4 tsp Bengal gram flour in one lit, butter milk. Heat 50-60 ml oil and add ½ tsp fenugreek seeds, a pinch of asafetida and 2-3 whole red chilli in it. Heat until spluttering. Add butter milk mixture and keep stirring. Add ½ tsp turmeric powder and 200g dehydrated sangria and boil it for 10-15 minutes.

Panchkutta vegetable

It is one of the most prestigious vegetable of western Rajasthan, which is prepared by mixing fruits of 5 traditional food plants viz., sangria, kumati, gonda, ker and kachari. Soak 50g sangria, 35g ker, 50g kumati, 25g gonda and 10 pieces of kachari (all dried) for overnight and boil with salt and turmeric powder for 10 minutes. Drain extra water and wash 4-5 times in tap water. Keep the boiled material under sun for 5-10 minutes. Heat 250 g oil in pan and add 2 pinches of asafetida, 2 tsp turmeric powder, 2 tsp coriander powder, and pachkutta vegetable along with salt and red chilli powder (to taste). Fry the whole material and cook it for 10-15 minutes. Turn it off from flame and cover with lid. By keeping it in dry place and turning up side down daily the vegetable can be stored for 7-8 days.

(ii) Products from green sangria

Sangri vegetable

To prepare vegetable from green *sangria*, select tender pods and blanch it for 4-5 minutes in boiling water. Mix 2 tsp each of red chilli and coriander powder and 1 tsp each of turmeric and green mango powder and salt (as per taste) with 150 g blanched *sangria*. Keep the material for 10 minutes. Heat 3 tsp oil in a saucepan and add a pinch of asafetida. Add the vegetable and stir until it is evenly fried. Turn off flame and cover the pan.

Sangri pickle

Sangri pickle is very common in Marwar region. During the month of April-May tender, sweet type pods are used for pickling. Cut 1 kg *sangria* into small pieces, blanch and dry for 1 hour. Heat 300 ml oil and add 30 g red chilli powder, 15 g turmeric, 75 g green mango powder, 35 g funnel, 40 g fenugreek seeds, 125 g salt into blanched sangria. Roast for 5 minutes and add 1 tsp acetic acid. Store it in a clean container.

(iii) Products from ripe khejri pods

Apart from the use of *sangria* at immature tender stage, the ripe khejri pods locally called *khokha* can also be utilized for various value added products. *Khokha* flour, which is prepared by grinding ripe pods in May-June, can be used for making chapatti, biscuits, confectionaries, etc. At CIAH, Bikaner attempts have been made to develop *khokha* powder and its use in various value added products. Cookies of different recipes using *khokha* powder (with or without seeds) have been prepared (Plate 18.7). The recipe consists of per cent replacement of wheat maida by *khokha* powder. 10, 20, 30 and 40 per cent maida was replaced with *khokha* powder and product was prepared in automatic bakery oven. The other ingredients included for making cookies were constant amount of ghee, water, etc. Amount of sugar was calculated by considering the available TSS in *khokha* powder. In general, the sugar was kept at lower side. The cookies were organeleptically analyzed (Amerine *et al.*, 1965). Scores clearly indicates the maximum acceptability of cookies prepared by replacing 10 per cent maida with *khokha* flour without seed among the treatments. The chemical composition of cookies has also been analyzed.

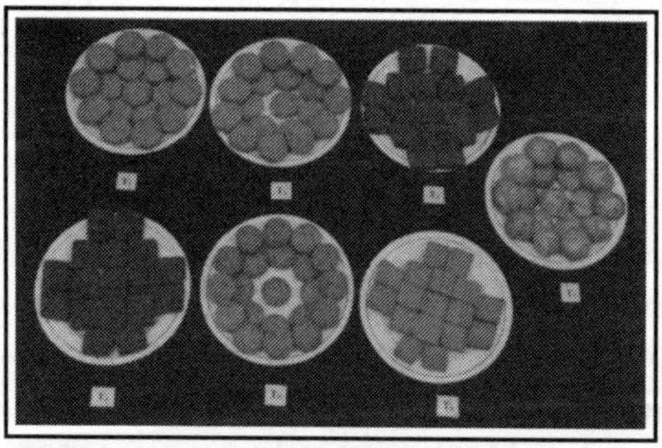

Plate 18.7. Cookies prepared from the ripe khejri pods (*Khokha*)

REFERENCES

Amerine, N. A., Pangborn, R.M. and Roester, E.B. (1965) Principal of sensory evaluation of foods. Academic ress. London

Anonymous (1970) 20 years of Agricultural Research in Rajasthan. Entomology. Department of Horticulture, Rajasthan : 171

Anonymous (1998) Annual Report, NRCAH, Bikaner

Arya, S., Toky, O. P., Tomar, R., Singh, L. and Harris, P. J. C. (1993) Seasonal variation in auxin induced rooting of *Prosopis cineraria* stem cuttings. *The Inter Tree crop Journal*, 7 : 249-59.

Arya, S., Tomar, R. and Toky, O. P. (1994) Effect of plant age and auxin treatment on rooting response in stem cuttings in *Prosopis cineraria. Journal of Arid Envinronment*, 27 : 99-103.

Arya, S., Toky, O. P., Bist, R. P. and Tomar, R. (1991) *Prosopis cinerania.* Promising multipurpose tree of arid lands. *Agroforestry Today*, 3(4) : 13-14.

Bhandari, D. S., Govil, H. N. and Hussian, A. (1979) Chemical composition and nutritive value of khejri (*P. cineraria*) tree leaves. *Annals of Arid Zone*, 18 : 170-173.

Bhandari, M. M. (1978) Flora of Indian Desert, Scientific Publishers, Jodhpur : 152.

Duhan, A., Chauhan, B. M., Punia, D. (1992) Nutritional value of some non-conventional plant foods of India. *Plant Foods and Human Nutrition*, 42 : 193-200.

Bhimya, C. P. Kaul, R. N. and Ganguli, B. M. (1969) Studies on lopping intensities of *Prosopis cinerania. Indian Forester*, 90 : 19-23.

Gogal, V. and Arya, H. C. (1984) Tissue culture in desert trees. I. Clonal multiplication of *Prospis cineraria* by bud culture. *Journal of Plant Physiology*, 115 : 183-189.

Kacker, N. L., Solanki, K. R., Singh, M., Vyas, S. C. and Singh, M. (1991) Micropropagation of *Propspis cineraria. Indian Journal Experimental Biology*, 29 : 165-167.

Kacker, N. L., Vyas, S. C., Singh, M., Solanki, K. R. and Singh, M. (1992) *In vitro* regeneration of *Prosopis cineraria* using root as explant. *Indian Journal Experimental Biology*, 30 : 429-430.

Kaul, R.N. and Jain, M.B. (1967) Growth attributes : Their relation to fuel yield in *P. cineraria* (Linn.) McBride (P. *specigera* Linn.). *Commonwealth Forestry Review*, 46(2): 128.

Lahiri, A. N. (1964) Moisture relation of some desert plant. Proceedings Problems of Indian Arid Zone. Ministry of Education, Government of India, New Delhi : 154-160.

Lahiri, A. N. (1977) Plants in relation to soil moisture conditions of arid areas. In desertification and its control, ICAR, New Delhi : 225-236.

Lahiri, A. N. (1980) *Prospis cinerania* in relation to soil water and other conditions of its habitat. In: Khejri in Indian Desert - its role in agro-forestry (Eds. Mann and Saxena) CAZRI, Monograph. 11 : 37-44

Macbride (1919) Contribution Gay. Herb n.s. LIX : 16

Mann, H. S. and Saxena, S. K. (1980) Khejri (*Prospis cinerania*) in India desert-its role in agro-forestry. CAZRI, Monograph. 11 : 1-85.

Mathur, C. S. (1976) Investigation on camel nutrition in Rajasthan desert (ICAR Project) : Final Report, University of Udaipur, Rajasthan.

Morsink, W. A. G. and Smith, V. G. (1974) Root and Shoot development of cuttings of bas wood as affected by auxin treatment and size of cuttings. *Canadian Journal of Forest Research*, 4 : 246 : 249.

Muthana, K. D. (1980) Silvicultural aspects of Khejri. In : Khejri in Indian Desert-its role in agro-forestry (Ed. Mann and Saxena), CAZRI, Monograph, 11 : 21-25.

Pal, S. K. (1977) White grubs and their management. Monograph No.5, CAZRI : 1-36.

Pareek, O. P., Vishal Nath, Singh, R. S. and Sharma, B. D. (1996) Regeration of *Prosopis cineraria* through cutting. Abst. National Seminar on Propagation and Nursery Management, G. A. U, Anand, Gujarat.

Pareek, O. P. and Vishal Nath (1997) Variability in horticultural traits of Khejri in thar desert. *International plant Genetic Resources Institute, Newsletter for Asia, Pacific and Oceania*, 24 : 23-25

Pareek, O. P. (1997) NRC for Arid Horticulture. In : 50 years of Horticultural Research, ICAR New Delhi : 95-99.

Pareek, O. P. and Purohit, A. K. (2002) Patch budding in Khejri (*Prosopis cineraria*). *Indian Journal of Horticulture*, 59 (1) : 89-94.

Parihar, D. R. (1978) Field observations on the nature and extent of damage by Indian desert termite and their control. *Annals of Arid Zone*, 17(2) : 192-199.

Parihar, D. R. (1980) Insect pests of Khejri. In : Khejri in Indian Desert - its role in agro-forestry (Ed. Mann and Saxena), CAZRI, Monograph-11 : 53-55

Puri, G. S., Jain, S. K., Mukherjee, S. K., Santisarup and Kotwal, M. M. (1964) Flora of Rajasthan. Recommendations of Botanical Survey of India. 19 : 1-159.

Purthi, H. S. (1969) Text book of Agricultural Entomology, ICAR : 1-977.

Rawat, T. S., Sharma, H. C. and Pundir, J. P. S. (1983) Air layering-a success in propagation of Khejri. *Udyanika*, 5 : 43-44.

Satyanarayan, Y. (1964) Habitat and Plant communities of India Desert. Proceedings : National Symposium on Problem of Indian Arid Zone : 59-67.

Saxena, S. K. (1978) Vegetation and its succession in Indian Desert. Desertification and its control (Ed. F. L. Jaiswal) ICAR, New Delhi : 176-192.

Saxena, S. K. (1980) Taxonomy, morphology, growth and reproduction of Khejri and its succession in North West India. In: Khejri in the Indian Desert - its role in agro-forestry (Eds. Mann and Saxena). CAZRI, Monogodd, 11 : 5-15.

Shankar, V. (1980) Distribution of Khejri (*Prosopis cinerania* (L.) Macbride) in Western Rajasthan. In : Khejri in the Indian Desert - its role in agro-forestry (Eds. Manu and Saxena), CAZRI, Monograph, 11 : 16-20.

Singh, K. S. and Lal, P. (1969) Effect of Khejri (*P. Cinerania* L.) and Babool (*A. arabica*) trees on the soil fertility and profile characterstics. *Annals of Arid Zone*, 8(1) : 33-36.

Singh, G. (1970) Palaeobotanical features of that desert. *Annals of Arid Zone*. 8(2) : 188-195.

Solanki, K. R., Kacker, M. L. and Jindal, S. K. (1986) Air layering in *Prosopis cineraria* (L.) Macbride. *Indian Forester,* 112 : 202-207.

Tiwari, P. (1993) *Prosopis cineraria* : Pods in the human diet. *Prosopis* species in arid and semi arid zones of India. (Eds. Tiwari *et al.,*). PSI & HRDA : 125-127.

Vishal Mittre (1975) Problems and Prospects of Paleobotanical approach towards the investigation of the history of Rajasthan Desert. Proc. Problems of desert in India. G S I, Jaipur.

Vishal Nath, Singh, R. S. and Dwivedi, N. K. (1993) Variation in some horticultural traits in *Prosopis cinerania* in Rajasthan. *Prosopis* species in arid and semi arid zones of India (Ed Tiwari *et al.,*), PSI & HRDA : 121-124.

Vishal Nath, Pareek, O. P., Saroj, P. L. and Sharma, B. D. (2000) Biodiversity of Khejri in arid region of Rajasthan. I. Screening of Khejri for culinary value. *Indian Journal of Soil Conservation*, 286 : 43-47.

Vishal Nath (2004) Value addition in arid horticultural crops. Advances in Arid Horticulture Vol. I. (Eds Saroj *et al.,*), IBDH, Lucknow : 483-504.

—x—

19

KHIRNI

Khirni (*Manilkara hexandra* Roxb) is one of the important under utilized fruit crop of tropical and sub tropical region of India. The tree is medium size, evergreen with spreading growth habit suitable for growing on waste and barren lands. The young immature fruits are boiled and eaten. The fresh ripe fruit or dried fruits are sweet and eaten by the local people in rural areas. It is a good source of minerals and vitamins with less fat content. The ripe fruit content 24.35-26.35 ^{0}Brix total soluble solids, 0.122-0.129 % acidity, 18.75 mg/100g vitamin C and 3.65 mg/100g of total carotenoids. The total sugar in the fruits ranges from 17.5 to 18.5 %. The seeds are fried or roasted, taste like peanuts but contain an alkaloid and are not to be eaten in excess. It also contains 24.6% oil known as Ryan oil which has got medicinal value. The wood is used for making agricultural implements and as a fuel wood in rural area.

Shah *et al.* (2004) studied the effects of the flavonoid rich fraction of the stem bark of *Manilkara hexandra* (Roxb.) Dubard, on gastric ulcers in animals. Oral administration of the ethyl acetate extract (extract A3) inhibited the formation of gastric lesions induced by ethanol. The protective effect of extract A3 against ethanol induced gastric lesions was not abolished by pretreatment with indomethacin (10 mg/kg). Further, extract A3 inhibited increase in vascular permeability due to ethanol administration. Extent of lipid peroxidation was significantly reduced in animals treated with extract. Extract A3 also inhibited the formation of gastric ulcers induced by pylorus ligation, when administered both orally and intraperitoneally. Moreover, pretreatment with extract A3 increased mucus production and glycoprotein content, which was evident from the rise in mucin activity and TC: PR ratio.

Origin and Taxonomy

Khirni is a native of South East Asia and found in Central and peninsular India. The tree is 3-12 m tall (Plate 19.1), bark gray, branch lets glabrous, leaves alternate, often closely clustered at end of branchlets with conspicuous scars; petiole 0.8 – 2 cm; leaf blade obovate to obovate-elliptic, 5 – 10 X 3 – 7 cm, both surfaces glabrous, base broadly cuneate to obtuse, apex retuse, midrib raised abaxially, lateral veins slender and parallel, veinlets dense. Flowers axillary, fascicled. Pedicel thick, 1 – 1.8 cm. Sepals ovate-triangular, 3 – 4 mm, outside yellowish gray tomentose. Corolla white or light yellow (4 mm), lobes oblong (3 mm). Stamens (5 mm); staminodes 2-parted, lobes linear, (3 mm). Ovary ovoid, (2 mm), 6-locular, tomentose. Berry obovoid-oblong to ellipsoid, 1-1.5 cm, 1- or 2-seeded. Seeds (8-10 mm).

Plate 19.1. Morphological Characteristics of khirni (i) A healthy plant (ii) Branches transformed into thorns to tolerate drought

Area and Distribution

Khirni plant are found naturally growing in the forest area and barren lands of tropical and subtropical region of the country. It is growing in Madhya Pradesh,Orissa,U.P.,West Bengal,Andhra Pradesh and Bihar.

Soil and Climate

Khirni is a hardy plant and performs well in wide range of soil. It is generally grown on degraded and barren soil having poor fertility status and low water holding capacity. It can also be cultivated on saline and sodic soil condition. The plant is well adapted to varying climatic condition. It can tolerate extreme hot during summer and extreme cold during winter.

Plant Genetic Resource Management and Varieties

Genetic variability in khirni is available in the country. It is growing naturally in the forest as well as degraded lands.There is variation in morphological character, flowering and fruiting behaviour of the tree.The quality parameters like TSS, Total sugar, acidity, and ascorbic acid are also variable according to genotype available in our country. There is no named variety available in India. The available genotypes are categorized as per the size, shape and quality of fruits. Analysis of physico-chemical cheracters of ten genotypes of khirni at Lucknow, Uttar Pradesh revealed variations in seed number (1-2 per fruit), fruit weight (1.25 g to 3.5 g), fruit length (1.8 to 1.9 cm), fruit diameter (1.11 to 1.94 cm), pulp (82.12 to 90.33%) and TSS (24.35 to 26.35 brix). Acid content in fruits did not very significantly. Maximum Vitamin-C and total carotenoides were recorded to the tune of 18.75 mg and 3.65 mg / 100 g edible portion of fruit, respectively. The tannin content in fruit ranged from 2.21-2.35 percent. Total sugar in the fruits ranged from 17.5 to 18.5 % in different accessions. On the basis of fruits analysis and tree characters, three promising accessions have been marked for multiplication and further evaluation (Singh *et al.*, 2006).

Plant Propagation and Nursery Management

Khirni is generally propagated by seed. It is extensively used as a root stock since it has been found to be the most promising among various rootstocks used for sapota propagation. However,poor germination, short viability and slow growth of seedlings are major problem in khirni thereby limiting its rapid multiplication. Experiments at the Indian Institute of Horticultural Research, Bangalore have revealed that under controlled conditions, seed grown at 32°C and or 25/32°C resulted in higher germination. The germination was 40, 75.6, 70 and 73.2% at 25, 32, 25/32°C (6^h+18^h) and ambient temperature, respectively. Whereas, under field grown condition (27.7°C+50.5% R.H.) only 50.5 % germination was noticed.

Studies on seed viability further revealed that khirni seeds have tendancy to loose viability quickly. When compared to fresh ,seed stored for one, three and six months showed loss in germination to the extent of 12,31 and 36% respectively. Seed priming using GA_3 1000 ppm for 24 hours restored germination significantly over all other treatments. Due to GA_3 priming, about 0,17 and 23% restoration in germination was noticed in one, three and six month old seeds, respectively. Although priming had no effect on days to 50% germination, increase in shoot vigour index to the tune of 35.9% which was noticed in primed seeds when compared unprimed seeds. The loss in khirni seeds germination during storage could be restored using seed priming. The uniform emergence and early vigour obtain due to priming could be commercially exploited to obtained graftable size of seedlings early, so that nursery costs can be reduced (Bhanuprakash, *et al.*, 2006).

Planting

Planting should be done in square system at 5 m x 5 m spacing. The pit size (50 x 50 x 50 cm) should be opened and filled with FYM and soil in equal proportion. The planting should be done in field during rainy season for better plant establishment.

Training and Pruning

At initial stage plant should be trained to provide strong frame work by allowing 4-6 scaf old branches from a height of one meter above the ground level. Other shoots emerging from the main stem during initial 1-2 years should be removed. It does not require regular pruning however,undesirable, week and diseased branches should be removed regularly.

Orchard Management

Khirni is a wild species which is mainly used as rootstock for sapota. As such, the orchards of khirni is not available however, the scattered population and or group of plants are being maintained for commercial production of fruits for rootstock. Therefore, the orchard management, as present, is not being practiced for khirni. Visualizing the potential of the species particularly, the antioxident property, khirni has bright future in marginal and waste land. The information, therefore on various orchard management techniques have been compiled.

Nutrient Management

Being under utilized fruit crop, very limited information is available on manure and fertilizer for khirni. For optimum growth and yield, chemical fertilizers in the ratio of 100:50:100 g NPKand 10 kg FYM/year /plant should be given. The doses can be increased upto five years and after wards constant dose of 500 g N:250 g P :500 g K and 50 kg FYM should be given every year till plant provides economical yield.

Water Management

Khirni is a hardy plant and therefore, very little irrigation is given to young growing plant especially during summer month. In arid and semi arid region,water harvesting structure should be created for run off collection near root zone of plant during rainy and post monsoon season at initial year of establishment. Application of grass mulch during summer season is beneficial for moisture conservation and suppression of weed growth in dry region.

Orchard Floor Management

The field should be free from weed during initial year of plant establishment. However, weeding is not required in full grown up tree. Summer ploughing will control the weed growth. During rainy season, seasonal vegetable crops can be taken up as an intercrop depending upon the soil and climatic condition.

Disease and Pest Management

Leaf webbers are serious pests of khirni especially during young stage of plants. Heavy infestation of plants by the larvae causes loss of leaves which restricts plant growth. The infestation of leaf webbers is serious during sprouting of new leaves and flower bud emergence. It can be managed by spraying with malathion 50 EC @ 2 ml / litre of water at 10 - 12 days interval during new vegetative growth.

Bark eating caterpillar also feed on bark of trees and bone into the trunks. Heavy infestation with bark eating caterpillars (*Inderbela* sp.) devitalizes the plants. The affectedportions should be cleaned and holes be injected with kerosene or 0.1% nuvansolution followed by plugging with clay mud.

Flowering, Fruiting and Yield

Flower are borne in cluster during spring season. The fruits are oval in shape, sweet edible with one or more seeds. The fruit mature during May-June (Plate 19.2). A full grown tree of khirni can produce 1-2 q fruits/tree/year.

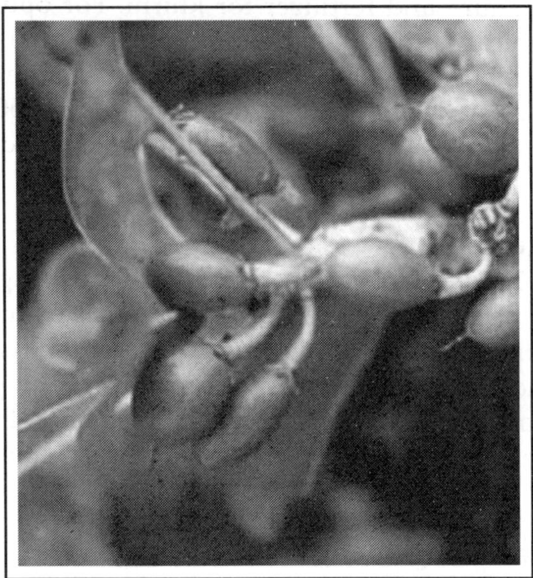

Plate 19.2. Fruiting in Khirni

Harvesting and Post Harvest Management

Harvesting is done when fruit become yellow in colour. The ripe fruit is consumed as fresh in rural area. The ripe fruit can also be dried and stored for longer period.

Processing and Value Addition

The fruit can be processed into various value added products. The ripe fruit can be used for making RTS and jam. It can also be used for making candy.

Future Research Needs

Khirni is an under utilized fruit crop but adopted well to adverse soil and climatic condition but tolerant to biotic and abiotic stresses. But much work has not been done for evaluation of varieties, agrotechniques and post harvest management. There is urgent need to take up work on priority basis on following aspect:

1. Identification and development of high yielding varieties with desired fruit quality.

2. Development of improved agro-technique for better crop production particularly in the wasteland.
3. Development of technology to manage pest and diseases.
4. Development of variety suitable for processing and value addition purpose.

REFERENCES

Bhanuprakash, B., Yogeesha H.S., Arun, M.N and Naik, L.B. (2006) Studies on seed germination improvement in Khirni (*Manilkara hexandra*). *National Symposium on under utilized Horticulural crops.* IIHR, Bangalore, June, 8-9, pp.66.

Guangxi, S. and Hainan, S.W.(1915) *Manilkara hexandra* (Roxburgh) Dubard, Ann. Inst. Bot.-Geol. Colon. Marseille, Ser. 3 (3): 9.

Shah, M.B.,Goswami, S.S. and Satani, D.D.(2004) Effect of *Manilkara hexandra* (Robx) Dubard against experimentally induced gastric ulcer. Department of pharmacognosy, Navarangapura, Ahmedabad.

Singh, A.K., Shukla, S.K., Bajpai, Anju, Singh, M.P. and Singh, A. (2006) Variability in qualitative characteristics of Khirni (*Manilkara hexandra*) accessions from Central Uttar Pradesh. *National Symposium on under utilized Horticulural crops.* IIHR, Bangalore, June, 8-9,pp.53.

20

ROSE APPLE

The rose apple (*Syzyzium jambos*) is an important under utilized fruit crop which belongs to family Myrtaceae. The tree is moderately resistant to winds and tolerates cool and coastal conditions. It is a highly decorative evergreen large shrub or small tree growing to about 5-6 m height with low spreading branches and pale-brown bark. It is wide spreading and often will be wider than its height. The fruits are very nutritious, mostly eaten out-of-hand by children. They are seldom marketed. But during May-June, rose apple fruits are sold sporadically in cities like Bangalore, Kolkata, Chennai, Bhubaneswar etc. In the home, they are sometimes stewed with some sugar and served as dessert. Culinary experimenters have devised other modes of using the cuplike halved fruits. One stuffs them with a rice-and-meat mixture, covers them with a tomato sauce seasoned with minced garlic and bakes them for about 20 minutes. Possible variations are many. The fruit is made into jam or jelly with lemon juice added, or more frequently preserved in combination with other fruits of more pronounced flavour. It is also made into a syrup for use as a sauce or to flavour cold drinks. In Jamaica, the halved or sliced fruits are candied by stewing them in very heavy sugar syrup with cinnamon. The nutritive value of rose apple fruits have been presented in Table 20.1:

In 1849, it was announced in Bengal that the ripe fruits, with seeds removed, could be distilled 4 times to make a "rosewater" equal to the best obtained from rose petals. The seeds are said to be poisonous. An unknown amount of hydrocyanic acid has been reported in the roots, stems and leaves. An alkaloid, jambosine, has been found in the bark of the tree and of the roots and the roots are considered poisonous.

The flexible branches have been employed in Puerto Rico to make hoops for large sugar casks and also are valued for weaving large baskets. The bark has been used for tanning and yields a brown dye. The sapwood is white. The heart wood is dark-red or brown, fibrous, close-grained, medium-heavy to heavy, strong and has been used to make furniture, spokes for wheels, arms for easy chairs, knees for all kinds of boats, beams for construction, frames for musical instruments (violins, guitars, etc.) and packing cases. It is also popular for general tannery. It is not durable in the ground and is prone to attack by dry wood termites.

Table 20.1. Food value per 100 g of edible portion of rose apple

Constituent	Content	Constituent	Content
Calories	56	Sodium	34.1 mg
Moisture	84.5-89.1 g	Potassium	50 mg
Protein	0.5-0.7 g	Copper	0.01 mg
Fat	0.2-0.3 g	Sulfur	13 mg
Carbohydrates	14.2 g	Chlorine	4 mg
Fiber	1.1-1.9 g	Carotene	123-235 I.U.
Ash	0.4-0.44 g	Thiamine	0.01-0.19 mg
Calcium	29-45.2 mg	Riboflavin	0.028-0.05 mg
Magnesium	4 mg	Niacin	0.521-0.8 mg
Phosphorus	11.7-30 mg	Ascorbic Acid	3-37 mg
Iron	0.45-1.2 mg		

The tree grows back rapidly after cutting to a stump and consequently yields a continuous supply of small wood for fuel. Rose apple wood makes very good charcoal. A yellow essential oil, distilled from the leaves, contains, among other properties, 26.84% *αa*-pinene and 23.84% *l*-limonene, and can be resorted to as a source of these elements for use in the perfume industry. The flowers are a rich source of nectar for honeybees and the honey is of good amber colour. In India, the fruit is regarded as a tonic for the brain and liver. An infusion of the fruit acts as a diuretic. A sweetened preparation of the flowers is believed to reduce fever. The seeds are employed against diarrhoea, dysentery and cataract. In Nicaragua, it has been claimed that an infusion of roasted, powdered seeds is beneficial to diabetics. In Colombia, the seeds are believed to have anesthetic property. The leaf decoction is applied to sore eyes, also serves as a diuretic and expectorant and treatment for rheumatism. The juice of macerated leaves is taken as a fabrifuge. Powdered leaves have been rubbed on the bodies of smallpox patients for the cooling effect. The bark contains 7-12.4% tannin. It is emetic and cathartic. The dececetion is administered to relieve asthma, bronchitis and hoarseness. Cuban people believe that the root is an effective remedy for epilepsy.

Origin and Taxonomy

The rose apple is native to the East Indies and Malaya and is cultivated and naturalized in many parts of India, southeast Asia and the Pacific Islands. It was introduced into Jamaica in 1762 and became well distributed in the West Indies, and at low elevations from southern Mexico to Peru. The tree was planted in Florida before 1877 and later in California. In Guatemala, the tree may be planted as a living fencepost or in hedgerows around coffee plantations. For this purpose, it is drastically pruned to promote dense growth. It grows wild abundantly, forming solid stands and thickets in Puerto Rico, the Virgin Islands, Guatemala, Honduras and Panama.

The Rose Apple is a tree, reaching 25 or even 40 ft (7.5-12 m) in height, and has a dense crown of slender, wide-spreading branches, often the overall width exceeding the height. The evergreen leaves are opposite, lanceolate or narrow-elliptic, tapering to a point 10-22 cm long, and from 2.5-6.25 cm wide, somewhat leathery, glossy, dark-green when mature, rosy when young. The flowers are creamy-white or greenish-white, 5-10 cm wide, consisting mostly of about 300 conspicuous stamens to 3-4 cm long, a 4-lobed calyx, and 4 greenish-white, concave petals. There are usually 4 or 5 flowers together in terminal clusters. Capped with the prominent, green, tough calyx, the fruit is nearly round, oval, or slightly pear-shaped (Plate 20.1), 4-5 cm long, with smooth, thin, pale-yellow or whitish skin, sometimes pink-blushed, covering a crisp, mealy, dry to juicy layer of yellowish flesh, sweet and

Plate 20.1. Developing fruits of Rose apple

resembling the scent of a rose in flavour. In the hollow center, there are 1 to 4 brown, rough-coated, medium-hard, more or less rounded seeds, 1-1.6 cm thick, which loosen from the inner wall and rattle when the fruit is shaken. Fragments of the seed coat may be found in the cavity.

Area and Distribution

There is no organized orcharding of rose apple. Therefore, informations on area and production of rose apple are not available. Plants are found naturally growing on degraded land and in forest. Rose Apple is native to the East Indies and Malaya and is cultivated and naturalized in many parts of India, Ceylon and former Indo-china and the Pacific Islands. It was introduced into Jamaica in 1762 and became well distributed in Bermuda, the Bahamas, the West Indies and, at low and medium elevations, from southern Mexico to Peru. (Morton, 1987).

In 1825, eight young trees were taken from Rio de Janeiro to Hawaii by ship, and, in 1853, a United States warship delivered avocado and rose apple trees from Central America to the Island of Hilo. The rose apple became naturalized on the islands of Kauai, Molokai, Oahu, Maui and Hawaii. In 1893, it was reported as already cultivated in Ghana. It is semi-naturalized in some areas of West Tropical Africa and on the islands of Zanzibar, Pemba and Reunion. It is believed to have been first planted in Queensland, Australia, about 1896. A tree obtained from an Italian nursery has grown and borne well on the coastal plain of Israel. However, it is not of interest there as a fruit tree but as an ornamental. The rose apple was introduced into Florida, at Jacksonville, before 1877, but, as a fruit tree, it is suited only to the central and southern parts of the state. In California, it is planted as far north as San Francisco for its ornamental foliage and flowers. Because the tree occupies considerable space and the fruit is little valued, the rose apple has not been planted in Florida in recent years, though there are quite a number of specimens remaining from former times.

Soil and Climate

A deep, loamy, well-drained soil is best for the rose apple, but it also flourishes on sand and limestone with very little organic matter. In India it grows along the streams and banks of canals and streams and tolerates semi-arid conditions. Prolonged dry spells, however, are detrimental. It is a favorite dooryard tree in the Peruvian part of the Amazon, where the trees are planted high enough to avoid the frequent floods.

The rose apple flourishes in the tropical and sub-tropical climates only. In Jamaica, it is naturalized from near sea-level up to an altitude of 3,000 ft (915 m); in Hawaii, from sea-level to 4,000 ft (1,200 m). In India, it ranges up to 4,400 ft (1,350 m); in Ecuador, to 7,500 ft (2,300 m). At the upper limits, as in California, the tree grows vigorously but will not bear fruit. Rose apples flourish in tropical or near-tropical climates, but the tree is proving to be hardy enough (to about 25° F) to be grown as an ornamental as far north in California as San Francisco. A beautiful specimen is thriving in the rather cold, windy rare fruit section of Quail Gardens in Encinitas. The rose apple is too large to make a suitable container plant.

The rose apple will tolerate several degrees of frost but does best when planted in a protected spot on the south side of a wall or building. Young plants can be given overhead protection and covered when heavy frosts are expected.

Plant Genetic Resource Management and Varieties

Variation in the seedling origin plants of rose apple in respect to fruit size, yield and quality in our country. There is need to collect and evaluate for its performance in arid, semi arid and coastal area. One genotype of rose apple (*Syzyzium jambos*) collected from local sources after propagating through air layering and planted at the research farms of this Station. The average fruit weight was observed to be 23.5g while TSS 8-9% and yielded 7-8kg fruits per tree during second fortnight of April. The fruits could be stored for 5-6 days in rose apple at room temperature under coastal Orissa conditions (Pandey *et al.*, 2006).

There are no known varieties available in our country. Varieties are known on the basis of size, shape and colour of fruit in different growing areas.

Plant Propagation and Nursery Management

Rose apple trees are grown from seeds, which are poly embryonic (producing 1 to 3 sprouts), but the seedlings are not uniform in character nor behaviour. In India, vegetative propagation has been undertaken with a view to standardizing the crop and also to select and perpetuate dwarf types. Using cuttings, it was found that hardwood does not root even treated with growth promoters. Treated semi hard wood cutting gave 20% success. Air-layers taken in the spring and treated with 1,000 ppm NAA gave 60% success. In budding experiments, neither chip nor "T" buds would take. Veneer grafting in July of spring-flush scions on 1-year-old rootstocks gave 31% success. In West Bengal, air-layering is commonly performed in July and the layers are planted in October and November. Fruiting can be expected within 4 years. Sometimes the rose apple is inarched onto its own seedlings.

Planting

Planting should be done in square system at 5 m x 5 m spacing. The pit size (50 x 50 x 50 cm) should be open and filled with FYM and soil in equal proportion. The planting should be done during rainy season for better plant establishment.

Training and Pruning

Training is essential to give proper shape of the plant at initial year of establishment. Pruning of rose apples is not usually necessary. In some countries it is pruned drastically to promote dense growth and used as hedgerows around coffee plantations.

Orchard Management

No systematic orchards of rose apple have been planted however, for the plantations under nutrition garden or arborata, plant management techniques should be ensured to harvest the optimum yield. From the available literature, informations have been compiled and synthesized.

Nutrient Management

Nutrient management plays an important role in rose apple production. Systematic work has not been done to standardize fertilizer requirement of rose apple. However,to achieve optimum growth and fruiting, an annual dose of 100:50:100 g NPK/plant/year should be applied. The doses should be increased upto five years and after wards constant dose of 500g N:250 gP :500 g K should be given every year till plant produces economical yield.

Water Management

The tree tolerate semi-arid conditions, but prolonged dry spell are detrimental. It should have frequent irrigation when the weather is warm, and kept on the dry side when it is cold. In arid and semi arid region, water harvesting structure should be created for run off collection near root zone of plant during rainy and post monsoon season at initial year of establishment. Application of grass mulch during summer season is beneficial for moisture conservation and weed control in dry region.

Orchard Floor Management

Inter cropping at initial year should be done for land utilization and additional income generation. In rainfed condition,moong and cluster bean should be taken during rainy season. If irrigation facilities are available, winter season vegetable crops can be grown. Inter cropping can also control the weed growth at initial stages of plant growth. Once plant is established,weeding is not needed.

Disease and Pest Management

The rose apple tree has few insect enemies. In humid climates, the leaves are often coated with sooty mold growing on the honeydew excreted by aphids. They are also prone to leaf spot caused by *Cercospora* sp., *Gloeosporium* sp., and *Phyllosticta eugeniae*; algal leaf spot (*Cephaleuros virescens*); black leaf spot (*Asterinella puiggarii*); and anthracnose (*Glomerella cingulata*). Root rot caused by *Fusarium* sp., and mushroom root rot (*Armillariella (Clitocybe) tabescens*) attack the tree.

Flowering, Fruiting and Yield

Rose apple flowers are large and showy, white to pale cream and sweetly scented. They are 2 - 4 inches wide and consist mostly of about 300 conspicuous stamens (Plate 20.2). There are usually 4 or 5 flowers together in terminal clusters. The flowers are a rich source of nectar for honeybees.

The fruits are 1 - 2 inches wide, almost round or a little longer than wide. When ripe they may be greenish or dull-yellow flushed with pink (Plate 20.3). The skin is smooth and thin, and the firm flesh yellowish, sweet and rose scented. The texture is crisp, almost crunchy when the fruit is ripe and freshly picked. They contains one to four medium hard, round seeds, which rattle around inside the fruit. The

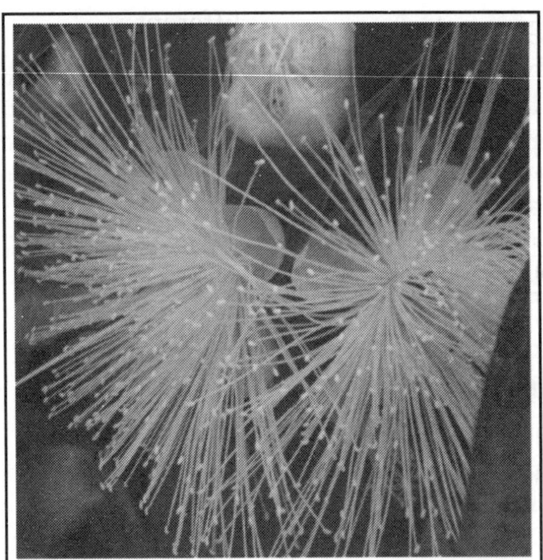

Plate 20.2. Flowers of Rose apple

Plate 20.3. Mature fruits of Rose apple

seed as well as the roots are regarded as poisonous. Seedless, thick-fleshed fruits have been experimentally produced by treating opened flowers with growth regulators such as naphthoxy acetic acid.

In Jamaica and Puerto Rico, the rose apple trees bloom and fruit sporadically nearly all year, though somewhat less in summer than at other times. The main season in the Bahamas and in Florida is May to

July. The fruiting period varies in different parts of India. In South India, blooming usually occurs in January, with fruit ripening in March and April, whereas in the Circars, ripening takes place in April and May. In the central part of the country, flowering occurs in February, March and April and the fruits ripen from June to July. Then again, it is reported that there are varieties that produce fruit in February and March.

A mature rose apple tree can yield 20-80 kg fruits under optimum management conditions. The fruits are, of course, very light in weight because they are hollow, but this is a very small return for a tree that occupies so much space.

Harvesting and Post Harvest Management

Rose apples bruise easily and are highly perishable. Since the fruits are non-climacteric, hence, these should be harvested carefully at crisp ripe stage. The fruits should be graded and packed in CFB boxes or bamboo baskets giving paper shreddings or paddy straw as cushioning material. Generally, fruits are sold locally but with proper packaging and faster transport, these can be sent to metropolitan cities for higher return.

Processing and Value Addition

The fruit is only moderately interesting, eaten out-of-hand and; is more often used in jellies and jams or preserved in combination with other fruits of more pronounced flavour. It is also cooked with sugar to make a dessert. When cooked with custards or puddings, they impart a rose flavour. The flowers can also be candid.

Future Research Needs

Rose apple is an important under utilized fruit crop and mostly grown on degraded land condition. For expansion of area of rose apple under crop diversification programme, attention must be given on following line of work:

1. Collection,conservation and identification of high yielding varieties with desired fruit quality for different agro-climatic condition.
2. Development of improved agro-technique for wasteland.
3. Development of integrated pest and disease management techniques.

4. Development of variety suitable for processing and value addition.

REFERENCES

Morton, J. (1987) Rose Apple (*Syzygium jambos*). In: Fruits of warm climates. pp. 383-386.

Pandey, V., Dinesh Kumar, Singh, H.S., Naik, G. and Bharthi, L.K. (2006). Observational Trial on Rose Apple (*Syzyzium jambos*) and Star Apple (*Chrysophyllum cainito*) in Costal regions of Eastern India. National symposium on under utilized fruit crop. IIHR, Bangalore.

—x—

21

CACTUS PEAR

Scientists with far reaching vision consider that the future of arid and semi-arid lands in entire world largely depends on the sustainable farming system using most suitable crops (Barbera *et al.*, 1995). In India, 12 percent of geographical area comes under such arid and 27.7 percent under semi-arid zone where extremes of temperatures, high wind velocity and solar radiation, scanty rainfall and moisture holding capacity of soil makes the crop production much difficult. The most suitable crops under such hostile climate and edaphic conditions are those that successfully cope with water shortage, high temperature and poor soils and require low energy inputs to provide food and forage for the subsistence agriculture. Cactus pear (*Opuntia ficus indica* Mill.), a zerophytic plant of Mexican origin has been identified and established as fruit, vegetable and fodder crop (Pimenta, 1994; Kaidanu, 1995; Barbera *et al.*, 1995). Owing to presence of pricks on its body, cactus pear is commonly known as prickly pear while its edible fruit is called "tuna". Tender and often spineless cladodes of "nopalea" (*Opuntia cochenilifera*) are used as vegetable salad while mature cladodes of both the types can be used as nutritive fodder for animals. Almost all parts of the plant have great industrial and commercial importance.

Cactus pear grows abundantly in western prairies of USA and is becoming a valuable livestock feed (Felker, 1995). In South Africa, spineless cacti were introduced for livestock feeding (De Kock, 1990). Similarly 3, 00,000 ha spineless cacti are being grown in north-eastern Brazil for livestock feeding (Domingues, 1963). Nutritional quality of *Opuntia* forage used in animal diet has been evaluated by number of workers time to time (Griffiths and Hare, 1906; Shoop *et al.*, 1977; Meyer and Brown, 1985; Ritamal *et al.*, 1987). In general, *Opuntia* fodder has 85 per cent moisture, 5-12 percent crude protein and good amount of minerals (P, K, Ca and Mg). *In vitro* dry matter digestibility of cactus fodder is around 75 percent (Felker, 1995). Under drought situation, cactus cladodes are an excellent feed for cattle and goat (Hansclka and Paschal, 1990). In India, Sirohi *et al.*, (1997) reported that cactus cladodes are highly palatable and can be used as nutritive feed in comparison with conventional material for cattle and sheep.

Cactus produces fruits of 100-200g weight which are highly priced in various developed countries. These fruits, after ripening, can be used as fresh as well as for processing. Fruits are processed into jam, syrup, juices, etc. in Mexico, Chile, Argentina, Brazil, South Africa, Isreal, Italy and USA. In general, a ripe fruit contains moisture (85%), carbohydrates (11%), crude fibre (1.8%), protein (0.5%), lipid (0.1%), acidity (0.18%), calcium (60mg /100mg), vitamin C (300mg /100g) and 150 IU vitamin A (Hernandez *et al.*, 1980; Askar and El-Samahy, 1981 and Sawaya *et al.*, 1983).

The young tender cladodes of vegetable type cactus (*Nopalea*) are used for culinary and salad purposes. Felker *et al.* (1997) noted the taste of prepared *nopalea* vegetable like okra. Due to its excellent nutritional qualities (Russel and Felker, 1987) and therapeutic properties against hyperglycaeamia, hypercholesterolasmia and many other ailments (Hegwood, 1990; Mulas *et al.*, 1993), this type has high demand in International Market for human consumption. Wessels (1994) mentioned that *nopalea* is a traditional vegetable in many parts of Mexico and Some Latin American countries. The young *nopalea* cladode, in general contains 3.7 percent carbohydrates, 1.0 percent protein, 1.3 percent mineral, 1.0 percent fibre, 18-23 mg/1000g β-carotene and 10-18 mg/100g vitamin C except moisture to the tune of 90 percent (Cantwell *et al.*, 1992).

Opuntia species are commercially used in cosmetic and dye industry. Shampoo, cream, body lotions, moisturizing agents etc. are some commercial products prepared from its various parts. Opuntia is one of the best rearing sources of the Cochineal insect at mass scale. This insect is used for the production of red dye 'Carminic acid' (Brutsch and Zimmermann, 1993). Cactus plant parts have been found cheap and economical source to generate non-conventional energy like biogas (Uribe *et. al.*, 1992; Anon., 1993).

Visualizing the horticultural, industrial and commercial potential of cactus pear, Food and Agricultural Organization (FAO) has established a "Cactus Pear Network" with the aim to promote cactus pear as an important fruit crop world wide (Pimienta *et al.*, 1993).

Origin and Taxonomy

Prickly pear is native of Central American and Mexican region (Pareek *et al.*, 1998). The earliest known record in this context is available in oldest edition of Oviedo-y-valdes, *La historia general* published in 1535, which is based on the plants growing in America only (Barbera, 1995). Although, introduction of plants to Europe and other parts of the world has not been properly documented, however, from the scanty literature it is evident that the introduction of plants in Europe took place before 1552 when *Lapec de Gomara* wrote about *Nopals*, a well known plant in Spain at that time (Donkin, 1977). The plants of genus *Opuntia* were reported and have been presented in Italy around 1560, in Germany and Netherland in 1583, and in England around 1556. Since the climatic conditions of Mediterranean region are exceptionally suitable for *Opuntias* therefore, the species engrossed in the countries around this region very soon. Consequently, it became the most common element of the flora in these countries. Plants of genus *Opuntia* have also been reported in Philippines (1695), China (1700), South Africa (1772), India (1780) and south-east Asia (1970), however, there are probabilities that the plants were introduced in these countries at an earlier dates (Donkin, 1977). At the moment, *Opuntias* are the integral part of the farming system and natural landscape in several regions of the world. Some species are even naturalized as weeds in certain countries (Wessels, 1988). In India, prickly pear is in wild state and thorny species are being planted as live fence to protect field crops in arid and semi-arid regions from antiquity.

Genus *Opuntia* belongs to family Cactaceae which has nearly 1600 species. These species have been grouped under 11 sub-genera : *Opuntia, Gnsolea, Austrocylindropuntia, Brasilopuntia, Corynopuntia, Cylindropuntia, Grusonia, Marenopuntia, Nopalea, Stenopuntia and Tephrocactus.* Out of 1600 species, about twenty species are being grown for fruits, vegetables and fodder in which *Opuntia ficus indica* is of prime importance. The prickly pear has the highest chromosome number (2n = 66, 2n = 44) found in cultivated conditions and changes in ploidy levels are commonly expressed by an increase in vegetative

(Cladode size) and reproductive (fruit size) vigour which perhaps played an important role in domestication of this crop (Pimienta, 1994).

Taxonomy of *Opuntia* sp. varies according to ecological conditions and level of polyploidy that exists in great number of population regenerated through vegetative and sexual means (Leuenberger, 1991; Scheinvar, 1995). Generally, the *Opuntia* plant has thick succulent, spiny or spineless leaves called 'cladodes'. These cladodes have numerous areoles which functions like meristematic buds. These buds develop into new cladodes and fruits (from areal portion) and roots (from under ground parts) with the passage of time. Plants are glabrous, joints obovate, 30-60 cm long, 20-40 cm wide and 19-28 mm thick with dark green colour which are covered by layers of wax. Areoles disposed in 8-9 spiral series, pyriform, 2-4.5mm long and approximately 3mm wide; 1-2mm long, caducus, light brown or yellow, glochids are arranged in the middle of areole (Lauenberger, 1991).

The flowers are 7-9 cm long, orange, yellow or red in colour, pericarple 2-2.5 times longer than the perianth and tuberculated with approximately eight spiral series of aeroles. Fruits (5-10cm long and 4-8cm wide) are sweet; juicy, edible; pyriform; yellow, orange, red and purple in colour with abundant freshly pulp are commonly called as Tuna (Plate 21.1). Tuna is very rich in sugars, minerals and vitamins.

Plate 21.1. Flowering and fruiting in cactus pear plant

Area and Distribution

The Opuntia species has been domesticated since prehistoric times (Scheinvar, 1995). In Mexico, the tender stems are used as a vegetable (*nopalitos*) and sweet fruits are in very much demand in International Market. Joints are also used as forage or to propagate new plants. The species is now being cultivated in many parts of world.

Opuntia has highest degree of genetic diversity in Mexico and hence these are utilized by the mankind in larger amount than anywhere else (Pimienta, 1993). The most common species of *Opuntia* i.e. *O. ficus indica* (Cactus pear) and its hybrids are distributed widely from North and South American continent to South-East Asia encompassing several regions of Europe, Africa and Asia. At present, commercial cultivation of Cactus pear is being done world wide for fruit, vegetable and fodder (Inglese *et al.*, 1995). Today, this crop covers around 50,000 ha area in Mexico and 1000 ha in Chile. In USA, cactus pear is cultivated for fruit production in state of California where it covers around 120 ha (Russel and Felker, 1987). Peru is the largest producer of Carmine dye (a product of *Opuntia sp.*). In Peru, wild plantation of Opuntia covers about 35,000 ha which provide 80% of natural yield. Intensive plantations of the species have recently been set up along the coasts. In Brazil, 40,000 ha is under wild and intensive plantation of the species and there, it serve as an important source of fodder. In Bolivia, approximately 630 ha is under this crop. *Opuntias*, especially *O. ficus indica* are the commercial fruit crop in Argentia and Colombia (Ochoa de corneli, 1993).

In Sicily province of Italy, *O. ficus indica* has been widely grown since eighteenth century (Barbera *et al.*, 1992) which accounts for 90% of the total production of the country. At the moment, there are 2500 ha of specialized orchards and 25000 ha of mixed plantations in Sicily (Basile *et al.*, 1996). Cactus pear is also grown in Spain, Portugal and Turkey but only on limited scale for domestic consumption.

Systematic plantation of cactus pear covers around 1500 ha area in South Africa. Once upon a time, wild *Opuntia* species had around 90,000 ha area in South Africa but now they have almost been eliminated with biological cowlort through an act applied to the spiny forms, prohibiting the uncontrolled diffusion of the plant. The spineless *O. ficus indica* was not subjected to this regulation (Brutsch and Zimmermann, 1993). In place of wild plantations, specialized plantations are gradually coming up (Brutsch, 1984). *O. ficus indica* has spread over an area of 2, 00,000ha in North Africa as whole and

60,000 to 80,000 ha in Tunisia alone. Besides, the species are also distributed and/or cultivated in Algeria. Ethiopia, Morocco and Egypt (Pimienta, 1994). In Middle East Asia, specialized plantations of the cactus pear are spread over an area of 400 ha. Out of which Israel alone has about 200 ha area with an estimated fruit yield of 18 t/ha (Nerd *et al.*, 1993). In India, cactus pear has been introduced from Texas, USA only in late eighties and thus, at present, the area under cultivation is negligible. Scattered plantation of cactus pear particularly that of wild types is existing in the country. The crop is now gradually becoming popular in arid and semi-arid regions because of its high yield and export potential from these fragile regions.

Soil and Climate

Cactus pear has great adaptability to various soil and climatic conditions and can produce economical yield under fragile environmental conditions of arid and semi-arid region in India. Due to xerophytic nature of plant, arid and semi-arid conditions are very much suitable for cactus pear cultivation. It can tolerate as high as 50-55^{0}C temperature (Mizrahi *et al.*, 1997). Owing to CAM (Cassulacean Acid Metabolism) path way, cactus pear can withstand high degree of moisture stress. Cactus pear puts forth new flush (both vegetative and reproductive) during March-April and produces yield during hottest season in arid and semi-arid zone. Besides, poor soils (sandy, gravely and stony) are not coming in the way of cactus pear production in these regions. Ferroyra *et al.* (1997) reported that cactus pear can do better even when they are irrigated with saline (8.2 dsm) and boron rich (17 mg/lit) water available in plenty in arid region.

Cactus pear has deep, fibrous root system which absorbs the moisture available in the lower strata of the soil. Thorny, wax coated cladodes and fruits allow least moisture loss through aerial portion. Under sandy soil conditions, cactus pear roots bind allow the soil particles and minimizes their dispersal by wind forces on one hand and keep the plant standing well under high wind velocity situation on the other hand. Thus, under fragile environmental conditions of Indian arid zone, cactus pear is a potential fruit crop which can bring revolution in years to come.

Plant Genetic Resource Management and Varieties

In recent years, interest in *Opuntia* species especially for fruit, vegetable and fodder types has remarkably increased, particularly in arid and semi-arid regions. Few years back, concerted efforts had been made on horticultural aspect of this genus which has taken a shape of Cactus Network all over the world in recent years. As a result, cultivation and commercial exploitation of cactus pear has been done in a number of countries (Pimienta *et al.*, 1993 Barbera *et al.*, 1995). With the advancement in scientific knowledge, cactus pear area is increasing day by day and now its various products are available in International Market. Owing to greater conversion of water to dry matter (than either C_3 or C_4 plants) through specialized photosynthetic mechanisms, known as Crassulacean Acid Metabolism (Felker *et al.*, 1997) and its remarkable ability to bind soil particles through fibrous roots, cactus pear is well suited crop for acid and semi-arid zone where low moisture index and sandy soils are the characteristic features. Visualizing the importance of crop, Indian Council of Agricultural Research has sanctioned an APCESS fund scheme to establish cactus pear germplasm at Nimbkar Agricultural Research Institute (NARI) Phaltan (MS) where 61 genotypes including 39 from Texas A & M University, Kingsville, Texas have been maintained (Anon., 1993). Five genotypes at Central Soil Salinity Research Institute (CSSRI), Karnal, three at Indo-Italian Project, Dhakrani, Dehradun, have also been maintained. During 1996-97, 51 genotypes have been introduced from Texas A & M University, Kingsville; Texas at National Research Centre for Arid Horticulture (now Central Institute of Arid Horticulture), Bikaner. At present, the Institute has a total collection of 109 genotypes (Table 21.1).

Table 21.1 : Total collection of germplasm in the world

Place of collection	Collected genotypes	Genotypes
NARI, Phaltan (MS)	41	Fruit vegetable, fodder type
Texas A & M University, Texas (USA)	51	Fruit vegetable, fodder type
CSSRI, Karnal (Haryana)	5	Fruit vegetable, fodder type
Indo-Italian Project, Dehradun	3	Fruiting type
CAZRI, Jodhpur, Rajasthan	1	Fruiting type
B.G.U., Beersheva, Israel	3	Fruit and vegetable
Regional Plant Resource Centre Bhubaneswar	1	Fruiting type
Fatehpur Shekhawati, Rajasthan	1	Fruiting type
Rahuri (Local), Maharashtra	1	Fruiting type
Jodhpur (Local), Rajasthan	1	Fruiting type
Jhunjhunu (Local), Rajasthan	1	Fruiting type

Source: Vishal Nath *et al.* (1999)

Cactus pear has wide range of cultivars. Depending upon the agro climatic adaptability, the cultivars can be classified into Mexican or American group, Italian group and Argentiean group apart from some cultivars from South Africa, Isreal and Malta. The important cultivars are Amarilla, Montesa, Fafayuco, Roja Pilona, Copena, Cardona, Burrona, Cristallina (Mexico); Bianca, Gialla, Rossa (Italy); Algerian, Direkteur, Fusicaulis, Nudosa, Roly-Poly (South Africa) and Ofer, BS-1 (Isreal). Wessels (1988) descrived the characteristics of 37 cultivars which have potential in Indian sub continent (Pareek *et. al.*, 2001). In Indian soil, however the performance evaluation of these cultivars is under progress.

Plant Propagation and Nursery Management

Cactus pear can be propagated through sexual and asexual means. Owing to hard, lignified seed coat which though protects the seeds from adverse environmental conditions prohibits the germination. Consequently, seed takes longer period to germinate. Being heterogeneity in seed lot the resulted seedlings are not true to type which is a limiting factor for commercial cultivation of the crop. Seed propagation also takes long time to select desirable plants. In view of these negative aspects of sexual propagation, vegetative propagation through cladodes have been attempted and found suitable for raising true to type cactus pear orchard (Pimienta, 1990).

Though, much literature is not available on vegetative propagation of cactus pear but for deciding suitable propagating material, cladode size appears to be an important factor (Barbera *et al.*, 1988; Wessels, 1988). A unit of two to three leaves was found to be highly suitable cladode size for planting (Wessels, 1988). Mondragon (1992) and Mondraon and Pimienta (1995) observed that clodode size is the most important factor for selecting planting material because it affects the number and size of shoots produced during the first year of growth in the field and inferred that bigger cladodes could be divided into two or even four pieces and small cladodes should be planted whole as planting material. Wessels *et al.*, (1997) has also emphasized that cladode size has positive correlation with growth of the plant.

The age of cladode is another important factor to decide the planting material as it is very much associated with the survival and growth of plants in field. Barbera *et al.*, (1988) observed that 2-3 one year old cladode, attached to a two year old cladode are succefully used in Multiple Cladode Cutting (MCC) method of plating by Italian, Mexican and Chilean growers since last four years. Mondragon and Pimienta (1995) reported that optimum age of cladode for planting is one year but who two year old cladodes can also be planted successfully. Plant growth regulators (PGR) particularly auxins have been found to improve the rooting in cactus pear (Mules *et al.*, 1992). Mulas *et al.*, (1992 a) reported that young cladodes of cultivar Gialla treated with 1000 ppm Naphthalene Acetic Acid (NAA) for 15 minutes showed the best performance in terms of number and weight of root per cladode. Treatment of cladodes with 0.2% Bavistine or 0.1% Ridomil before planting in nursery improves survival and reduces rotting of cut ends (Nallathambi *et. al.*, 1998).

Planting

Cactus pear is a succulent plant in which modified leaves called cladodes are the main planting material. Workers have contradictory opinion regarding the time of planting in cactus pear. Bonifacio (1961) mentioned that autumn or spring is the appropriate time for planting of cladodes whereas, Damigella (1958) considered autumn to be the best season. According to Barbera *et al.*, (1992), late spring is the best time for planting in Italy. However, in South African conditions, August-September is considered to be the best time for planting of cladodes (Wessels, 1988). Recently, Inglese (1995) observed that late spring is the best time for planting of cladodes.

Orientation of cladodes during planting is another important aspect for successful production of cactus pear. To receive optimum photosynthetically active radiation (PAR) correct row orientation is necessary for better growth and production (Nobel, 1982). At nearly 27^0 latitude, east-west face of cladodes gave better results in terms of survival, growth and production of cactus pear orchards. However, with change in latitude and topographical features as well, the orientation of cladode can be changed. According to Mondragon and Pimienta (1995), cladodes placed facing east-west improves the productivity. In Semi-arid region of India, Cladodes when planted on ridges under ridge-furrow system, exhibited appreciably better growth and production (Anon., 1993). Inglese (1995) observed that planting of cladodes at an angle of 30^0 resulted in better growth and production but flat planting by placing planner surface of cladode on ground is also practiced in many European countries because it saves the cost of pit digging. Singh *et. al.*, (1999) reported that under Bikaner condition planting of cladode in East-West direction and by putting 1/3rd lower portion in the soil have been found to produce more growth in Genotype 1270.

Training and Pruning

Cactus pear is a typical plant which produces number of cladodes from the mother cladode. These initial cladodes make the branches in long run. Therefore, to develop well spaced branches and ultimately the strong frame of a plant, initial training of cladodes is very much essential. To develop a strong frame, 2-3 cladodes are allowed to develop on each of which 2-3 new cladodes are permitted. In this way, the plant is trained. Besides strong frame, the main objective of training is to allow penetration of more light in plant canopy and to ease plant protection, pruning and harvesting operations.

The main objective of pruning is to make a balance between the vegetative and reproductive growth of plant which also facilitates the penetration of light and thus, improves the yield and quality of fruits. Pruned cladodes can also be used as source of planting material to establish new orchards. Mulas and D'Hallewin (1992) found positive results of pruning in cactus pear cultivar Gialla. They reported that pruning promotes the vigorous shoot growth. They also mentioned that during initial years, the yield in terms of cladodes number per plant and cladode number/m^3 is reduced by pruning. Inglese *et al.* (1955a) also emphasized that cladode load and time of thinning maximizes the fruit size and flesh content in *Opuntia ficus indica* cultivar Gialla. Singh and Felker (1988) emphasized that pruning should be done when the crop has been reaped and the plant is in non-active growth stage. They further pointed out that never prune a cactus pear during active growth phase under Indian condition.

Orchard Management

Cactus pear is an emerging fruit crop particularly in the most neglected and resource poor regions of the country. Visualizing the importance of the crops orchard management techniques including raising of saplings for commercial plantation, planting methods, time and direction, water and nutrient management, plant architecture and canopy management in Indian context are very much important.

Nutrient Management

Although manures and fertilizers are frequently used in cactus pear (Russel and Felker, 1987) but the doses of nutrients, their time of application in different soil and climatic conditions has not been clearly delineated. Application of 10kg manure along with 250g ammonium sulphate, 200g single super phosphate and 100g potassium sulphate per plant per year during winter season is a common practice in Chile. It was observed that nutrients applied during this period improved fruit yield and quality. Further, Nerd *et al.,* (1989) found that any interruption in fertilizer application during winter reduced the number of flower buds and delayed their emergence. Nerd *et al.,* (1991) further clarified that continuous fertigation reduced the flower bud in winter than in summer crop. They found that increase in flower bud production in fertigated plants were associated with the increase of NO_3 content in the cladodes. In semi-arid condition of India, application of 200g poultry manure/plant at the time of planting with the provision of sufficient moisture produced good vegetative growth (Anon., 1993) whereas, Nerd *et al.,* (1993) reported that NPK fertigated young plantation of *Opuntia ficus indica* produced 18 tons/ha fruits after 2.5 years of planting in Negev desert of Israel.

Influence of nitrogen and phosphorus on biomass productivity, fruiting and crude protein have been examined by Gathaara *et al.,* (1989). Results reveal that the highest dry fruit production (734 kg/ha/year) was obtained by the application of 100 kg nitrogen and 80kg phosphorus/ha. Further, it was pointed out that soil application of N and P has significant correlation with fruiting whereas, total biomass productivity for 3 years growth does not have any correlation with soil application of nitrogenous and phosphatic fertilizers.

Water Management

Cactus pear is a drought hardly fruit crop however, it responds to irrigation. Singh and Felkar, (1998) mentioned that 2-3 irrigation at 20-25 days interval during fruit formation and development enhances the yield and checks fruit cracking. Since the root system in cactus pear is fibrous and confined to upper 20-30cm layer of soil, light irrigation is most beneficial. Under semi-arid conditions of Maharashtra, irrigation at monthly interval except during monsoon was found most useful for plant growth and production (Anon., 1993). Barbera (1984) reported that 2-3 irrigations during fruit development stage increases yield, fruit size and flesh percentage. Areas receiving little higher rainfall (400-600 cm), can produce cactus pear under rainfed condition. Wessels (1988) considered that 300-600cm well distributed rainfall is sufficient to ensure good yield, however, light irrigation during fruit swelling is essential particularly in light soils to regulate the soil moisture balance and ultimate fruit cracking. Rainfed production can be more profitable through moisture management techniques such as *in situ* and *ex situ* water harvesting and mulching. At critical stage of fruit development, high irrigation can be provided from *ex situ* stored water whereas, soil moisture status can be improved by providing 5 percent slope towards plant from all the direction and provision of organic (local weed, crop wastes, etc.) and inorganic (black polythene) mulching. For *nopalea* production, however, assured irrigation facility is essential particularly during summer months.

For efficient use of irrigation water and to minimize the cost of water application, suitable method of irrigation should be followed in cactus pear. Traditional basin system is consuming a lot of labour therefore, for better establishment of plants in arid region furrow system is an easy and economical method of irrigation during initial year. Once the plant established, drip system has been found much effective. For establishment of cactus pear in hyper arid situation at Bikaner, *Pitcher* system has been found much useful. Pareek *et al.,* (1998) reported that by fixing an earthen pitcher of 10-12 lit capacity to each plant improves the initial cladode development by 2-3 times in fruiting type cactus as compared to basin system. They further suggested that by mixing inorganic fertilizer in pitcher water, an effective integration of water and nutrient application can be done.

Orchard Floor Management

Weeds compete for water and nutrients leading to poor plant and cladode growth. To enhance the productivity, weed management is a must (Monjauze and Le Howeron, 1965). However, weeding in *Opuntia* field is difficult, tedious and expensive owing to spiny and tender vegetative parts. Felker and Russell (1988) found that application of Glyphosate is most suitable herbicide to control both monocot and dicot weeds without any phytoxicity to *Opuntia* plants. Soil sterilants like Bromocil, Hexazinon and Tabuthiuron when applied @ 4.5 kg/ ha in soil have also been found very effective to check weed population. Cactus pear is generally not inter cropped with seasonal crops however, for soil enrichment short duration cover crops can be allowed.

Disease and Pest Management

Pest and disease problem in cactus pear is not serious but its control is essential when required. At the time of planting, cladodes should be treated with fungicide (Bavistin 0.2%) to control rotting of cladodes. Cochineal is a type of wooly aphid, which makes white colonies on the leaf and stem surface. The pest has negative effect on fruit yield. To control the insect pests spray of methidithion @ 0.1% is found more effective. The larval stage of pear moth is destructive which can be controlled by using Decis (15ml/100 lit water) as spray.

Flowering, Fruiting and Yield

Flowering in cactus pear is influenced by age of plant and environmental and nutritional factors. Fruiting generally starts after 3 years of planting but with better management practices (fertigation) it can be induced earlier also (Nerd *et al.*, 1993). Flowering generally takes place on terminal cladodes of one year age but subterminal cladodes can also produce flowers (Wessels, 1988). Flower bud initiation generally takes place on areoles located along the edges of cladode. Spring-summer is the peak time of flowering. Under Indian condition, early flower bud initiation have been observed in clones 1118, 1270 and 1271. Detailed study on flowering has also been conducted. Nieddu and Spano (1992) reported that flowering in cactus pear cultivar Gialla occurred mainly on one year old cladode. They found that 74 percent flowers were arising from the upper margin and 17 percent from the central part of the cladodes. Regarding the time, they found that the maximum flowers (76%) emerged during June while 23% appeared during May. The flowers took 25-37 days to come into full bloom whereas, the fruit development period was 56-76 days. Similarly, Gutterman (1995) also reported that flowering fruiting and yield in cactus pear is influenced by temperature not by light. They also mentioned that young cladodes show significantly lower fertility. Cactus pear blooms during day and most of the flowers open late in the morning and close by the evening. Perfect pollination is essential for fruit and seed set in cactus pear. Number of seeds/fruit ranged from 80-300 (Wessels, 1988). After pollination, growth and development of fruit starts which take 80-90 days for ripening. Parthenocarpy has also been observed in cactus pear.

Botanically, fruit of cactus pear is a berry in which exocarp and endocarp are not separable and thus, it has delicate skin. Therefore, during harvesting and handling it requires utmost care. Harvesting of fruits should be done at proper maturity (Plate 21.2). Fruits have small pricks on its outer surface; therefore, these pricks should be removed by rubbing the fruit or by propane torches. Fruiting in cactus pear

Plate 21.2. Fruits of edible type cactus pear

generally starts after 3-4 years of planting and 10-15 t/ha yield can be obtained from a well managed plantation. A yield of 80-90 t/ha immature cladodes can be harvested from vegetable types whereas, fodder type yields 100-150 t/ha forage. Wessels (1994) reported that a yield of 21.35 t/ha immature cladode (fresh weight) has been obtained from ultra high density planting system (40,000 plants/ha) which could supply the annual protein requirement to 12.2 peoples.

Harvesting and Post Harvest Management

Harvesting time and yield in cactus pear vary according to purpose for which it is being grown. For vegetable purpose, Flores (1995) reported that when cladodes attain optimum size (12-15cm) and remain tender, should be harvested preferably in morning hours. After sprouting, a cladode takes 30-60 days to attain vegetable maturity. A recent report from Israel suggested that marketable size (17-20 cm long) cladode is obtained within 20-30 days after sprouting and new vegetative buds start developing after cladode reaches to its maximum (130 cm^2) size (Anon., 1996). Cactus pear produces fruits covered with small pricks which need to be removed at the time of harvesting. For this, purpose propane torch is used in western countries. Fruits are harvested individually and can be stored for a considerable period under normal condition. Waxing of fruits improves the shelf life.

Processing and Value Addition

Fruits are consumed as fresh but they can also be processed in to quality products like squash, RTS, Jam etc. A famous wine called *Taquilas* is produce by fermenting the fruit pulp in Mexico.

Future Research Needs

In Indian arid and semi-arid region, cactus pear has immense potential. Research efforts are on but they needs to be more intensified in following areas.

1. Collection, conservation and screening of germplasm for Indian environmental condition.
2. Crop improvement programmes to develop high yielding cultivars having tolerance to biotic and abiotic stresses.
3. More attention is needed to standardize the agro-techniques particularly for time and methods of planting and water, nutrient, weed and canopy management.
4. Emphasis should also be given to develop maturity indices, harvesting time and techniques and handling of the produce.
5. Medicinal value of various parts of cactus pear needs to be examined in context to major human health hazards.
6. Plant protection measures needs to be developed.
7. Work should be undertaken to minimize the post harvest losses in fruit and vegetable type.
8. To popularize the crop and its importance in Indian farming community, extension programmes should also be under taken up at various level.

REFERENCES

Anonymous, (1993) Final report of the project "Genetic selection and improvement of *Opuntia* cultivars for human and animal food on semi arid lands" submitted to ICAR by Nimbkar Agricultural Research Institute, Phaltan, Maharashtra, 89p.

Anonymous, (1996) Scientific Activities 1992-95. IAR. Ben Gurion Univ. of Negev., ISRAEL.

Askar, A. and Samahy S. K. E., (1981) Dtsch. Lebensm. Rdsch. 77 : 279-281.

Barbera, G. (1995) History, Economic and agro-ecological importance In. Agro-ecology, cultivation and uses of cactus pear. FAO, Rome.

Barbera, G.; Inglese P. and Pimienta B.E., (1995) Agroecology, cultivation and uses of cactus pear. FAO, Rome.

Barbera, G.; Carimi F. and Inglese P., (1992). Past and present role of the Indian - Fig Prickly pear coputing ficurindica (L.) Miller, Cactaceal) in the Agriculture of sicily. *Economic Botany*, 461 : 10-22.

Barbera, G.; Carimi F. and Inglese P., (1988) *Fruitticoltura*, 10 : 37-43.

Barbera, G., (1984) Ricerche sull' irrigazoine del ficodindia. *Fruitticoltura*, 46 : 49-55.

Basile, F. V., Foti T.; Inglese P. and Brutsch M.O., (1997) *Acta Horticulture*, 438 : 139-150.

Bonifacio, G., (1961) Ficod, Inida: Una piñata preziosa per le regioni meridionali. *Fruitticolture*, 23 : 49-55.

Brustsch, M.O., (1984) Prickly pear (*Opuntia ficus-indica*) cultivation in Southern Africa. In : Symposim of agricultural use of the Cactaceae. Prospects and problems. International Organization for Succulent Plant study, Frankfurth.

Brutsch, M. O. and Zimmermann H.G., (1993) The prickly pear (*Opuntia ficus-indica* (Cactaceae) in South Africa : utilization of the naturalized weed, and of the cultivated plants. *Economic Botany*, 47(2): 154-162.

Cantwell, M.A., Rodriguez F. and Robles F., (1992) Post physiology of prickly pear cactus stem. *Science Horticulture*, 50 : 1-9.

Damigella, P., (1958) *Tecnica Agricola*, 10 : 474-502.

De Kock, G. C., (1980) Browse in Africa. ILCA. Ethiopia : 399-408.

Domingues, O., (1963) Origen e introducao dap alma forrageira no nordeste. IJN PS. *Pemambuco* : 73.

Donkin, R.A., (1977) Trans of American Philos. Soc. Vol. 67, part 5.

Ferroyra, R.E., Aljaro A.U.; Ruiz R. S.; Rojas L. P. and Oster J.D., (1997) *Water Management*, 34(2): 111-124.

Felker, P., Singh G. B. and Pareek O.P., (1997) Opportunities for development of cactus (*Opuntia* spp.) in arid and semi arid regions. *Annals of Arid Zone*, 36(3) : 267-278.

Felkar, P., (1995) FAO Text Book, Rome.

Flores-Valdez.C.A., (1995) Agroecology, cultivation and uses of cactus pear. FAO, Rome : 92-99.

Gathara, N.G., Felker P. and Margaret Land (1989) Influence of nitrogen and phosphorus application on *Opuntia angelmanii* tissue N and P concentrations, biomass production and fruit yields. *Journal of Arid Environment*, 16 : 337-346.

Griffiths, D. and Hare R.F., (1906) Agriculture Experiment Station Bulletin (Mexico). 60 : 125.

Guttermann, Y. (1995) Environmental factors affecting flowering and fruit development of *Opuntia ficus indica* cuttings during the three weeks before planting. *Israel Journal.Plant Science*, 43(2) : 151-15.

Hasnclka, C. W. and Paschal J.C. (eds.) (1989) Proc. of Conference Tesax A & Uni : 63

Hernandez, X.E., (1970) American Associan for Advancement of Sciences Publication 90. Washington D.C.

Hegwood, D.A., (1990) Human health discoveries with *Opuntia* sp. (prickly pear). *Horticulture Science,* 25(12) : 1515-1516.

Inglese, P., (1995 a) Agroecology, cultivation and uses of cactus pear. FAO, Rome : 78-92.

Inglese, P.G. Barbera; T.L.Mantia and S. Protolano, (1995a). *Horticulture Science,* 30(2) : 227-230.

Inglese, P., Barbera G., La Mantia T. and Portolano S., (1994a). The effect of thinning on growth and ultimate size of cactus pear (*Opuntia ficus-indica* (L.) Mill.) fruits. *Horticulture Science.*

Inglese, P., Barbera G. and La Mantia T. (1995) Research strategies for the improvement of cactus pear (*Opuntia ficus indica*) fruit quality and production. *Journal of Arid Environment.* 29 : 455-468.

Kaidanu, A., (1995) IAR Newsletter of Agric. Research 10 : 2-3.

Leuenberger, B., (1991) Cactaceae. Taxon. 40 : 621-627.

Meyer, M.M. and Brown R.D., (1985) *Journal of Range Manage,* 38 : 154-157.

Mizrahi, Y., Nerd A. and Nobel P.S. (1997) Cacti as crops. *Horticultural Review,* 18 : 291-321.

Mondragon, J.C. and Primienta B.E., (1995) Agroecology, cultivation and uses of cactus pear. FAO, Rome : 64-73.

Mondragon-Jacobo, C., (1992) Prueba de germinacion de semillas de nopal tunero de variedades comerciales y semicomerciales, Informe Tecnico. Campo Experimental Norte de Gto. INIFAP. SARH. Mexico.

Mulas, M. D'Hallewin G. and Canu D. (1992) Rivista di Fruitcaltura,. 10 : 67-70.

Mulas, M. and D'hallewin G. , (1992). Improvement pruning effects on vegetative and yielding behaviour in prickly pear (*Opuntia ficus indica* Mill.) cultivar "Gialla". *Acta Horticulture,* 296 : 139-146.

Mulas, M., (1992) Medicinal properties and yield possibilities of the prickly pear (*Opuntia* spp.) in the Mediteranean environment. *Acta Horticulture,* 331 : 79-84.

Mulas, M., D'Jallewin G. and Davide Conu (1992) Rooting Experiments on one year old cladodes of *Opuntia ficus indica* Mill. *Revista di fruitcoltura,* 10 : 67-70.

Mulas, M.Sapno D.; Pellizzaro G. and D'Hallewin G., (1922a). *Advances Horticulture Science*, 6 : 44-46.

Nallathambi, P., Umamaheshwari C. , Singh R. S. and Pareek O.P. (1998). Foot rot disease in cactus (*Opuntia sp.*) and its management. In. Nat. Sym. On pests and disease management, 11-13 Dec., 98 at C.S.A.U.A. & T., Kanpur, India.

Nerd, A.; Karadi A. and Mizrahi Y., (1991) *Horticulture Science,* 26 : 527-529.

Nerd, A.; Karadi A. and Mizrahi Y., (1989) *Horticulture Science,* 24 : 773-775.

Nerd, A.; Mesika R. and Mizrahi Y. , (1993) *Journal Horticulture Science,* 24 : 773-775.

Nerd, A. and Mizrahi Y., (1995). *Advances in Horticultural Sciences,* 9(4) : 188-191.

Neiddu, J. and Spano D., (1992) *Acta Horticulture,* 296 : 153-159.

Nobel, P.S., (1982) Orientations of terminal cladodes of planty opuntias. *Botanical Gazzett,* 143 : 219-224.

Pareek, O. P.; Singh R. S., Vishal Nath and Vashishtha B. B. (2001) The Prickly Pear. Agro's Gardening Series. Agrobios (India): 1-76.

Pimenta, B.E., Barbera G. and Inglese P., (1993). Cactus pear (*Opuntia* spp. Cactaceae) International Nerwork : An effort for productivity and environmental conservation for arid and semi arid lands. *Cactus and Succulent Journal,* 65 : 225-229.

Pimienta, B.E., (1994) Prickly pear (Opuntia spp.) : a valuable fruit crop for the semi arid lands of Mexico. *Journal of Arid Environments,* 28 : 1-11.

Ochoa de Cornelli, J., (1993) Abstract and Final Report of round table Conf. for the establishment of Int. Tech. cooperation Network on cactus pear. August 18-20. 1993. Mexico.

Pareek, O.P., Sharma B.D. and Singh R.S., (1998) Abstract: National Seminar on Micro irrigation Research in India : status and perspective for 21st Centure. WTCER, Bhubaneswar. 26-28 July, 1998.

Pareek, O.P.; Sharma S. and Arora R.K., (1998) Under utilized fruits and nuts. IPGRI office of South Asia, New Delhi : 10.

Pimienta, B.E., (1990) Editorial Universidad de Guadalajara : 246.

Pimienta, B.E., (1993) Ciencia. 44 : 345-356.

Ritmal, N.; Duran J.M. and Fernandez J., (1987) *Journal of Science food Agriculture*, 38 : 303-311.

Russell, C.E. and Felker P., (1987) The prickly pear (*Optuntia* spp. Cactaceale); a source of human and animal food in remiand regions. *Economical Botany*, 41 : 433-445.

Sawaya, W. N., Khatachadowian H. A.; Safi W. M. and Al-Muhamad H. M., (1983) *Journal of Food Technology*, 181 : 183-193.

Scheinvar, L., (1995) Agroecology, cultivation and uses of cactus pear, FAO, Rome : 12-19.

Shoop, M.C.; Alford E.J.and Mayland H.F., (1997) *Journal of Rangeland Management*, 30 : 225-420.

Sirohi, S.K., Karim S.A. and Misra A.K., (1997) *Journal of Arid Environment*, 36(1) : 161-166.

Singh, G.B. and Felker P. (1998) Catus : A new world food. *Indian Horticulture*, 26-31.

Singh, R. S., Bhargava, R., Sharma B.D., Vishal Nath and Pareek O.P. (1999) Effect of planting depth and direction of cladode on growth behaviour in cactus pear (*O. ficus indica* (L.) Mill.). *Indian Journal Horticulture.* (Accepted).

Uribe, J.M., Varneo M.T. and Benavides C., (1992) Biomass de tune (*Opuntia ficus-indica* (L.) Mill.) como acelerador de la digestion anaerobica de guano de bovino. *Simiente*, 62(1) : 14-18.

Vishal Nath, Singh, R. S., Shukla, A. K. and Vashishtha, B. B. (1999) Cactus pear (*Opuntia ficus indica* Mill.) as an emerging fruit crop for arid and semi arid regions of India. *Current Agriculture*, 23 (1-2):49-58.

Wessels, A.B. (1988) Spineless prickly pear. Perskor, Johannesburg, South Africa. 61p.

Wessels, A.B., (1994) *Applied Plant Science*, 8(2) : 64-66.

Wessels, A.B., Van L.L. Der Merwe and Du-Plessis H., (1997). *Acta Horticulture*, 438 : 73-76.

—x—

22

PALMYRA PALM

Palmyra palm (*Borassus flabellifer* Linn. syn. *B. flabelliformis* Roxb. or *Borassus flabelliformis* Murr.) is an important palm species of tropical and sub tropical parts of the world. It is known by different vernacular names in various continents and languages. In tropical Africa, where it has its natural abode, it is known as Palmyra palm, toddy or wine palm, siwalan, rontal, lontar, talauriksha palm and Borassus palm. Palmyra palm is known as Daleib (Sudan) in ARABIC, Chibangalala, Kakoma, Kambili in BEMBA, African fan palm, African palmyra palm, Black-rum palm, Borassus palm, Deleb palm, Selati palm in ENGLISH, Rondier, Rônier in FRENCH, Delebpalme in GERMAN, Lontar in MALAY, Palmeira-abânico in PORTUGUESE, Mchapa, Mtappa, Mvomo in SWAHILI and Abgon eye in YORUBA. In CHINESE it is called as Shan ye shu tou zong or Shan ye tang or Shan ye zi (Taiwan) zong whereas in DANISH and DUTCH it is known as Lontar or Lontarpalm or Palmyra palm. In Malagasy, it is called as Befelatanana, Dimaka, Marandravina whereas in NEPALESE, it is called as Taadii. In Indian regional languages it is recognized in the name of Taala in BENGALI, Taada in GUJARATI, Taad (Tad), Tad mar, Tala in HINDI, Taad, Tadh in MARATHI, Karimabana, Pannei in TAMIL, Karatalamu, Tatichettu in TELUGU.

Different parts of the Palmyra palm have economical value. Palm sap is stimulant, antiphlegmatic and cures dropsy. The sap can be concentrated to prepare jaggery. Its leaves and stalks are used for extraction of fibre and for brush making. Bienvenu Sambou *et. al.*, (2000) reported that the palm tree *Borassus aethiopum* Mart. is used by the Bassari ethnic group in the Republic of Guinea in various ways. The Bassari use this palm mostly as a palm wine resource. The study conducted in the north-western part of the Republic of Guinea by using 'Rapid Rural Appraisal' with palm wine harvesters, collectors, carriers and sellers revealed that a woody ladder fixed to the stem from where the Bassari extract the sap by cutting a hole into the palm heart and by removing a small part of the palm heart two times per day. The palm wine extraction of *B. aethiopum* is the main activity of the Bassari for 7 months of the year. For harvesters, collectors, carriers and sellers, palm wine extraction provides one of the most important annual incomes.

It is a common tree in a larger part of Africa particularly in south of the Sahara and in tropical eastern Asia. The fruits and the young seedlings are important as an article of food. The fibrous pulp around the large nuts is of a sweet, fruity taste and is eaten. The natives bury the nuts until the kernels begin to sprout; when dug up and broken; the inside resembles coarse potatoes and is prized in times of scarcity as nutritious food. During several months of the year, palm wine, or sura, is obtained in large quantities and when fresh is a pleasant drink, somewhat like champagne and not at all intoxicating, though, after standing a few hours, it becomes highly so. According to Grant, it is called as the doub palm by the negroes in the Upper Nile, who eats the boiled roots during famines.

The Palmyra palm is cultivated in India and Sri Lanka. 'Neera' a special kind of liquor is prepared from the adult palm by tapping the sap from new inflorescence in Asian countries. It is collected by hanging earthen pots below the stabbed inflorescence having maximum sap flow. The fresh 'Neera' has large population of beneficial bacteria and yeast, and therefore considered good for health. The pulp of the fruit is also eaten raw or after roasting in India, where as a preserve is made from it in Ceylon. The unripe seeds and particularly the two to three months old young plant are an important article of food during famine in many countries. The fruit and fusiform roots are used as food by the poorer classes in the Northern Circars. Firminger says that the insipid, gelatinous, pellucid pulp of the fruit is eaten by the natives but is not relished by Europeans. A good preserve may, however, be made from it and is often used for pickling. But the most valuable product of the tree is the sweet sap which runs from the peduncles, cut before flowering and is collected in bamboo tubes or in earthen pots tied to the cut peduncle. Nearly all of the sugar made in Burma and a large proportion of that made in south India is the produce of this palm. The sap is also fermented into toddy and distilled.

In the rural areas of developing countries palms are major sources of many products. One of these is palm wine, which has an important role for nutrition, income and social life. In West Africa, palm wine is extracted from three different species: *Borassus aethiopum* Mart., *Raphia hookeri* Mann & Wendland and *Elaeis guineensis* Jacq. As the extraction methods are destructive, palm stands are locally threatened by extinction. This could have serious consequences not only for the nutritional situation of the population but also for their income.

The chief product of the palmyra is the sweet sap (toddy) obtained by tapping the tip of the inflorescence, as is done with the other sugar palms and, to a lesser extent, with the coconut. The sap flows for 5-6 months - (200 days in Ceylon) and each male spadix producing 4-5 lit per day; the female gives 50% more than the male. The toddy ferments naturally within a few hours after sunrise and is locally popular as a beverage. It is distilled to produce the alcoholic liquor called palm wine, arrack, or arak. Rubbing the inside of the toddy-collecting receptacle with lime paste prevents fermentation, and thereafter the sap is referred to as sweet toddy, which yields concentrated or crude sugar (gur in India; jaggery in Ceylon); molasses, palm candy and vinegar.

Palmyra palm jaggery (gur) is much more nutritious than crude cane sugar, containing 1.04% protein, 0.19% fat, 76.86% sucrose, 1.66% glucose, 3.15% total minerals, 0.861 % calcium, 0.052% phosphorus. also contains 11.01 mg iron and 0.767 mg copper per 100 *gur*. The fresh sap is reportedly a good source of vitamin B complex.

The peeled seedlings are eaten fresh or sun-dried, raw, or cooked in various ways. They also yield starch, which is locally made into gruel with rice, herbs, chilli peppers, fish or other ingradients added. It has been proposed for commercial starch production.

Small fruits are pickled in vinegar. During April and May, in India, the shell of the seed can be punctured with a finger and the sweetish liquid sucked out for refreshment like coconut water. Immature seeds are often sold in the markets. The kernels of such young seeds are obtained by roasting the seeds and then breaking them open. The half-grown, soft-shelled seeds for the hollow jelly-like kernels are sliced longitudinally to form attractive loops, or rings and these, as well as the whole kernels, are canned in clear, mildly-sweetened water and exported. Tender fruits that fall prematurely are fed to cattle.

The pulp of mature fruits is sucked directly from the wiry fibers of roasted and then peeled fruits. It is also extracted to prepare a product called 'Punatoo' in Ceylon. It is eaten alone or with the starch from the palmyra seedlings. The fresh pulp is reportedly rich in vitamins A and C. The values of proximate analysis of leaves, fruit, seedlings, immature seed and "seed" of *B. flabellifer* have been assembled from various sources by Atchley (1984).

There are innumerable medicinal uses for all parts of the palmyra palm. Briefly, the young plant is said to relieve biliousness, dysentery, and gonorrhea. Young roots are diuretic and anthelmintic and a decoction is given in certain respiratory diseases. The ash of the spadix is taken to relieve heart burn and enlarged spleen and liver. The bark decoction with salt is used as a mouth wash and charcoal made of the bark serves as a dentifrice. Sap from the flower stalk is prized as a tonic, diuretic, stimulant, laxative and anti phlegmatic and amebicide. Efforts have also been made to explore the aerobiologic and allergenic significance of the pollen of palmyra palm (*Borassus flabellifer* Linn.), one of the important and common palm trees in India, with a view to isolate and purify its major allergenic components partcularly from the pollen. Sugar made from this sap is said to counteract poisoning and it is prescribed in liver disorders. Candies of palmyra palm are remedy for coughs and various pulmonary complaints. Fresh toddy, heated to promote fermentation, is bandaged onto all kinds of ulcers. The cabbage, leaf petioles and dried male flower spikes all have diuretic activity. The pulp of the mature fruit relieves dermatitis. The chemical constituent of palmyra palm on dry matter basis has been given in (Table-22.1).

Table 22.1 : Nutritive value of Palmyra palm

Plant part	Crude Protein	Crude Fiber	Ash	Mineral matter	Moisture	Ca	P
Leaves	13.3	38.0	7.4	4.6	74.7	-	-
Fruit	6.5	16.1	4.8	0.8	87.9	0.22	0.24
Immature seed	5.1	7.9	1.7	0.6	92.7	0.17	0.18
Shoot	8.9	7.2	3.3	0.7	87.2	0.06	0.46
Fruit	3.1	-	3.1	0.9	93.4	-	-
Seed	8.1	-	3.5	1.4	85.1	-	-

Source: Atchley (1984)

Origin and Taxonomy

Palmyra palm (*Borassus flabellifer* Linn.) is native of tropical Africa. It is an important member of family Palmeae or Arecaceae. There are many species of palmyra palm being used by the habitants' world wide. These are *Borassus aethiopium* Mart. or *Borassus aethiopum* Mart, *Borassus flabellifer* L. *Borassus flabellifer* L. var. *aethiopum* Warb. *Borassus flabelliformis* Murr. Syn. *Borassus flabellifer* L., *Borassus madagascariensis* Bojer & Bec. *Borassus sambiranensis* Jum. & Perrier, *Borassus sonneratii* Giseke.

Morton, (1988) mentioned that the palmyra palm is a large tree up to 30 m high and the trunk may have a circumference of 1.7 m at the base. There may be 25-40 fresh leaves (Plate 22.1). They are leathery, gray green, fan-shaped, 1-3 m wide, folded along the midrib; are divided to the center into 60-80 linear- lanceolate, 0.6-1.2 m long, marginally spiny segments. Their strong, grooved petioles, 1-1.2 m long, black at the base and black-margined when young, are edged with hard spines. The coconut-like fruits are three-sided when young, becoming rounded or more or less oval, 12-15 cm wide and capped at the base with overlapping sepals (Plate 22.2). The outer covering is smooth, thin, leathery, and brown, turning nearly black after harvest. Inside is a juicy mass of long, tough, coarse, white fibres coated with yellow or orange pulp. Within the mature seed is a solid white kernel which resembles coconut meat but is much harder. When the fruit is very young, this kernel is hollow, soft as jelly and translucent like ice and is accompanied by a watery liquid, sweetish and potable.

Plate 22.1. Plant of palmyra palm (i) Sketch of the palm and inflorescence (ii) A healthy palm with functional leaves (iii) Fruiting in the palmyra palm

Plate 22.2. Fruits of Palmyra palm

Area and Distribution

Palmyra palm is a tropical tender perennials palms found in India, Burma, Sri Lanka, Malaysia, New Guinea, Nepals, Florida and South east Asia. It has a very good likley hood of growing in California. In India, Palmyra palm grows in southern and western coastal areas, Bihar, Orissa and West Bengal. Different parts of the palm have economical value. Palm sap is stimulant, anti-phlegmatic and cures dropsy. The sap can be concentrated to prepare jaggery. Its leaves and stalks are used to extract fibre and for brush making. *B. aethiopum* is a typical savanna tree (> 20 palms per hectare) and is almost absent in other vegetation types. In 75% of the villages, the exploitation of *B. aethiopum* is not sustainable. The preferred vegetation type for *R. hookeri* is the forest along the rivers (> 33 palms per hectare) compared to the other vegetation types of the forest zone (< 5 palms per hectare). Due to the high preference for the palm wine of this species its exploitation is prevalent in many areas. The distribution of *E. guineensis* depends mostly on human activities, the main vegetation types being plantations (34 palms per hectare), fallow land and cultivated fields (15 palms per hectare) and the forest along the rivers (31 palms per hectare). The exploitation of this species is in general sustainable throughout all eco-zones with some rare exceptions. It grows wild from the Persian Gulf to the Cambodian-Vietnamese border whereas it is commonly cultivated in India, South-east Asia, Malaysia and occasionally in other warmer regions including Hawaii and southern Florida. In India, it is planted as a wind break on the plains. It is also used as a natural shelter by birds, bats and wild animals.

Soil and Climate

It prefers dry climate and grows on a variety of soils. Soil pH requirements for Palmyra palm is 6.6 to 7.5 (neutral), however it can also be grown in soils having pH 7.6 to 7.8 (mildly alkaline). Great looking palm from Africa can't grow in southern California (U.S.) because of too cool climate. It can not bear persistently cool climate but it does well in central Florida and south. It has quite a bit of cold tolerance as it handles frosts and even freezing temperatures. They do suffer some leaf damage from such an event, but seem to quickly grow out of it. As a young plant it looks a bit like a *Bismarckia* (related genus), but very green. Seeds are massive (almost as large as a coconut) but a bit tricky to germinate. Develops wonderful black coloration along the petioles and is extremely ornamental as a larger seedling.

Plant Genetic Resources Management and Varieties

Due to the wrong method of tapping tree sap by some communities in Africa, the fast decline of this important resource has recently gained attention at governmental level. In order to conserve the populations of this palm, it is essential to understand whether the Bassari use a lethal technique because of socio-economic reasons or due to a lack of knowledge. There is no standard variety of palmyra palm. In coastal region of India, there is considerable variability with respect to vigour of palm and in fruit characters. Some high yielding hybrids have been developed by crossing *Borassus aethiopum* X *B. flabellifer*, recently.

Plant Propagation and Nursery Management

Mainly the Palmyra palm is propagated through seed. Although the seed germinates naturally in the soil below the palm, it can also be regenerated *in vitro* in gelatin, agar or other medium. Bergman (1998) mentioned that naturally germinating seed produce deep roots penetrating in the soil which often fails to establish in the field. Therefore, the seed should be allowed to germinate in bed of chicken wire with about 6 inches of germinating medium over the wire. To keep the medium from falling through the chicken wire holes, put a 1/2 inch layer of sphagnum moss down. On top of this bed, the seeds germinate quickly. The radicals worked their way through the six inch germinating bed and then protruded through the bottom. When they were down about 4-6 inches below the chicken wire, remove them with intact radical. With this technique, the leaves opened up correctly, above the soil level. Care must be taken to deal with the hanging radicals quickly when they penetrate the initial chicken wire or the radicals turn dark brown and then rot.

Planting

Although no systematic work has been done to standardize spacing and method of planting palmyra palm, however, Pareek and Vishal Nath (2006) mentioned that 8-10 m spacing is considered optimum. A spacing of 30-40 ft. (9-12 m) has been found optimum for planting palms in tropical climate of Africa and Sri Lanka. Pits of 1x1x1 m size should be dug and filled with soil mixed with 20-25 kg well rotten FYM and leaf mould just before planting. While planting, crown of the sucker/plant should be kept well above the ground level to avoid rotting of the growing tip due to deposition of sand and water.

Training and Pruning

Palmyra palm does not require training and pruning. However, diseased leaves and arial suckers should be pruned to improve the growth and productivity of the palm. While removing the leaves at the time of pruning, care should be taken to maintain optimum number of photosynthetically active leaves (15-20) on the palm.

Orchard Management

As such, there in no recommended management practices for palmyra palm plantations in India because of no systematic planting of palmyra palm in orchard form, however the natural grooves and boundary plantations should be given due care particularly when they are young. In other countries where commercial utilization of palms are being done, proper management of plantation is practiced.

Nutrient Management

Palmyra palm is seldom manured in natural condition. In systemic plantation, however, Pareek and Vishal Nath (2006) advised a yearly dose of 50-60 kg well rotten FYM to each palm during the vegetative growth period. Manure and fertilizer requirement for commercial plantations needs to be standardized.

Water Management

Palmyra palm is a drought tolerant species suitable for xeric escaping. For establishment of newly planted palm, regular watering is essential. Plant needs regular watering during initial phase of growth if the weather conditions are dry. Over watering is harmful. While watering, the crown portion of the palm should not submerged in water for longer period. Mulching with palm leaves conserves soil moisture. Palmyra palm plantation can also be raised under drip irrigation system.

Orchard Floor Management

Owing to lack of systematic cultivation, orchard floor management in palmyra palm has not been worked out. The palms however, growing at the bunds of agricultural fields provide some shade to the crop in the initial years (Plate 22.3). After attaining the sufficient height, palms do not produce shade effect. Similar is the situation with the palms growing in the middle of the field in upland of Eastern India but at later stage each palm occupies 3-5 square meter area. As a principle, leguminous crop species and some partial shade loving crops can be grown to utilize the inter space between the palms.

Plate 22.3. Plant of Palmyra palm at the bund as wind break

Disease and Pest Management

No serious pest and diseases has been observed on the scattered populations of Palmyra palm. However, in the areas where concentrated population of palmyra palm exist particularly in the natural grooves, it is attacked by insects like rhinoceros beetle, red palm weevil, back headed caterpillar, etc. The fungus, *Pythium palmivorum* also attacks the tree causing bud rot destroying the growing point and consequently killing the palm. Efforts are needed to survey and identify the important pests and diseases causing severe damage to the palmyra palm trees and their appropriate control measures needs to be standardized.

Flowering, Fruiting and Yield

Palmyra palm comes in floriferous phase after about 10 years. It produces white or near white bloom on a medium size flowering stalk called rachis. The flowering in palmyra palm takes place during spring-summer in eastern Indian condition. Pollen grains of *B. flabellifer* were found to be present in air from February to May and June. Pollination takes place through air. Palmyra palm pollens are reported to cause allergic to susceptible persons. In a medical study, it has been pastulated that pollen grain of *B. flabellifer* contributing more than 7% of the total airborne pollen load of the study area in West Bengal (Chakraborty, *et al*, 1998). The pollen elicited maximum concentration at 19 to 20 hours in the evening and at a height of 4 m from ground level. Skin prick testing of 455 respiratory allergic patients with whole pollen extract exhibited 31.64% positive response. Each palm may bear 6-12 bunches of about 50 fruits per year. An average crop of *B. flabellifer* in Ceylon is 350 fruits.

Palm sap (*neera*) is a commercial product of palmyra palm. A healthy palm yields about 0.5 gallon sap daily giving a total yield of 40 gallons *neera* per season. Individual trees produce on an average 10 litters per day. After 35–45 days, the sap production ceases and the palm dies. A harvester kills on average 56 individuals of *B. aethiopum* per year in Africa where destructive method is used for extraction of palm sap. In India, however, the *neera* is extracted from the inflorescence.

Harvesting and Post Harvest Management

Neera is extracted by cutting the inflorescence and trapping the sap in earthen pots during spring season. The ripe fruits of palmyra palm, known as 'Koya' in eastern India, is harvested during May-June. The fresh fruits as well as fresh '*Neera*' is consumed immediately after harvest.

Processing and Value Addition

Apart from use of fresh *neera*, jaggery is prepared by concentrating the juice by boiling immediately after its tapping from the palm. A brown coloured sugar candy is also prepared from the juice. Squeezed juice is dried into thick layers (sheets) called *punatoo* which is used as home made confectionery. The soft, yellow and pulpy tissue under the outer skin of seed is edible when fresh. The soft, translucent kernel containing a sweet milky liquid (*tedgole*) is nutritious and edible.

Future Research Needs

Palmyra palm is a hardy species naturalized in the waste and marginal lands in India and many other countries. Although, it is not much popular for its fruits it has the commercial importance and socio-economic impact on rural community. As a matter of fact, there is no harm in promoting this species in waste land and at the farm boundary. If at all it is promoted, research thrust pertaining to development of improved varieties with fast growth, high sap flow and/or desired fruit (*koya*) traits, agro-techniques, plant protection and post harvest management aspects needs to be concentrated in future. It will be amply rewarded if some branded product from palmyra palm is developed and popularized through research and development.

REFERENCES

Atchley, A. A. (1984) Nutritional value of palms. *Principles*, 28(3):138-143.

Bienvenu, S., Assane G., Finn E., Daouda D. and Mamadou C. C. (2002) Palm wine harvesting by the Bassari threatens *Borassus aethiopum* populations in north-western Guinea. Biomedical and Life Sciences and Earth and Environmental Science Biodiversity and Conservation. Springer. Netherlands 11 (7): 1149-1161.

Bergman, P. (1998) Germination techniques in Palmyra palm seeds. *Jungle Music.*

Bittle B. (1999) Common names of common Palms, *Virtual Palm Encyclopedia*, Palm & Cycad Societies of Florida. USA.

Chakraborty, P.; Chowdhury, I.; Gupta, S.; Roy, I.; Chatterjee, S. and Chanda, S. (1998) Aerobiologic and Immunochemical studies on *Borassus flabellifer* Pollen: Evidence for a 90-kD Allergen. *Annals of Allergy, Asthma and Immunology*, 80 (4): 311-317.

EURODICAUTOM, - On-line Technical Dictionary in 12 languages provided by the European Commission < http://eurodic.echo.lu/cgi-in/edicbin/EuroDic>(Danish, Dutch, English, Finnish, French, German, Italian, Portuguese, Spanish, Swedish, Latin with authority names).

Hedrick, U. P. (1919). Sturtevant's Edible Plants of the World. *Borassus flabellifer*- Sustainable Exploitation of *Borassus aethiopum, Elaeis guineensis* and *Raphia hookeri* for the Extraction of Palm Wine in Cote- d'Ivoire.

Mabberley D.J. (1997) The Plant-Book : a Portable Dictionary of the Vascular Plants, 2d Edition, Cambridge University Press, Cambridge, U.K.

Mollet M.; Herzog F.; Behi Y.E.N. and Farah Z. (2000). The Palmyra or Toddy Palm (*Borassus flabellifer* L.). *Environment, Development and Sustainability*, 2 (1):45-59.

Morton, J.F. (1988). Notes on Distribution, Propagation, and Products of *Borassus* Palms (Arecaceae). *Economic Botany*, 42(3): 420-441.

Pareek O. P. and Vishal Nath (2006) Arid and minor fruits. In: Text of Horticulture (Ed. K.L. Chadha), MPH, New Delhi (In Press).

USDA, ARS, (2000) National Genetic Resources Program. Germplasm Resources Information Network - (GRIN). National Germplasm Resources Laboratory, Beltsville, Maryland. www.ars-grin.gov/cgi-bin/npgs/html/taxdump.

Wiersema J. H. and Leon B. (1999) World Economic Plants: A Standard Reference. CRC Press LLC.

Xie Zongwa and Yu Youqin, (1996) Materia Medica, Chinese Terminology, Vol. 1, Renminweisheng Publishing House Beijing, China.

—x—